Computational Mechanics

Computational Mechanics

Edited by Wallace Sanders

CLANRYE
INTERNATIONAL
www.clanryeinternational.com

Clanrye International,
750 Third Avenue, 9th Floor,
New York, NY 10017, USA

ISBN: 978-1-63240-603-3

Cataloging-in-Publication Data

Computational mechanics / edited by Wallace Sanders.
   p. cm.
Includes bibliographical references and index.
ISBN 978-1-63240-603-3
1. Mechanics, Applied. 2. Mechanics--Data processing. 3. Engineering--Data processing.
I. Sanders, Wallace.
TA350 .C66 2017
620.1--dc23

For information on all Clanrye International publications
visit our website at www.clanryeinternational.com

Printed in the United States of America.

Contents

Preface

Computational mechanics is the study of mechanics through computational methods. This book on computational mechanics deals with numerical and computational modeling and their applications that can solve problems in mechanical engineering. This field has three main sub-fields which are mathematics, computer science and mechanics. Different approaches, evaluations, methodologies and advanced studies on this field have been included in this book. It elucidates concepts and innovative models around prospective developments with respect to computational mechanics. The book is appropriate for students seeking detailed information in this area as well as for experts. It will help the readers in keeping pace with the rapid changes in this field.

After months of intensive research and writing, this book is the end result of all who devoted their time and efforts in the initiation and progress of this book. It will surely be a source of reference in enhancing the required knowledge of the new developments in the area. During the course of developing this book, certain measures such as accuracy, authenticity and research focused analytical studies were given preference in order to produce a comprehensive book in the area of study.

This book would not have been possible without the efforts of the authors and the publisher. I extend my sincere thanks to them. Secondly, I express my gratitude to my family and well-wishers. And most importantly, I thank my students for constantly expressing their willingness and curiosity in enhancing their knowledge in the field, which encourages me to take up further research projects for the advancement of the area.

Editor

Vibration of imperfect rotating disk

L. Půst[a,*], L. Pešek[a]

[a]Institute of Thermomechanics AS CR, v. v. i., Dolejškova 5, 182 00 Praha 8, Czech Republic

Abstract

This study is concerned with the theoretical and numerical calculations of the flexural vibrations of a bladed disk. The main focus of this study is to elaborate the basic background for diagnostic and identification methods for ascertaining the main properties of the real structure or an experimental model of turbine disks. The reduction of undesirable vibrations of blades is proposed by using damping heads, which on the experimental model of turbine disk are applied only on a limited number of blades. This partial setting of damping heads introduces imperfection in mass, stiffness and damping distribution on the periphery and leads to more complicated dynamic properties than those of a perfect disk. Calculation of FEM model and analytic — numerical solution of disk behaviour in the limited (two modes) frequency range shows the splitting of resonance with an increasing speed of disk rotation. The spectrum of resonance is twice denser than that of a perfect disk.

Keywords: bladed disk, damping, imperfect disk, travelling waves

1. Introduction

The vibration of a turbine bladed disk is undesirable and highly dangerous, as it can lead to a failure by fatigue and may cause serious accidents. In order to quench these vibrations, various types of dampers are used. Even a small imperfection makes the vibration analysis more complicated. Large-scale research of these problems has been carried out in many institutes and laboratories in the world (e.g. [2–5,9–11]). Experimental model investigated in the laboratories of the Institute of Thermomechanics (IT AS CR) consists of a steel disk with prismatic models of blades fastened on the perimeter of the disk. The disk is fixed in its centre either to the steel plate, or it is overhung on the rotating shaft. On the opposite ends of the disk's diameter, there are several blades equipped with damping heads.

2. FEM model

For a theoretical solution of deformations and stress, a three dimensional FE-model has been developed. The mesh structure is shown in Fig. 1. The eight-node hexagonal elements were used. Numerical method LANCZOS was applied for calculations of eigen-values and modes of vibrations. For modal analysis [6,7], the damping elements were fixed to the ends of the blades.

Due to the added masses on the ends of selected blades, the bladed disk losses its perfect circular properties having infinite number of symmetry axes and therefore, becomes an imperfect disk with limited $2n$ number of axes. FEM solution enables to calculate all interested eigenfrequencies and modes of vibrations in very wide range of frequencies. This paper investigates the

---

*Corresponding author. e-mail: pust@it.cas.cz.

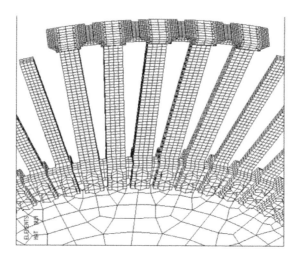

Fig. 1. Mesh structure of FE-model

eigenfrequencies of the two lowest modes of nodal diameters ($n = 1$ and $n = 2$) and no nodal circle ($l = 0$).

FEM calculated the lowest pairs of eigenfrequencies of a perfect, non-rotating bladed disk without any damping elements. The results are: $f_1 = 59.21$ Hz and $f_2 = 78.60$ Hz. Due to small masses added to the ends of five blades of selected groups, the bladed disk lost its perfect circular properties and became imperfect with a countable number of axes in this case two or four symmetry axes. The perfect disk has double eigenfrequencies f_1, f_2, which split into two pairs of close eigenfrequencies at the imperfect disks: $f_{1a} = 59.02$ Hz and $f_{1b} = 45.81$ Hz for $n = 1$ and $f_{2a} = 77.98$ Hz and $f_{2b} = 72.70$ Hz for $n = 2$.

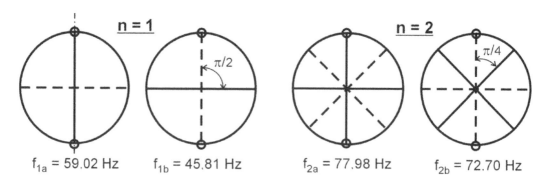

Fig. 2. Split modes with 1 and 2 nodal diameters

The imperfection of the bladed disk, caused by the addition of damper heads, results in free oscillations of two different orthogonal shape modes with the same number of nodal diameters, first ones (f_{1a}, f_{2a}) with one node line going through this imperfection (see full diameters in Fig. 2) and other ones (f_{1b}, f_{2b}) with this imperfection lying on one anti-node line (see dashed diameters in Fig. 2). For FEM modal analysis, the damping elements were stiff connected with the blade ends and created elastic shroud over five blades bundles. Therefore, no damping is present in the system calculated by FEM.

3. Experimental bladed disk model

The damping elements were attached to the real experimental bladed disk. These damping elements were formed as small masses among top-heads and were connected with them only by friction contacts. They allow the axial relative motion of heads, but suppress their relative turning and increase the bending stiffness, similar to the elastic shroud. Therefore these dampers are modelled for computation procedure not only by added masses Δm at the ends of the corresponding diameters, but also by increased damping Δb and stiffness Δc. These additional magnitudes must be multiplied twice, as there are two groups of damping heads.

3.1. Free vibrations of non-rotating disk

Free vibrations of the real experimental bladed disk can be investigated by means of four differential equations of the imperfect non-rotatig disk in the excitation frequency range ω near to the lowest eigenfrequencies f_1, f_2, i.e. approximately in $f \in (0, 100)$ Hz, or $\Omega = 2\pi f \in (0, 630)$ 1/s. These equations are in generalized coordinates q_{an}, q_{bn} ($n = 1, 2$) as follows:

$$
\begin{aligned}
m_{redn}\ddot{q}_{an} + (b_{redn} + 2\Delta b)\dot{q}_{an} + c_{redn}q_{an} &= 0, \\
(m_{redn} + 2\Delta m)\ddot{q}_{bn} + b_{redn}\dot{q}_{bn} + (c_{redn} + 2\Delta c)q_{bn} &= 0, \qquad n = 1, 2,
\end{aligned} \tag{1}
$$

where m_{redn}, b_{redn}, c_{redn} are reduced values of perfect disk masses, damping and stiffness for the first ($n = 1$) or second ($n = 2$) mode of vibrations and are ascertained from the kinetic, potential energy and Rayleigh dissipative function of deformed bladed disk at corresponding modes [1,8,9].These values can be ascertained either from drawings by numerical calculation by means of some FE packages (e.g. ANSYS, CONSOL, etc.) or from data gained by measurements on a real structure or on its physical model.

Values q_{an}, q_{bn} are generalized coordinates belonging to the a – sine modes, b – cosine modes of stationary or rotating disk vibrations with n nodal diameters $n = 1, n = 2$, see Fig. 2. General coordinate system is connected and rotates together with the disk.

Additional damping $2\Delta b$ is added to the sine modes q_{an}, because this modes have the greatest shear deformations on the nodal diameter, i.e. in the damping heads position. Added mass and stiffness $2\Delta m$, $2\Delta c$ is zero in this position. On the contrary, the additional mass $2\Delta m$ and stiffness $2\Delta c$ strongly influence the vibrations of cosine modes q_{bn}, as they are in the anti-mode position.

3.2. Free un-damped vibrations

The various forms of vibrations of an un-damped disk $b_{red1} = b_{red2} = \Delta b = 0$, with amplitudes a_{na}, a_{nb} are described by

$$
q_{an} = z_{na}(r, \phi) = a_{na}g_0(r)\sin n\phi, \qquad q_{bn} = z_{nb}(r, \phi) = a_{nb}g_0(r)\cos n\phi, \tag{2}
$$

where φ is the circumferential angle, connection of radius r and rectangular coordinates x, y is $r = \sqrt{x^2 + y^2}$, $g_0(r)$ denotes the form of vibration in the radial direction; it depends on the structure and mass distribution of the disk vibrating with the mode of one or two nodal diameter and without any nodal circle. This function $g_0(r)$, which can be gained for given type of bladed disk either by FE solution, or by a measurement on real structures, is here supposed for simplicity to be equal for both $n = 1, 2$.

Indexes a, b ascertain the position of nodal diameters in relation to the position of the imperfection. Sinus forms correspond to the nodal diameters drawn by full lines at frequencies

f_{1a}, f_{2a}, in Fig. 2, cosine forms correspond to positions of nodal lines at frequencies f_{1b}, f_{2b}, in Fig. 2. The proper initial conditions in time $t = 0$ produce the general vibrations as combination of eigenmodes with common eigenfrequencies $\Omega_{na} = 2\pi f_{na}$, $\Omega_{nb} = 2\pi f_{nb}$ of the un-damped disk:

$$z(r, \phi, t) = \sum_{n=1}^{2} \left(a_{na} g_0(r) \sin n\phi \cos(\Omega_{na} t + \phi_{na}) + a_{nb} g_0(r) \cos n\phi \cos(\Omega_{nb} t + \phi_{nb}) \right). \quad (3)$$

This expression contains eight free constants a_{1a}, ϕ_{1a}, a_{1b}, ϕ_{1b}, a_{2a}, ϕ_{2a}, a_{2b}, ϕ_{2b}, which can be determined by proper initial conditions. If the imperfection limits to zero, $\Omega_{na} \approx \Omega_{nb}$ and travelling waves which rotate on the disk by angular frequencies

$$\frac{d\phi}{dt} = \frac{\Omega_{nl}}{n} \quad \text{or} \quad \frac{d\phi}{dt} = -\frac{\Omega_{nl}}{n} \quad \text{for} \quad n = 1, 2 \quad (4)$$

can arise at appropriate initial conditions. However, damping in a real disk makes a quick decay of these free travelling wave oscillations.

The travelling waves exist in a non-rotating damped disk with imperfections when it is excited by a non-rotating harmonic force $F_0 \cos \omega t$ acting in a point between nodal (a) and anti-nodal (b) diameters as shown in Fig. 3. The strongest travelling waves appear if the frequency of exciting force lies in one of these intervals: $\omega \in (\Omega_{1a}, \Omega_{1b})$ or $\omega \in (\Omega_{2a}, \Omega_{2b})$ corresponding to the inter-resonance ranges. Differential equations of such system are derived in [6,7,9,10] and are:

$$m_{redn} \ddot{q}_{an} + (b_{redn} + 2\Delta b)\dot{q}_{an} + c_{redn} q_{an} = g_0(r_F), \sin n\phi F_0 \cos \omega t \sin n\lambda,$$
$$(m_{redn} + 2\Delta m)\ddot{q}_{bn} + b_{redn}\dot{q}_{bn} + (c_{redn} + 2\Delta c)q_{bn} = g_0(r_F) \cos n\phi F_0 \cos \omega t \cos n\lambda, \quad (5)$$
$$n = 1, 2.$$

Here the left sides are identical with equations (1), q_n are generalized coordinates belonging to the eigen-forms at $n = 1$, $n = 2$. Radius r_F corresponds to the position of excitation force.

More detailed analysis of the stationary vibrations and travelling waves is presented in the publication [7].

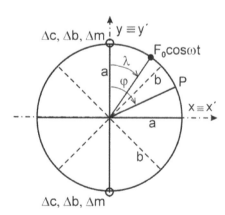

Fig. 3. Harmonic excitation of standing imperfect disk

4. Rotating imperfect disk

Dynamical properties of rotating imperfect disk with constant angular speed ν and modes of vibrations having n nodal diameters and no nodal circles are similar to those of stationary disk. However, one of the main differences observed are eigenfrequencies of the rotating disk which are increasing with increasing angular velocity ν:

$$\Omega_n = \Omega_{n,0} + c_n \nu^2.$$

Because revolutions of experimental equipment in IT ASCR are comparatively low, this change of frequency can be neglected in the following analysis.

Two coordinate systems are used to describe the investigated system:

1. A fixed one r, ϕ_a connected with the standing space,

2. Another coordinates r, ϕ connected to the rotating disk. Relation $\phi = \phi_a - \nu t$ is valid for the initial situation $\phi = \phi_a$ at $t = 0$.

Let in the fixed coordinate system 1) a harmonic force $F_0 \cos \omega t$ at $\phi_a = \lambda$ and at radius r_F near to the ends of blade acts on an imperfect disk, rotating with speed ν. In the disk coordinates 2), this force moves linearly with time in the negative direction ϕ and excites all modes of vibrations. Point of action of $F_0 \cos \omega t$ is given in rotating coordinates x', y' (see Fig. 4) by angle $\phi_F = \lambda - \nu t$, and a general point P of the disk has in the non-rotating coordinates increasing angle position $\phi_a = \phi + \nu t$. Vibrations of the rotating imperfect disk with one or two nodal diameters ($n = 1, 2$) are described by equations

$$m_{redn} \ddot{q}_{an} + (b_{redn} + 2\Delta b)\dot{q}_{an} + c_{redn} q_{an} = g_0(r_F) \sin n\phi F_0 \cos \omega t \sin n(\lambda - \nu t),$$
$$(m_{redn} + 2\Delta m)\ddot{q}_{bn} + b_{redn} dot q_{bn} + (c_{redn} + 2\Delta c)q_{bn} = g_0(r_F) \cos n\phi F_0 \cos \omega t \cos n(\lambda - \nu t), \quad (6)$$
$$n = 1, 2.$$

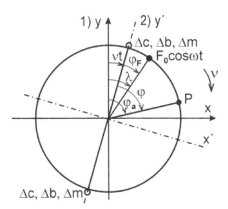

Fig. 4. Harmonic excitation of rotating imperfect disk

Simple arrangement of equations (6) by introducing expressions

$$u_{an} = \frac{q_{an} m_{redn}}{g_0(r_F)F_0}, \qquad u_{bn} = \frac{q_{bn}, m_{redn}}{g_0(r_F)F_0}, \qquad \Omega_{an}^2 = \frac{c_{redn}}{m_{redn}}, \qquad \Omega_{bn}^2 = \frac{c_{redn} + 2\Delta c_b}{m_{redn} + 2\Delta m},$$

$$\beta_{an} = \frac{b_{redn} + 2\Delta b}{m_{redn}}, \qquad \beta_{bn} = \frac{b_{redn}}{m_{redn} + 2\Delta m}, \qquad n = 1, 2 \qquad (7)$$

gets equations

$$\ddot{u}_{an} + \beta_{an}\dot{u}_{an} + \Omega_{an}^2 u_{an} = \cos\omega t \sin n\phi \sin n\nu t,$$
$$\ddot{u}_{bn} + \beta_{bn}\dot{u}_{bn} + \Omega_{bn}^2 u_{bn} = \cos\omega t \cos n\phi \cos n\nu t, \qquad n = 1, 2. \tag{8}$$

The dynamical parameters $\Omega_{an}, \Omega_{bn}, \ldots$ etc. of bladed disk can be ascertained either from working drawings by numerical calculation using some professional FE packages (e.g. ANSYS, COMSOL, etc.) or by measurements on a real structure or on its physical model. Values used in examples in this article are based on data gained from experiments carried out on simplified model of bladed disk in Dynamical Laboratory of Institute of Thermomechanics, ASCR.

The functions on the right sides of equations (8) are complicated functions of time — product of two harmonic functions. For a simple solution of forced vibrations it is convenient to transform these expressions into a sum of terms containing no more than one harmonic function of time:

$$\cos\omega t \sin n\nu t = [\sin(\omega + n\nu)t - \sin(\omega - n\nu)t]/2,$$
$$\cos\omega t \cos n\nu t = [\cos(\omega + n\nu)t + \cos(\omega - n\nu)t]/2, \qquad n = 1, 2. \tag{9}$$

It is shown that one in space-fixed harmonic force $F_0 \cos\omega t$ acts on the rotating disk as two harmonic forces with different frequencies $(\omega + n\nu)$ and $(\omega - n\nu)$. These force-decompositions at appropriate values ω and ν cause travelling waves. Results of our analysis show that the response curves of a disk rotating at constant speed ν can have twice as many resonance peaks as the non-rotating disk.

Solution (8) after introduction (5) gives

$$u_{an} = \frac{\sin n\phi \sin[(\omega + n\nu)t - \psi_{an1}]}{2\sqrt{(\Omega_{an}^2 - (\omega + n\nu)^2)^2 + \beta_{an}^2(\omega + n\nu)^2}} + \frac{\sin n\phi \sin[(\omega - n\nu)t - \psi_{an2}]}{2\sqrt{(\Omega_{an}^2 - (\omega - n\nu)^2)^2 + \beta_{an}^2(\omega - n\nu)^2}},$$

$$u_{bn} = \frac{\cos n\phi \cos[(\omega + n\nu)t - \psi_{bn1}]}{2\sqrt{(\Omega_{bn}^2 - (\omega + n\nu)^2)^2 + \beta_{bn}^2(\omega + n\nu)^2}} + \frac{\cos n\phi \cos[(\omega - n\nu)t - \psi_{bn2}]}{2\sqrt{(\Omega_{bn}^2 - (\omega - n\nu)^2)^2 + \beta_{bn}^2(\omega - 2\nu)^2}},$$
$$n = 1, 2. \tag{10}$$

The corresponding phase angles are

$$\psi_{an1} = \text{arctg}\frac{\beta_{an}(\omega + n\nu)}{\Omega_{an}^2 - (\omega + n\nu)^2}, \qquad \psi_{bn1} = \text{arctg}\frac{\beta_{bn}(\omega + n\nu)}{\Omega_{bn}^2 - (\omega + n\nu)^2},$$
$$\psi_{an2} = \text{arctg}\frac{\beta_{an}(\omega - n\nu)}{\Omega_{an}^2 - (\omega - n\nu)^2}, \qquad \psi_{bn2} = \text{arctg}\frac{\beta_{bn}(\omega - n\nu)}{\Omega_{bn}^2 - (\omega - n\nu)^2}, \qquad n = 1, 2. \tag{11}$$

Dynamical coefficients in expressions (10)

$$\frac{1}{\sqrt{(\Omega_{an}^2 - (\omega + n\nu)^2)^2 + \beta_{an}^2(\omega + n\nu)^2}}, \qquad \frac{1}{\sqrt{(\Omega_{an}^2 - (\omega - n\nu)^2)^2 + \beta_{an}^2(\omega - n\nu)^2}},$$
$$\frac{1}{\sqrt{(\Omega_{bn}^2 - (\omega + n\nu)^2)^2 + \beta_{bn}^2(\omega + n\nu)^2}}, \qquad \frac{1}{\sqrt{(\Omega_{bn}^2 - (\omega - n\nu)^2)^2 + \beta_{bn}^2(\omega - 2\nu)^2}} \tag{12}$$

reach their maximum values at excitation frequencies

$$\omega = \Omega_{a1} + \nu, \quad \omega = \Omega_{a1} - \nu, \quad \omega = \Omega_{b1} + \nu, \quad \omega = \Omega_{b1} - \nu,$$
$$\omega = \Omega_{a2} + 2\nu, \quad \omega = \Omega_{a2} - 2\nu, \quad \omega = \Omega_{b2} + 2\nu, \quad \omega = \Omega_{b2} - 2\nu, \tag{13}$$

at which the resonance peaks in response curves occur.

5. Response curves of rotating imperfect disk

The graphical representation of these properties for the first mode with one nodal diameter ($n = 1, l = 0$) is shown in Fig. 5 in the form of frequency-speed diagram [9].

The response curve of a stationary disk (disk speed $v = 0$) has two resonance peaks at $\omega = \Omega_{b1}$ and $\omega = \Omega_{a1}$, determined by the intersections of a vertical line A with the oblique thick lines. The thick lines correspond to equations in the first row (13). The response curve is drawn in Fig. 6.

Rotation at low revolutions, as observed in Fig. 5 line B at 200 rpm, causes that the two resonance peaks split into four, as seen in Fig. 7. The split resonances have lower height (ap-

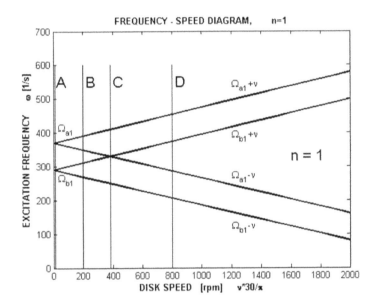

Fig. 5. Frequency-speed diagram of imperfect disk

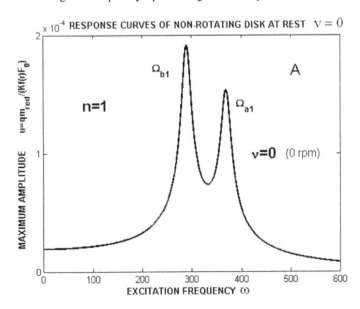

Fig. 6. Response curve of standing imperfect disk

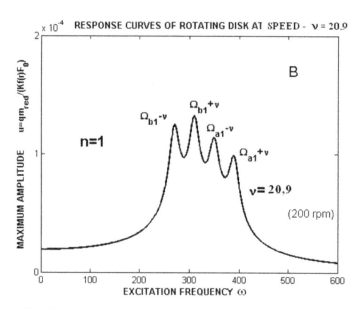

Fig. 7. Response curve of rotating (200 rpm) imperfect disk

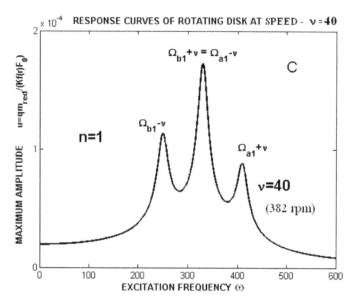

Fig. 8. Response curve of rotating (382 rpm) imperfect disk

proximately half) then those in Fig. 6. In both figures are dimension u [s^2] and ω [rad/s]. The same dimensions are used in Fig. 8, 9, 11, 12.

Line C in Fig. 5 at $\nu = 40$ rad/s goes through the point of intersection of lines given by equations $\omega = \Omega_{a1} - \nu$ and $\omega = \Omega_{b1} + \nu$.

Response curve of this case (shown in Fig. 8) has only three peaks. The combined resonance is higher then the neighbouring peaks at $\omega = \Omega_{b1} - \nu$ and $\omega = \Omega_{a1} + \nu$.

At a higher disk speed up to 800 rpm ($\nu = 83.81$ rad/s), represented by the line D in Fig. 5, the number of interactions increased to four again. Corresponding response curve is shown in Fig. 9. It is similar to Fig. 7 (200 rpm) except that the sequences of resonance peaks are different; the peaks with the mode b alter with a at increasing frequency of excitation ω.

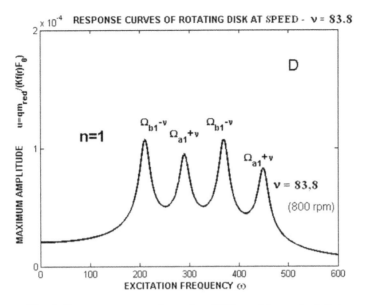

Fig. 9. Response curve of rotating (800 rpm) imperfect disk

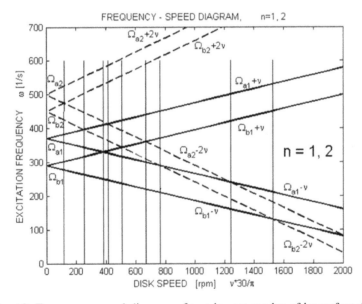

Fig. 10. Frequency-speed diagram of two lowest modes of imperfect disk

Results based on the analysis of disk vibrations when considering only one isolated mode ($n = 1$) cannot describe the full dynamic behaviour of the real rotating imperfect disk as it presents the rich spectrum of modes with various nodal diameters and nodal circles.

As an example of the increasing complications due to mutual interactions of disk modes with different number n of circumferential waves Fig. 10 shows frequency-speed diagram of rotating imperfect disk, where two basic modes $n = 1$ and $n = 2$ are considered. The thick full inclined lines correspond to the equations in the first row of (13) and illustrate the dynamic behaviour of $n = 1$ mode of disk vibrations. The thick dashed inclined lines belong to the second row of equations (13) and graphically demonstrate the dynamical properties of $n = 2$ mode of vibrations. Due to different gradient of lines for $n = 1$ and $n = 2$, ten points of

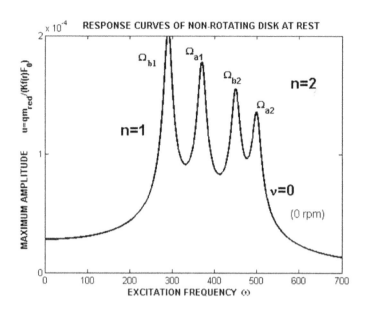

Fig. 11. Four resonance peaks of standing imperfect disk

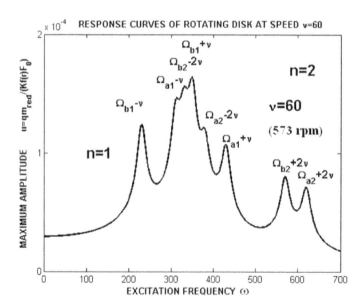

Fig. 12. Eight resonance peaks of rotating (573 rpm) imperfect disk

interaction are present in this diagram. They are marked by vertical lines, which, in general, intersect the thick inclined lines in points, giving the positions of resonance peaks of response curves.

The simplest response curve belonging to the standing non-rotating disk ($\upsilon = 0$) is shown in Fig. 11. There are four resonance peaks at $\omega = \Omega_{b1}$; Ω_{a1}; Ω_{b2}; Ω_{a2}.

Rotation of disk ($\upsilon > 0$) causes the split of resonances again. Example of such response curve is shown in Fig. 12 for $\upsilon = 60$ rad/s. There are eight resonance peaks, but four of them are very close and form in the range $\omega \in (300, 400)$ rad/s a strong common resonance peak with four local peaks.

Similar, but in sequence of modes very different response curves are obtained at various speeds of a disk rotation. It is evident that the analysis of experimentally gained resonance curves of a rotating imperfect disk is extremely difficult without any detailed theoretical investigation.

6. Conclusion

This paper presents computational methods for the analysis of dynamic properties of a non-rotating or rotating disk. The circular disk fixed in its centre has many double eigenfrequencies, each one described by different numbers of nodal diameters and of nodal circles. Positions of nodal diameters are arbitrary at perfect disks, but they are fixed with respect to the position of imperfection at imperfect disks.

Investigated imperfection was formed by the attachment of two groups of vibration-damping heads on simplified model of an experimental bladed disk. Detail analysis was focused on the modes with one and two modal diameters and on the frequency range containing corresponding split resonance peaks. Analysis of dynamic properties of a rotating imperfect disk excited by an external single in space-fixed harmonic transversal force $F_0 \cos \omega t$ shows the existence of travelling waves, and split of eigenfrequencies depending on the rotation speed.

The response curves of the imperfect disk rotating with speed ν have twice as many resonance peaks than the same non-rotating disk.

This study is a contribution to the theoretical support of experimental research of a rotating model of a bladed disk carried out in Institute of Thermomechanics AS CR with the aim to investigate the influence of elastic, mass and damping imperfections on the dynamic behavior of turbine disks.

Acknowledgements

This work has been supported by the grant project GA CR 101/09/1166 "Research of dynamic behaviour and optimisation of complex rotating system with non-linear couplings and high damping materials".

References

[1] Brepta, R., Půst, L., Turek, F., Mechanical vibrations, Technical guide No. 71, Sobotáles, Praha, 1994. (in Czech)

[2] Bucher, I., Feldman, M., Minikes, A., Gabay, R., Real-time travelling waves and whirl decomposition, Proceedings of the 8th International Conference on Vibrations in Rotating Machinery), IMechE 2004, Wiltshire, 2004, pp. 261–267.

[3] Pešek, L., et al., Dynamics of rotating blade disk identifical by magneto-kinematic measuring system, Proceedings of ISMA 2008, Leuven, Belgium, 2008, pp. 1 097–1 111.

[4] Půst, L., Applied continuum mechanics II: Continuum dynamics, ČVUT-FJFI, Praha, 1986. (in Czech)

[5] Půst, L., Effect of imperfection of rotary bodies on their vibrations, Proceedings of the V. National Congress on Theoretical and Aplied Mechanics, Vol. 3, Sofia, Bulgaria, 1985, pp. 131–136. (in Russian)

[6] Půst, L., Pešek, L., Traveling waves in circular disk with imperfections, Proceedings of the 10th Conference on Dynamical Systems – Theory and Applications, Vol. 1, Lodž, Poland, 2009, pp. 345–352.

[7] Půst, L., Pešek, L., Traveling waves in rotational machines, Proceedings of the conference Engineering Mechanics 2009, Svratka, Czech Republic, 2009, pp. 1 065–1 078.

[8] Slavík, J., Stejskal, V., Zeman, V., Basics of machine dynamics, ČVUT, Praha, 1997. (in Czech)

[9] Tobias, S. A., Arnold, R. N., The Influence of dynamical imperfection on the vibration of rotating disks, Proceedings of the Institution of Mechanical Engineers 171(1) (1957) 669–690.

[10] Torii, T., Yasuda, K., Nonlinear oscillation of a rotating disk subject to a transverse load at a space-fixed point, Proceedings of the X. World Congress on the Theory of Machine and Mechanisms, Oulu, Finland, 1999, pp. 1 752–1 757.

[11] Yan, L.-T., Li, Q.-H., Investigation of travelling wave vibration for bladed disk in turbomachinery, Proceedings of the 3rd International Conference on Rotordynamics – IFToMM, Lyon, France, 1990, pp. 133–135.

2

Periodically stimulated remodelling of a muscle fibre: perturbation analysis of a simple system of first-order ODEs

J. Rosenberg[a,*], M. Byrtus[a]

[a] *Faculty of Applied Sciences, UWB in Pilsen, Univerzitní 22, 306 14 Plzeň, Czech Republic*

Abstract

The paper deals with the dynamical analysis of the system of first-order ODE's describing the isometric stimulation of the muscle fibre. This system is considered to be a non-autonomous one having the periodical excitation. For the analysis of dynamical behaviour the system the multiple scale method (MSM) is employed. The main goal of this contribution is to show the application of MSM to the non-autonomous dynamical system using the first order approximation of the solution. The existence of the degenerated Hopf's bifurcation of the gained solution is presented.

Keywords: dynamical non-autonomous system, muscle fibre, multi-scale method, Hopf's bifurcation, chaos

1. Introduction

The paper follows the authors previous papers [6] dealing with the application of the non-irreversible thermodynamics and the growth and remodelling theory (GRT) [1] to the muscle fibre modelling. The approach allows taking into account also the change of the muscle fibre stiffness during time. The effect change of the muscle fibre stiffness during time was experimentally approved and modelled [2]. The same approach can be used to model the piezo-electric stack time evolution. The final simplified dimensionless formulation has the form of the dynamical system with two degrees of freedom. The numerical experiments have shown the interesting behaviour of this system, e.g. the existence of bifurcations. These effects correspond to some medical recognition like vasomotion or myogenic response. This contribution is devoted to the analysis of these properties using the multi-scale method (MSM) [3] which is kind of the perturbation method. Here, MSM is used to model the behaviour of mentioned system close to the Hopf's bifurcation point leading to the periodical or even chaotic motion.

2. Problem setting

In GRT the starting point is the initial configuration B_0 that "growths" and "remodels", i.e. it changes its volume ("growth"), anisotropy ("geometrical remodelling") or material parameters ("material remodelling"). This process is expressed by the tensor $\mathbf{P}$ (further growth tensor) firstly that relates the initial configuration to the relaxed one B_r with zero inner stress to the real configuration B_t where the inner stress invoked by growth, geometrical remodelling and external loading can already exists. It is stated by the deformation tensor $\mathbf{F}$ (see Fig. 1).

---

*Corresponding author. e-mail: rosen@kme.zcu.cz.

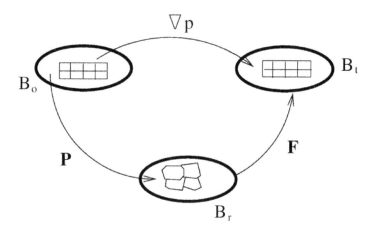

Fig. 1. Initial, relaxed and current configurations

The deformation gradient between the configurations B_0 a B_t is

$$\nabla p = \mathbf{F}\,\mathbf{P}. \tag{1}$$

Let the 1D continuum have the initial length l_0. Its actual length after growth, remodelling and loading will be l. The relaxed length (it means after growth and remodelling) is l_r. For the corresponding deformation gradients we can write

$$P = \frac{l_r}{l_0}, \qquad F = \frac{l}{l_r}, \qquad \nabla p = \frac{l}{l_0}. \tag{2}$$

In the isometric case, the actual length is constant, i.e. it holds $l = \mathrm{const}$.

In [6], the set of the ODE's describing in dimensionless form the behavior of the muscle fiber during the isometric excitation was derived according the work of DiCarlo and Quiligotti [1]

$$\dot{x} = -x\left\{C + \frac{y}{\lambda}e^{\frac{\lambda}{2}(x-1)^2}\left[\lambda x(x-1) - 1\right] + \frac{y}{\lambda}\right\}, \tag{3}$$

$$\dot{y} = \mathrm{sgn}\,m\left[-\frac{1}{\lambda}\left(e^{\frac{\lambda}{2}(x-1)^2} - 1\right)\right]. \tag{4}$$

The meaning of used variables is following

$$x = \frac{1}{l_r'}, \qquad y = k', \qquad k' = k\sqrt{\frac{|m|}{g}}, \qquad l_r' = \frac{l_r}{l} = \frac{1}{F}, \qquad t' = \frac{t}{\sqrt{g|m|}}, \tag{5}$$

where l, l_r are the lengths of the continuum in actual and relaxed configurations and k' is the dimensionless stiffness parameter. The parameters C, D, λ, m, g are the parameters.

Eqs. (3) and (4) are based on the Fung's form of the free energy

$$\psi = \frac{k}{\lambda}\left(e^{\frac{\lambda}{2}(F-1)^2} - 1\right), \tag{6}$$

where k is the stiffness of the muscle fiber.

Setting $\lambda \to 0$ we obtain more simple form

$$\psi = \frac{1}{2}k(F-1)^2, \tag{7}$$

where C is the control parameter depending on the calcium concentration inside the muscle cell. Without using the complex model of the calcium concentration evolution we will suppose, that C is either constant or a periodical function of time and can be approximated in this form

$$C \to C + D\sin\omega t. \tag{8}$$

Using (7) and (8) we obtain the more simple non-autonomous system

$$\dot{x} = -x\left[C + D\sin\omega t + \frac{y}{2}(x^2 - 1)\right], \tag{9}$$

$$\dot{y} = \operatorname{sgn} m\left[r - \frac{1}{2}(x-1)^2\right]. \tag{10}$$

According to some numerical experiments we can see that both models have qualitatively the same properties. Therefore, we will further focus our attention on the simpler one (9) and (10).

3. Dynamical analysis for $D = 0$

This analysis was published in [5]. The existence of the degenerated Hopf's bifurcation was proved for $C = 0$ and $\operatorname{sgn} m = -1$. The situation is shown on Fig. 2 in the right corner. Depending on the sign of C there exists one stable and one unstable equilibrium point and a stable limit cycle around these points for $C = 0$.

4. Dynamical analysis using multiple scale method

We start with the dynamical system defined by the Eqs. (9) and (10). The steady solution of this dynamical system is

$$
\begin{aligned}
x_0 &= 1 \pm \Theta, \qquad \Theta = \sqrt{2r}, \\
y_0 &= 0, \\
C_0 &= D_0 = 0.
\end{aligned} \tag{11}
$$

Now we follow the procedure suggested by [4], but generalized for non-autonomous systems. Let us consider the solution

$$
\begin{aligned}
x &= x_0 + \varepsilon\xi, \\
y &= y_0 + \varepsilon\eta, \\
C + D\cos\omega t &= C_0 + D_0\cos\omega t + \varepsilon^2 C_2 + \varepsilon^3 D_3\cos\omega t.
\end{aligned} \tag{12}
$$

With this very specific approximation we restrict our analysis on the case of small C and even smaller D. We must not forget it when we will do some numerical experiments in order to validate this analysis.

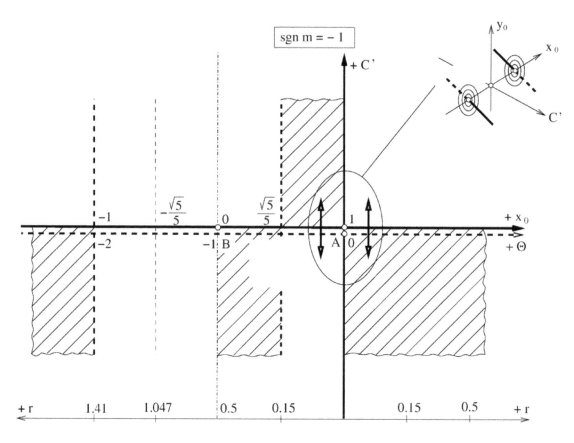

Fig. 2. Stability domains in the parameter space

We put (12) into (11) and while for x_0 and y_0 the RHSs are zero and neglecting the terms of order 4 and higher, we obtain the following non-autonomous dynamical system

$$
\begin{aligned}
\dot{\xi} &= -\frac{1}{2}x_0 \left(x_0^2 - 1\right) \eta - \varepsilon\frac{1}{2} \left(3x_0^2 - 1\right) \xi\eta - \varepsilon^2\xi C_2 - \\
&\quad \varepsilon x_0 C_2 - \varepsilon^2\frac{3}{2}x_0\xi^2\eta - \frac{1}{2}\varepsilon^3\eta\xi^3 - \\
&\quad \varepsilon^3\xi D_3 \cos\omega t - \varepsilon^2 x_0 D_3 \cos\omega t, \\
\dot{\eta} &= (x_0 - 1)\xi + \frac{1}{2}\varepsilon\xi^2.
\end{aligned}
\tag{13}
$$

This system will be solved using multiple-scale method. Let us assume

$$
\begin{aligned}
\xi &= \xi_1 + \varepsilon\xi_2 + \varepsilon^2\xi_3, \\
\eta &= \eta_1 + \varepsilon\eta_2 + \varepsilon^2\eta_3,
\end{aligned}
\tag{14}
$$

where $\xi_i(T_0, T_2), \eta_i(T_0, T_2); i = 1, 2, 3$ and $T_0 = t, T_2 = \varepsilon^2 t$. Limitation of our analysis will be the restriction only on the first approximation of the solution ($x = x_0 + \varepsilon\xi_1, y = y_0 + \varepsilon\eta_1$).

Inserting this approximation into (13) we obtain (again by neglecting terms with the order higher than 3)

$$\frac{\partial \xi_1}{\partial T_0} + \varepsilon^2 \frac{\partial \xi_1}{\partial T_2} + \varepsilon \frac{\partial \xi_2}{\partial T_0} + \varepsilon^3 \frac{\partial \xi_2}{\partial T_2} + \varepsilon^2 \frac{\partial \xi_3}{\partial T_0} + \varepsilon^4 \frac{\partial \xi_3}{\partial T_2} =$$

$$= -\frac{x_0}{2} \left(x_0^2 - 1 \right) \left(\eta_1 + \varepsilon \eta_2 + \varepsilon^2 \eta_3 \right) -$$

$$\frac{1}{2} \left(3x_0^2 - 1 \right) \left(\varepsilon \xi_1 \eta_1 + \varepsilon^2 \eta_1 \xi_2 + \varepsilon^3 \eta_1 \xi_3 + \varepsilon^2 \eta_2 \xi_1 + \varepsilon^3 \eta_2 \xi_2 + \varepsilon^3 \eta_3 \xi_1 \right) -$$

$$\varepsilon x_0 C_2 - \frac{3}{2} x_0 \left(\varepsilon^2 \eta_1 \xi_1^2 + 2\varepsilon^3 \eta_1 \xi_1 \xi_2 + \varepsilon^3 \eta_2 \xi_1^2 \right) - C_2 \left(\varepsilon^2 \xi_1 + \varepsilon^3 \xi_2 \right) -$$

$$\varepsilon^2 x_0 D_3 \cos \omega T_0 - \varepsilon^3 \xi_1 D_3 \cos \omega t, \tag{15}$$

$$\frac{\partial \eta_1}{\partial T_0} + \varepsilon^2 \frac{\partial \eta_1}{\partial T_2} + \varepsilon \frac{\partial \eta_2}{\partial T_0} + \varepsilon^3 \frac{\partial \eta_2}{\partial T_2} + \varepsilon^2 \frac{\partial \eta_3}{\partial T_0} + \varepsilon^4 \frac{\partial \eta_3}{\partial T_2} =$$

$$= (x_0 - 1) \left(\xi_1 + \varepsilon \xi_2 + \varepsilon^2 \xi_3 \right) + \frac{1}{2} \left(\varepsilon \xi_1^2 + \varepsilon^3 \xi_2^2 + 2\varepsilon^2 \xi_1 \xi_2 + 2\varepsilon^3 \xi_1 \xi_3 \right). \tag{16}$$

If we compare the terms with the same order of ε, we obtain the following systems of equations

$$\varepsilon^0: \qquad \begin{aligned} \frac{\partial \xi_1}{\partial T_0} &= -\frac{1}{2} x_0 \left(x_0^2 - 1 \right) \eta_1, \\ \frac{\partial \eta_1}{\partial T_0} &= (x_0 - 1)\xi_1, \end{aligned} \tag{17}$$

$$\varepsilon^1: \qquad \begin{aligned} \frac{\partial \xi_2}{\partial T_0} &= -\frac{1}{2} x_0 \left(x_0^2 - 1 \right) \eta_2 - \frac{1}{2} \left(3x_0^2 - 1 \right) \xi_1 \eta_1 - x_0 C_2, \\ \frac{\partial \eta_2}{\partial T_0} &= (x_0 - 1)\xi_2 + \frac{1}{2}\xi_1^2, \end{aligned} \tag{18}$$

$$\varepsilon^2: \qquad \begin{aligned} \frac{\partial \xi_1}{\partial T_2} + \frac{\partial \xi_3}{\partial T_0} &= -\frac{1}{2} x_0 \left(x_0^2 - 1 \right) \eta_3 - \frac{1}{2} \left(3x_0^2 - 1 \right) \left(\eta_1 \xi_2 + \eta_2 \xi_1 \right) - \\ & \quad \frac{3}{2} x_0 \eta_1 \xi_1^2 - C_2 \xi_1 - x_0 D_3 \cos \omega T_0, \\ \frac{\partial \eta_1}{\partial T_2} + \frac{\partial \eta_3}{\partial T_0} &= (x_0 - 1)\xi_3 + \xi_1 \xi_2. \end{aligned} \tag{19}$$

System (17) can be rewritten into form

$$\frac{\partial^2 \xi_1}{\partial T_0^2} + \Omega^2 \xi_1 = 0,$$

$$\eta_1 = -\frac{2}{x_0 \left(x_0^2 - 1 \right)} \frac{\partial \xi_1}{\partial T_0}, \tag{20}$$

where

$$\Omega^2 = \frac{1}{2} x_0 \left(x_0^2 - 1 \right) (x_0 - 1). \tag{21}$$

Solution of Eq. (20) can be written as

$$\begin{aligned} \xi_1 &= A(T_2)e^{i\Omega T_0} + \bar{A}(T_2)e^{-i\Omega T_0}, \\ \eta_1 &= -\Phi i \left(Ae^{i\Omega T_0} - \bar{A}e^{-i\Omega T_0} \right), \end{aligned} \tag{22}$$

where

$$\Phi = \frac{2\Omega}{x_0 (x_0^2 - 1)}. \tag{23}$$

Now we transform the system (18) into the form

$$\frac{\partial^2 \xi_2}{\partial T_0^2} + \Omega^2 \xi_2 = -\frac{1}{4} x_0 (x_0^2 - 1) \xi_1^2 - \frac{1}{2} (3x_0^2 - 1) \frac{\partial \xi_1}{\partial T_O} \eta_1 - \frac{1}{2} (3x_0^2 - 1) \xi_1 \frac{\partial \eta_1}{\partial T_O},$$

$$\eta_2 = -\frac{2}{x_0 (x_0^2 - 1)} \left[\frac{\partial \xi_2}{\partial T_O} + \frac{1}{2} (3x_0^2 - 1) \xi_1 \eta_1 + x_0 C_2 \right]. \tag{24}$$

After inserting from (22) we obtain the equation

$$\frac{\partial^2 \xi_2}{\partial T_0^2} + \Omega^2 \xi_2 = -\frac{1}{2} x_0 (x_0^2 - 1) A\bar{A} - Q \left[A^2 e^{i2\Omega T_0} + \bar{A}^2 e^{-i2\Omega T_0} \right], \tag{25}$$

where

$$Q = \frac{1}{4} x_0 (x_0^2 - 1) + (3x_0^2 - 1) \Phi\Omega \tag{26}$$

is constant. Particular solution of this equation with the frequency equal to 2Ω is

$$\xi_2 = -\frac{x_0}{x_0 - 1} A\bar{A} + \frac{Q}{3\Omega^2} \left(A^2 e^{i2\Omega T_0} + \bar{A}^2 e^{-i2\Omega T_0} \right),$$

$$\eta_2 = -\frac{2}{x_0 (x_0^2 - 1)} \left[i\frac{2}{3} \frac{Q}{\Omega} \left(A^2 e^{i2\Omega T_0} - \bar{A}^2 e^{-i2\Omega T_0} \right) - \right.$$

$$\left. i\frac{1}{2} (3x_0^2 - 1) \Phi \left(A^2 e^{i2\Omega T_0} - \bar{A}^2 e^{-i2\Omega T_0} \right) + x_0 C_2 \right]. \tag{27}$$

The last step is to transform the system of first order ODEs (19) into the second order ODE

$$\frac{\partial^2 \xi_3}{\partial T_0^2} + \Omega^2 \xi_3 = -\frac{1}{2} x_0 (x_0^2 - 1) \xi_1 \xi_2 + \frac{1}{2} x_0 (x_0^2 - 1) \frac{\partial \eta_1}{\partial T_2} - \frac{\partial}{\partial T_0} \left(\frac{\partial \xi_1}{\partial T_2} \right) -$$

$$\frac{1}{2} (3x_0^2 - 1) \left(\frac{\partial \eta_1}{\partial T_0} \xi_2 + \eta_1 \frac{\partial \xi_2}{\partial T_0} + \frac{\partial \eta_2}{\partial T_0} \xi_1 + \eta_2 \frac{\partial \xi_1}{\partial T_0} \right) -$$

$$\frac{3}{2} x_0 \frac{\partial \eta_1}{\partial T_0} \xi_1^2 - 3x_0 \eta_1 \xi_1 \frac{\partial \xi_1}{\partial T_0} - C_2 \frac{\partial \xi_1}{\partial T_0} - x_0 D_3 \frac{d}{dT_0} (\cos \omega T_0). \tag{28}$$

Now we put the previous solutions into this equation. The result is

$$\frac{\partial^2 \xi_3}{\partial T_0^2} + \Omega^2 \xi_3 = -\frac{1}{2} x_0 (x_0^2 - 1) \left(Ae^{i\Omega T_0} + \bar{A}e^{-i\Omega T_0} \right) \cdot$$

$$\left[-\frac{x_0}{x_0 - 1} A\bar{A} + \frac{Q}{3\Omega^2} (A^2 e^{i2\Omega T_0} + \bar{A}^2 e^{-i2\Omega T_0}) \right] -$$

$$\frac{1}{2} x_0 (x_0^2 - 1) \Phi i \left(A' e^{i\Omega T_0} - \bar{A}' e^{-i\Omega T_0} \right) - i\Omega \left(A' e^{i\Omega T_0} - \bar{A}' e^{-i\Omega T_0} \right) -$$

$$\frac{1}{2} (3x_0^2 - 1) \left\{ \begin{array}{l} \Phi\Omega \left(Ae^{i\Omega T_0} + \bar{A}e^{-i\Omega T_0} \right) \left[-\frac{x_0}{x_0-1} A\bar{A} + \frac{Q}{3\Omega^2} (A^2 e^{i2\Omega T_0} + \bar{A}^2 e^{-i2\Omega T_0}) \right] + \\ \Phi \left(Ae^{i\Omega T_0} - \bar{A}e^{-i\Omega T_0} \right) \frac{2Q}{3\Omega} (A^2 e^{i2\Omega T_0} - \bar{A}^2 e^{-i2\Omega T_0}) + ...NST.. - \\ \frac{2}{x_0(x_0^2-1)} (NST + x_0 C_2) i\Omega \left(Ae^{i\Omega T_0} - \bar{A}e^{-i\Omega T_0} \right) \end{array} \right\} -$$

$$..NST.. - iC_2\Omega \left(Ae^{i\Omega T_0} - \bar{A}e^{-i\Omega T_0} \right) - x_0 D_3 \frac{d}{dT_0} (\cos \omega T_0), \tag{29}$$

where NST means "non-secular terms" and A' designates derivative of A according to T_2.

Further, we will distinguish two cases:

1. Autonomous case – $D_3 = 0$,

2. Non-autonomous case corresponding to soft resonant stimulation – $D_3 \neq 0$;
$\omega = \Omega + \varepsilon^2 \nu_2$.

4.1. Autonomous case – $D_3 = 0$

The conditions for secular terms (periodical with the frequency Ω and therefore leading to the resonance) from (29) is

$$
e^{i\Omega T_0} : \quad
\begin{bmatrix}
\frac{1}{2}\frac{x_0^2(x_0^2-1)}{x_0-1} A^2 \bar{A} - \frac{1}{2}x_0 (x_0^2 - 1) \Phi i A' - \\[4pt]
i\Omega A' + \frac{1}{2} (3x_0^2 - 1) \frac{x_0}{x_0-1} \Phi \Omega A^2 \bar{A} + \\[4pt]
\frac{3x_0^2-1}{(x_0^2-1)} \Omega i C_2 A - i C_2 A \Omega
\end{bmatrix}
= 0,
\tag{30}
$$

$$
e^{-i\Omega T_0} : \quad
\begin{bmatrix}
\frac{1}{2}\frac{x_0^2(x_0^2-1)}{x_0-1} \bar{A}^2 A + \frac{1}{2}x_0 (x_0^2 - 1) \Phi i \bar{A}' + \\[4pt]
i\Omega \bar{A}' + \frac{1}{2} (3x_0^2 - 1) \frac{x_0}{x_0-1} \Phi \Omega \bar{A}^2 A - \\[4pt]
\frac{3x_0^2-1}{(x_0^2-1)} \Omega i C_2 \bar{A} - i C_2 \bar{A} \Omega
\end{bmatrix}
= 0.
\tag{31}
$$

It can be shown easily that both conditions are identical and in the next we will work with the first one. This condition can be simplified into the form

$$
\alpha A^2 \bar{A} - A'\beta i - A\gamma i = 0,
\tag{32}
$$

where

$$
\alpha = \frac{x_0}{2}\left(4x_0^2 + x_0 - 1\right), \qquad \beta = 2\Omega, \qquad \gamma = -C_2\Omega\frac{2x_0^2}{x_0^2 - 1},
\tag{33}
$$

where α, β and γ are constants. Now we insert

$$
A = a e^{i\phi}, \qquad \bar{A} = a e^{-i\phi}
\tag{34}
$$

and comparing the real and imaginary parts after reducing the equation by $e^{-i\phi}$ and $e^{i\phi}$ (they have unit absolute value) we obtain the following equations

$$
\begin{aligned}
\text{Re}: &\quad \phi'\beta + \alpha a^2 = 0 \Rightarrow \phi' = -\frac{\alpha}{\beta}a^2, \\
\text{Im}: &\quad a'\beta + \gamma a = 0 \Rightarrow a' = -\frac{\gamma}{\beta}a.
\end{aligned}
\tag{35}
$$

Solution for a is

$$
a = \text{const.}\, e^{-\frac{\gamma}{\beta}T_2}.
\tag{36}
$$

Only for $\gamma = 0 \Rightarrow C_2 = 0$ it exists $a = $ const. corresponding to the periodical motion. For $C_2 \neq 0$ depending on its sign we can observe either convergence or divergence of the solution. This result fully corresponds with previous analysis [5] for $C = 0$ where the Hopf's degenerated bifurcation exists. On Figs. 3, 4 and 5 the results of the numerical experiments are shown. The limit cycle for $C = 0$ is presented on Fig. 3 and the convergence to the point with the coordinate $x_0 = 1 + \Theta$ for $C < 0$ and divergence from this point and convergence to the point with the coordinate $x_0 = 1 + \Theta$ for $C > 0$ on Figs. 4 and 5.

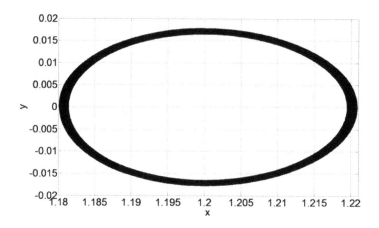

Fig. 3. Phase portrait for $r = 0.02$, $x_0 = 1 + \Theta$, $y_0 = 0.0177$, $C = 0$, $D = 0$

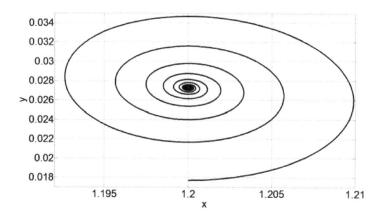

Fig. 4. Phase portrait for $r = 0.02$, $x_0 = 1 + \Theta$, $y_0 = 0.0177$, $C = -0.006$, $D = 0$

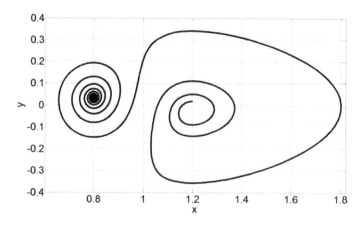

Fig. 5. Phase portrait for $r = 0.02$, $x_0 = 1 + \Theta$, $y_0 = 0.0177$, $C = 0.006$, $D = 0$

4.2. *Non-autonomous case – soft resonant stimulation – $D_3 \neq 0$; $\omega = \Omega + \varepsilon^2 \nu_2$*

Here, after setting we obtain

$$\omega T_0 = \Omega T_0 + \varepsilon^2 \nu_2 T_0 = \Omega T_0 + \nu_2 T_2. \tag{37}$$

In this case the Eq. (29) will have the form

$$\frac{\partial^2 \xi_3}{\partial T_0^2} + \Omega^2 \xi_3 = -\frac{1}{2} x_0 \left(x_0^2 - 1 \right) \cdot$$

$$\left(A e^{i\Omega T_0} + \bar{A} e^{-i\Omega T_0} \right) \left[-\frac{x_0}{x_0 - 1} A\bar{A} + \frac{Q}{3\Omega^2} \left(A^2 e^{i2\Omega T_0} + \bar{A}^2 e^{-i2\Omega T_0} \right) \right] -$$

$$\frac{1}{2} x_0 \left(x_0^2 - 1 \right) \Phi i \left(A' e^{i\Omega T_0} - \bar{A}' e^{-i\Omega T_0} \right) - i\Omega \left(A' e^{i\Omega T_0} - \bar{A}' e^{-i\Omega T_0} \right) -$$

$$\frac{1}{2} \left(3x_0^2 - 1 \right) \left\{ \begin{array}{l} \Phi\Omega \left(A e^{i\Omega T_0} + \bar{A} e^{-i\Omega T_0} \right) \left[-\frac{x_0}{x_0-1} A\bar{A} + \frac{Q}{3\Omega^2} \left(A^2 e^{i2\Omega T_0} + \bar{A}^2 e^{-i2\Omega T_0} \right) \right] + \\ \Phi \left(A e^{i\Omega T_0} - \bar{A} e^{-i\Omega T_0} \right) \frac{2Q}{3\Omega} \left(A^2 e^{i2\Omega T_0} - \bar{A}^2 e^{-i2\Omega T_0} \right) + ...NST... - \\ \frac{2}{x_0 \left(x_0^2 - 1 \right)} \left(NST + x_0 C_2 \right) i\Omega \left(A e^{i\Omega T_0} - \bar{A} e^{-i\Omega T_0} \right) \end{array} \right\} -$$

$$...NST... - iC_2\Omega \left(A e^{i\Omega T_0} - \bar{A} e^{-i\Omega T_0} \right) - \frac{1}{2} x_0 D_3 \Omega i \left(e^{i(\Omega T_0 + \nu_2 T_2)} - e^{-i(\Omega T_0 + \nu_2 T_2)} \right). \tag{38}$$

The condition for the secular terms (30) will have the form

$$e^{i\Omega T_0} : \quad \begin{bmatrix} \frac{1}{2} \frac{x_0^2 \left(x_0^2 - 1 \right)}{x_0 - 1} A^2 \bar{A} - \frac{1}{2} x_0 \left(x_0^2 - 1 \right) \Phi i A' - \\ i\Omega A' + \frac{1}{2} \left(3x_0^2 - 1 \right) \frac{x_0}{x_0 - 1} \Phi\Omega A^2 \bar{A} + \\ \frac{3x_0^2 - 1}{\left(x_0^2 - 1 \right)} \Omega i C_2 A - iC_2 A\Omega - \frac{1}{2} i x_0 D_3 \Omega e^{i\nu_2 T_2} \end{bmatrix} = 0 \tag{39}$$

and further

$$\alpha A^2 \bar{A} - A' \beta i - A\gamma i - i\delta e^{i\nu_2 T_2} = 0, \tag{40}$$

where

$$\delta = \frac{1}{2} x_0 D_3 \Omega. \tag{41}$$

After inserting from (34) to (40) we obtain

$$\begin{array}{ll} \text{Re :} & a\phi' \beta + \alpha a^3 + \delta \sin(\nu_2 T_2 - \phi) = 0, \\ \text{Im :} & a'\beta + \gamma a + \delta \cos(\nu_2 T_2 - \phi) = 0. \end{array} \tag{42}$$

To obtain the autonomous system we provide the substitution

$$\nu_2 T_2 - \phi = \psi(T_2). \tag{43}$$

And then

$$\begin{array}{l} \psi' = \nu_2 + \frac{\alpha}{\beta} a^2 + \frac{\delta}{a\beta} \sin \psi, \\ a' = -\frac{\gamma}{\beta} a - \frac{\delta}{\beta} \cos \psi. \end{array} \tag{44}$$

Let the stationary solution of this dynamical system be $\tilde{a}, \tilde{\psi}$. If we eliminate $\tilde{\psi}$ from the corresponding equations, we obtain the following frequency response equation

$$\tilde{a}^2 \left[\left(\beta\nu_2 + \alpha\tilde{a}^2 \right)^2 + \gamma^2 \right] = \delta^2. \tag{45}$$

Equations in variations of (44) are

$$\eta_a' = -\frac{\gamma}{\beta} \eta_a - \tilde{a} \left(\nu_2 + \frac{\alpha}{\beta} \tilde{a}^2 \right) \eta_\psi,$$

$$\eta_\psi' = \left[2\frac{\alpha}{\beta} \tilde{a} + \frac{1}{\tilde{a}} \left(\nu_2 + \frac{\alpha}{\beta} \tilde{a}^2 \right) \right] \eta_a - \frac{\gamma}{\beta} \eta_\psi. \tag{46}$$

The eigenvalues are

$$\lambda_{1,2} = -\frac{\gamma}{\beta} \pm \sqrt{\left(\frac{\gamma}{\beta}\right)^2 - \left(\nu_2 + \frac{\alpha}{\beta}\tilde{a}^2\right)\left(\nu_2 + 3\frac{\alpha}{\beta}\tilde{a}^2\right)}. \tag{47}$$

For the stability reason it is necessary that the square root is either zero or pure imaginary one

$$-\left(\nu_2 + \frac{\alpha}{\beta}\tilde{a}^2\right)\left(\nu_2 + 3\frac{\alpha}{\beta}\tilde{a}^2\right) \le 0. \tag{48}$$

If this condition is fulfilled, the system is asymptotically stable for $\frac{-\gamma}{\beta} < 0$. Putting from (33) into this condition, we obtain the form

$$\frac{1}{2}C_2\frac{x_0^2}{x_0^2 - 1} < 0. \tag{49}$$

For $x_0 = 1 + \Theta$, the system is stable if $C_2 < 0$ and for $x_0 = 1 - \Theta$ it is stable if $C_2 > 0$. The first approximation of the solution is

$$x = x_0 + \varepsilon 2\tilde{a}\cos\Omega t,$$
$$y = \varepsilon\frac{4\Omega}{x_0\left(x_0^2 - 1\right)}\tilde{a}\sin\Omega t. \tag{50}$$

The further approximations contain the multiple of Ω and the frequency of stimulation ω.

Again, now we will show some numerical examples. On Figs. 6 and 7, there is shown the solution of (46) together with the area of instability according (49) for different values of D.

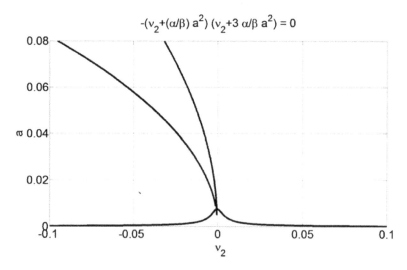

Fig. 6. Amplitude vs. frequency for $r = 0.02$, $x_0 = 1 + \Theta$, $y_0 = 0.017\,7$, $C = -0.004$, $D = 0.000\,06$

In Fig. 8, the solution of the original dynamical system for the corresponding data is stated. The correspondence for chosen small values of C and D is very good. The cause is the approximation (13).

Poincare mappings on Figs. 9 and 11 correspond to the quasi-periodical motion with two frequencies Ω and $\omega = \Omega + \varepsilon^2\nu_2$. These frequencies would occur in the second and further approximations that have not been solved in this contribution.

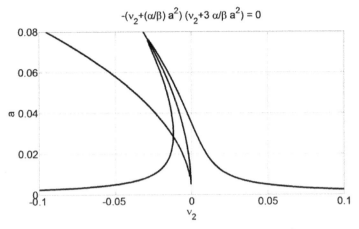

$$-(v_2+(\alpha/\beta)\, a^2)\, (v_2+3\,\alpha/\beta\, a^2) = 0$$

Fig. 7. Area of instability for $r = 0.02$, $x_0 = 1 + \Theta$, $y_0 = 0.0177$, $C = -0.004$, $D = 0.0006$

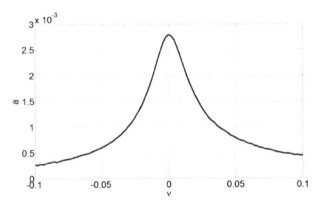

Fig. 8. Amplitude vs. frequency for $r = 0.02$, $x_0 = 1 + \Theta$, $y_0 = 0.0177$, $C = -0.004$, $D = 0.00006$

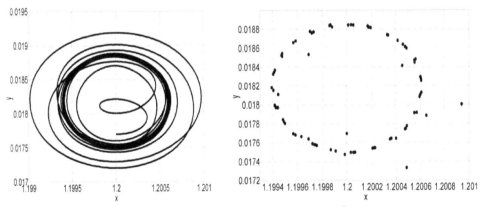

Fig. 9. Phase portrait and Poincare mapping with the period $T = \frac{2\pi}{\Omega}$ for $r = 0.02$, $x_0 = 1 + \Theta$, $y_0 = 0.0177$, $C = -0.004$, $D = 0.00006$, $v_2 = -0.05$

5. Conclusion

The paper shows the possibility of MSM usage for the analysis of non-autonomous dynamical system. This system is transformed into the autonomous one and the basic dynamical properties are examined using the common method. From given examples, it is obvious that the crucial point is the choice of the approximation (13) which influences the range where the approximated

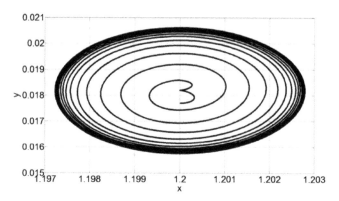

Fig. 10. Phase portrait for $r = 0.02$, $x_0 = 1 + \Theta$, $y_0 = 0.0177$, $C = -0.004$, $D = 0.00006$, $\nu_2 = 0$

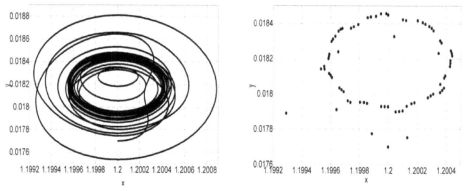

Fig. 11. Phase portrait and Poincare mapping with the period $T = \frac{2\pi}{\Omega}$ for $r = 0.02$, $x_0 = 1 + \Theta$, $y_0 = 0.0177$, $C = -0.004$, $D = 0.00006$, $\nu_2 = 0.1$

solution is valid. That is the reason why the transition to chaos slightly evident on Figs. 9 and 11 is not seen from the MSM analysis. The future analysis of this dependence effects seems to be necessary using the second order approximation.

Acknowledgements

This work was supported by the research project MSM 4977751303 of the Ministry of Education, Youth and Sports of the Czech Republic.

References

[1] DiCarlo, S., Quiligotti, Growth and balance. Mechanics Research Communications 29, Pergamon Press 2002, 449–456.

[2] Herzog, W., Force enhancement following stretch of activated muscle: Critical review and proposal for mechanisms, Medical & Biological Engineering & Computing 43 (2005) 173–180.

[3] Nayfeh, A. H., Perturbation Methods. John Wiley, New York, 1973.

[4] Nayfeh, A. H., Balachandran, B., Applied Nonlinear Dynamics, John Wiley, New York, 1995.

[5] Rosenberg, J. Hynčík, L., Dynamical properties of the growing continuum using multiple-scale method. Applied and Computational Mechanics 2 (2) (2008) 357–368.

[6] Rosenberg, J., Svobodová, M., Comments on the thermodynamical background to the growth and remodelling theory applied to a model of muscle fibre contraction. Applied and Computational Mechanics 4 (1) (2010) 101–112.

3

Determination of Lamb wave dispersion curves by means of Fourier transform

P. Hora[a,*], O. Červená[a]

[a] *Institute of Thermomechanics of the ASCR, v.v.i., Veleslavínova 11, 301 14 Plzeň, Czech Republic*

Abstract

This work reports on methods for determination of Lamb wave dispersion curves by means of Fourier transform (FT). Propagating Lamb waves are sinusoidal in both the frequency domain and the spatial domain. Therefore, the temporal FT may be carried out to go from the time to the frequency domain, and then the spatial FT may be carried out to go to the *frequency–wave-number* domain, where the amplitudes and the wave-numbers of individual modes may be measured. The result of this transform will be a 2D array of amplitudes at discrete frequencies and wave-numbers. Other method of determination of dispersion curves is based on the solution of FEM task in frequency domain. The frequency spectrum of a time-dependent excitation is defined using FT. Since a small number of frequencies are sufficient to achieve a correct representation of a wide variety of temporal excitations, this approach considerably speeds up the computation by avoiding the temporal FT and by decreasing the number of calculation steps.

Keywords: Lamb wave, dispersion curves, Fourier transform, FEM

1. Introduction

The application of the conventional ultrasonic methods, such as pulse-echo, has been limited to testing relatively simple geometries or interrogating the region in the immediate vicinity of the transducer. A new ultrasonic methodology uses guided waves to examine structural components [10]. The advantages of this technique include: its ability to test the entire structure in a single measurement; and its capability to test inaccessible regions of complex components.

The propagation of guided waves in a complex structure is a complicated process that is difficult to understand and interpret. The current research develops the mechanics fundamentals that model this propagation.

One approach to the modeling of guided wave propagation phenomena is to solve the governing differential equations of motion and their associated boundary conditions analytically. This procedure has been already done for simple geometries and perfect specimens without defects (see [7] and [11]). However, these equations become intractable for more complicated geometries or for non-perfect specimens.

A numerical solution is another approach to this problem. A number of different numerical computational techniques can be used for the analysis of wave propagation. These include finite difference equations, finite element methods (FEM), finite strip elements [4], boundary element methods, global matrix approaches [9], spectral element methods [8], mass-spring lattice models [13] and the local interaction simulation approach [6]. The primary advantage of

---

*Corresponding author. e-mail: hora@cdm.it.cas.cz.

the FEM is that there are numerous available commercial FEM codes, which eliminates any need to develop actual code.

Comparison between dispersion curves of a perfect structure and dispersion curves of a defect structure is main feature of the guided wave method. This work reports on methods for determination of Lamb wave dispersion curves by means of Fourier transform.

2. Theory

Lamb waves are two-dimensional vibrations propagating in plates with free boundary conditions. Their displacements may be symmetric or antisymmetric with respect to the middle plane of the plate. Sometimes they are referred to as free or normal modes, because they are the eigensolutions of characteristic equations. The velocities of all Lamb waves are dispersive. The frequency equation for the propagation of symmetric waves is

$$\frac{\tan \beta h}{\tan \alpha h} = -\frac{4\alpha\beta k^2}{(k^2 - \beta^2)^2} \tag{1}$$

and the frequency equation for the propagation of antisymmetric waves is

$$\frac{\tan \beta h}{\tan \alpha h} = -\frac{(k^2 - \beta^2)^2}{4\alpha\beta k^2}, \tag{2}$$

where $\alpha^2 = \omega^2/c_1^2 - k^2$, $\beta^2 = \omega^2/c_2^2 - k^2$, k is wavenumber, ω is angular frequency, c_1, c_2 are phase velocities (subscripts refer to wave mode: 1-longitudinal, 2-shear) and h is half-thickness of the plate [7]. Dispersion curves for symmetric (1) and antisymmetric (2) modes for phase velocities are shown in Fig. 1 – left and for group velocities you can see in Fig. 1 – right.

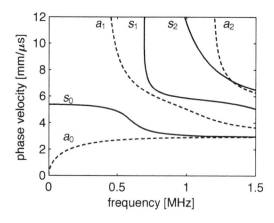 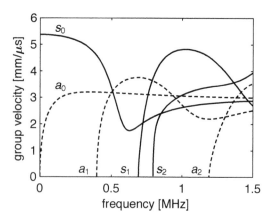

Fig. 1. Velocity dispersion curves for steel, where $c_1 = 5\,950$ m/s and $c_2 = 3\,180$ m/s; symmetric modes $s_0 \ldots s_2$ (solid lines), antisymmetric modes $a_0 \ldots a_2$ (dashed lines)

The displacements corresponding to the symmetric modes are given by

$$u_x = A\,i\beta \left(\frac{\cos \beta y}{\cos \beta h} - \frac{2k^2 \cos \alpha y}{(k^2 - \beta^2) \cos \alpha h} \right) e^{i(kx - \omega t)},$$

$$u_y = A\,k \left(\frac{\sin \beta y}{\cos \beta h} + \frac{2\alpha\beta \sin \alpha y}{(k^2 - \beta^2) \cos \alpha h} \right) e^{i(kx - \omega t)} \tag{3}$$

and the displacements corresponding to the antisymmetric modes are

$$
u_x = A\,i\beta \left(\frac{\sin \beta y}{\sin \beta h} - \frac{2k^2 \sin \alpha y}{(k^2 - \beta^2) \sin \alpha h} \right) e^{i(kx - \omega t)},
$$

$$
u_y = A\,k \left(\frac{\cos \beta y}{\sin \beta h} + \frac{2\alpha\beta \cos \alpha y}{(k^2 - \beta^2) \sin \alpha h} \right) e^{i(kx - \omega t)}, \tag{4}
$$

where x, y are spatial coordinates, t is time, i is imaginary unit and A is constant [7].

Since the propagating Lamb waves are sinusoidal in both the frequency domain and the spatial domain, the temporal FT may be carried out to go from the time to the frequency domain, and then the spatial FT may be carried out to go to the *frequency–wave-number* domain, where the amplitudes and the wave-numbers of individual modes may be measured.

Applying the spatial Fourier methods in practice to data gained experimentally or numerically requires us to carry out a two-dimensional FT giving

$$
H(k, f) = \int_{-\infty}^{+\infty} \int_{-\infty}^{+\infty} u(x, t) e^{-i(kx + \omega t)} \, dx \, dt. \tag{5}
$$

The discrete 2D FT may be defined similarly to the 1D FT. The result of this transform will be a 2D array of amplitudes at discrete frequencies and wave-numbers. As in the one-dimensional case, aliasing must be avoided by sampling the data at a sufficiently high frequency both in time and in wave number space. Usually, the signal will be not periodic within the temporal and spatial sampling windows and leakage will occur. Window functions such as the Hanning window may be used to reduce this leakage, and zeros may be padded to the end of the signal to enable the frequency and wave number of the maximum amplitude to be determined more accurately.

The algorithm of time analysis is following [1]:

1. Create the array (in column order) from experimentally or numerically gained the time histories of the waves received at a series of equally spaced positions along the propagation path.
2. Carry out the temporal FT of each column to obtain a frequency spectrum for each position. At this stage, an array with the spectral information for each position in its respective column is obtained.
3. Carry out the spatial FT of each row formed by the components at a given frequency to obtain the *frequency–wave-number* information.

In the case of frequency analysis the frequency spectra in particular places on the surface of the plate are available. Therefore it is sufficient to carry out only the spatial FT

$$
H(k, f) = \int_{-\infty}^{+\infty} u(x, f) e^{-ikx} \, dx \tag{6}
$$

for obtaining *frequency–wave-number* information.

Then the algorithm is following:

1. Create the array (in column order) from gained the complex displacement for each frequency received at a series of equally spaced positions along the propagation path.
2. Carry out the spatial FT of each row to obtain the *frequency–wave-number* diagram.

3. Numerical simulations

FEM calculations are performed in the commercial environment COMSOL Multiphysics with the Structural Mechanics Module [5]. The plane strain is used as an application mode. The analyzed steel plate is $2h = 4\,\text{mm}$ thick and $L = 120\,\text{mm}$ long. Young's modulus $E = 205\,\text{GPa}$, Poisson's ratio $\nu = 0.30$ and density $\rho = 7\,800\,\text{kg/m}^3$. The plate geometry and the used coordinate system are shown in Fig. 2. The mapped squared mesh with size of elements $0.5 \times 0.5\,\text{mm}$ is created. Elements are of the Lagrange–Quadratic type. The exciting constraint is set at the left edge of this plate. The others edges are free.

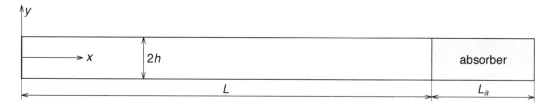

Fig. 2. The plate geometry and the used coordinate system

Two kinds of problems are realized: firstly, the time dependent analysis, and secondly, the frequency response analysis. The time stepping parameters for time dependent analysis are: times (from 0 to 100 with step $0.01\,\mu\text{s}$), relative tolerance (10^{-5}) and absolute tolerance (10^{-10}). For frequency analysis, $L_a = 20\,\text{mm}$ long absorbing part is appended to the main plate [3]. The Young's modulus of this absorbing part is complex, given by function $E_A = E\,(1 + i\,1\,500(x - L)^2)$, where i is imaginary unit and x is the position in x direction.

Exciting pulse

An individual Lamb wave is temporary launched from one end of the plate sinusoidal tone burst, modulated temporary by a Hanning window function to limit bandwidth. This time pulse is given by relation

$$f(t) = \frac{1}{2}\left(1 - \cos(\omega_0 t/N_c)\right)\sin(\omega_0 t), \qquad t \in \langle 0, N_c/f_0\rangle, \tag{7}$$

where f_0 is center frequency, ω_0 is center angular frequency, N_c is the number of cycles and t is time. The spectrum of pulse envelope is

$$\int_0^{N_c/f_0} \left(1 - \cos(\omega_0 t/N_c)\right)\,e^{-i\omega t}\,\mathrm{d}t = \frac{4i\pi^2 f_0^2\left(e^{i\omega\frac{N_c}{f_0}} - 1\right)e^{-i\omega\frac{N_c}{f_0}}}{\omega\left(\omega^2 N_c^2 - 4\pi^2 f_0^2\right)}. \tag{8}$$

The numerical tests are done for the number of cycles 3, 5 and 7 and for the center frequencies from $0.1\,\text{MHz}$ to $1\,\text{MHz}$ with step $0.1\,\text{MHz}$. Presented results are for $N_c = 5$ and $f_0 = 0.5\,\text{MHz}$. Fig. 3 shows the time history of such pulse and the amplitude spectrum of its envelope.

The amplitude of u_x and u_y displacement of exciting pulse at each node at the plate edge are given by the through thickness deflected shape calculated from (3) or (4) for center frequency. The examples of the u_x and u_y displacements for the first antisymmetric mode (a_0) and the first symmetric mode (s_0) and frequencies $0.31, 0.5$ and $0.69\,\text{MHz}$ are shown in Fig. 4.

Note that in this way of exciting the modes are pure only at the center frequency of the imposed excitation signal. This is to be expected as the imposed displacement deflected shapes

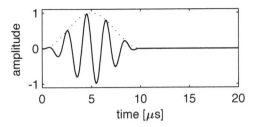
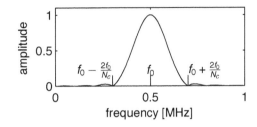

Fig. 3. Time history of pulse used in numerical tests (left) and amplitude spectrum of its envelope (right)

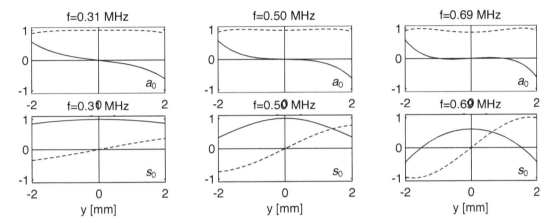

Fig. 4. The deflected mode shapes of Lamb waves a_0 and s_0 for the lowest, central and the highest frequency; u_x displacement (solid line), u_y displacement (dashed line)

shown in Fig. 4 – center were correct only at 0.5 MHz and as may be deduced from (3) and (4) the mode shapes are slightly different either side of this frequency. Hence, the frequency dependence of the mode shapes means that excitation signals of the form shown in Fig. 3 – left will generally excite more than one mode away from their center frequencies. Pure a_0 and s_0 modes may be excited at a frequency below the frequency of the first non-zero propagating modes a_1 and s_1 respectively if the input signal is symmetric in the x direction and antisymmetric in the y direction for s_0 and vice versa for a_0.

In order to excite a pure Lamb wave two conditions have to be simultaneously satisfied [2]. Firstly, the frequency of the harmonic excitation signal must be identical to the Lamb wave frequency being excited and secondly, the variation of the excitation with y at the excitation position ($x = 0$ in the test reported here) must correspond to the exact mode shape of the Lamb wave being excited. Assuming a single frequency input, the required excitation $\boldsymbol{f}(y, t)$ is of the form

$$\boldsymbol{f}(y, t) = \boldsymbol{\Phi}(y)\, e^{i\omega t} \tag{9}$$

and $\boldsymbol{\Phi}(y) = [u_x, u_y]^\top$ may be calculated from (3) and (4).

However, in almost all modeling applications single frequency excitation is not desirable or possible, for example in explicit time marching FE methods the duration of the input signal has to be finite. It is therefore very advantageous to be able to excite single modes with a wideband excitation signal $f(t)$. This can be achieved by summing the required inputs over a range of frequencies. For single frequency component ω_j, the required input is

$$\boldsymbol{f}_j(y, t) = \boldsymbol{\Phi}_j(y) A(\omega_j)\, e^{i\omega_j t}. \tag{10}$$

If all the significant energy components in the excitation signal are over a range of frequencies from $j = 1$ to k then

$$f(y,t) = \sum_{j=1}^{k} \boldsymbol{\Phi}_j(y) A(\omega_j) \, e^{i\omega_j t}. \tag{11}$$

Here, $A(\omega)$ is the complex amplitude of the Fourier transform of the excitation signal $f(t)$, where $f(t)$ is a tone burst modified by Hanning window in the tests reported here. This method of exciting Lamb waves in the numerical investigation is hence adopted in all the temporal FE simulations that follow.

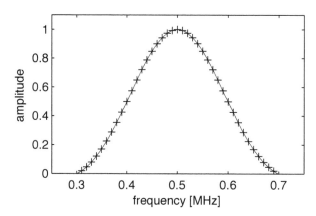

Fig. 5. Frequency spectrum

The frequency spectrum of this temporal waveform is quite narrow, see Fig. 5. The crosses in this figure indicate 39 discrete frequency components used for the FE analysis in frequency domain. The u_x and u_y displacement of excitation at each node at the left edge of the plate were given by the through thickness deflected shape calculated from (3) or (4) for given frequency. Then the displacements were multiplied by the corresponding amplitude of frequency spectrum.

Tests of the velocity predictions

A set of simple tests was carried out to velocity predictions obtained using COMSOL Multiphysics. The tests were proceeded on an extremely simple system. This was $2h = 0.5\,\text{mm}$ thick and $L = 300\,\text{mm}$ long steel plate, see Fig. 2. The phase velocities for steel were: $c_1 = 5\,950\,\text{m/s}$ and $c_2 = 3\,180\,\text{m/s}$.

All surface nodes were pinned in the y direction. The plate was therefore effectively infinite in the y and z directions, so simple plane wave propagation could be expected. A 0.5 mm square mesh was used. The plate was excited by applying the signal shown in Fig. 3 – left in the x direction to all the nodes at $x = 0$. This excited the bulk longitudinal wave. Fig. 6 – left shows the response of the top surface of the plate in the x direction at distance $x = 100\,\text{mm}$. The shape of the waveform is the same as the input shown in the Fig. 3 – left indicating that the numerical scheme has not introduced dispersion. Using the time flight between the two monitored points the velocity was calculated as $5\,917.2\,\text{m/s}$. The model was then modified. The surface nodes were pinned in the x direction now. This allowed a pure shear wave to propagate along the plate. A y direction displacement of the form shown in Fig. 3 – left was applied to each of the nodes at $x = 0$ and the resulting predicted waveform at $x = 100\,\text{mm}$ is shown in Fig. 6 – right. Again, no dispersion is evident and using the time flight between the two monitored points the

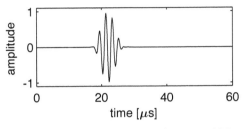 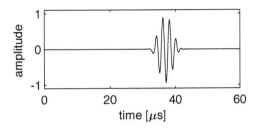

Fig. 6. Predicted time history at time $x = 100\,\mathrm{mm}$, when the input at $x = 0$ was excited: the longitudinal wave c_1 only (left), the shear wave c_2 only (right)

velocity was calculated as $3\,159.6\,\mathrm{m/s}$. In all the above cases, the FE predictions are excellent and the velocities differed from the theoretically predicted ones by less than 1 % confirming the applicability of the finite element program COMSOL Multiphysics.

Results from temporal analysis

As mentioned above, a wideband excitation signal $f(y, t)$ given by (11) was applied for center frequency $f_0 = 0.5\,\mathrm{MHz}$. Amplitude of vertical displacements in dependence on space and time is illustrated by pseudocolor plot in Fig. 7. The case when the left edge of the plate was loaded by displacements equivalent to the first antisymmetric mode (a_0) (see Fig. 4 – central above) is shown at the left side and the case when the left edge of the plate was loaded by displacements equivalent to the first symmetric mode (s_0) (see Fig. 4 – central bottom) is shown on the right side. These results are obtained by FEM calculations. The Fig. 8 shows the pseudocolor plot of *frequency–wave-number* spectrum, which was computed by 2D FT in MATLAB [12]. The dispersion curves obtained analytically are plotted by solid lines (symmetric mode) and dashed lines (antisymmetric mode). Negative values of wave-number correspond to the waves propagating from left to right edge of the plate and positive values to the waves propagating in opposite direction. Note that two antisymmetric modes (a_0, a_1) exist at given frequency (0.5 MHz).

Results from frequency analysis

Other method of dispersion curves determination is based on the solution of FEM task in frequency domain. The frequency spectrum of a time-dependent excitation was defined using FT. Then, the structure response (complex displacements and stresses at any locations) was calculated at each frequency of this load, i.e. the FE-code supplies stationary solution for each frequency component of the temporal excitation. The Fig. 9 shows the pseudocolor plot of *frequency–wave-number* spectrum, which was computed by 1D FT in MATLAB. The solid lines refer to symmetric modes and the dashed lines refer to antisymmetric modes of the analytically obtained dispersion curves. Negative values of wave-number correspond to the waves propagating from left to right edge of the plate again. The reflected waves practically do not exist because of presence of the absorber. Almost no spectral values exist in the area of positive wave-number values.

4. Experiment

The 2D FT algorithm was verified by the experimentally gained data. The experiment was performed at the Ultrasonic methods department of the Institute of Thermomechanics AS CR. The measurement of vertical displacements on surface of the steel plate was the object of the experiment.

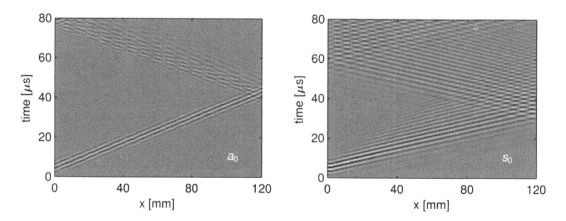

Fig. 7. *Time-spatial* distribution of vertical displacement when input at $x = 0$ was designed a_0 (left) or s_0 (right) only

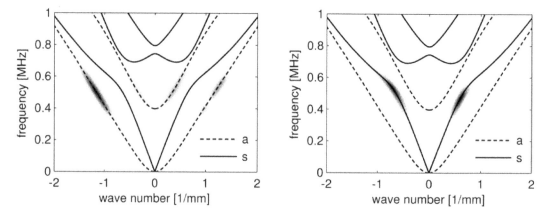

Fig. 8. *Frequency–wave-number* distribution after 2D FT when input at $x = 0$ was designed a_0 (left) or s_0 (right) only

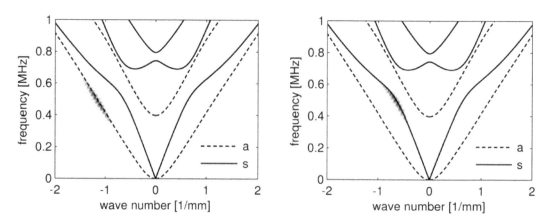

Fig. 9. *Frequency–wave-number* distribution after 1D FT when input at $x = 0$ was designed a_0 (left) or s_0 (right) only

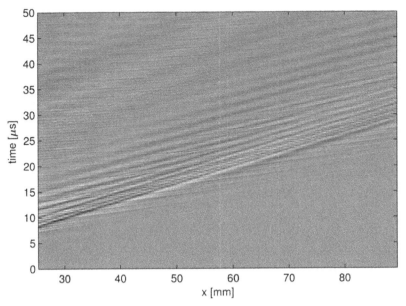

Fig. 10. *Time-spatial* distribution of vertical displacement on surface of the steel plate for the laser excitation

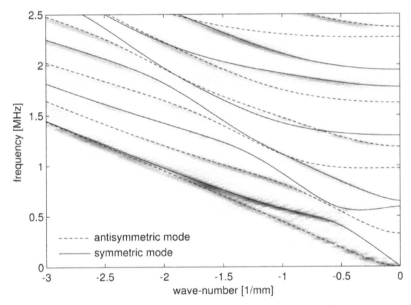

Fig. 11. *Frequency–wave-number* distribution after 2D FT of experimental data on surface of the steel plate

The steel plate size was $280.2 \times 280.4 \times 5.04$ mm. The measured mechanical properties of the plate were $c_1 = 5\,973.3$ m/s, $c_2 = 3\,270.6$ m/s, Poisson's ratio $\nu = 0.285\,9$ and density $\rho = 7\,793$ kg/m^3. The laser source of excitation was placed in the center of the plate. The displacements were scanned on the plate surface in distance 25.25 to 89.4 mm (with step 1.002 3 mm) from the center of the plate by the miniature transducers (the VP-1093 Pinducer). The digital oscilloscope LeCroy 9304AM was used for recording of the transducer signals.

The vertical displacements in dependence on space and time are illustrated by pseudocolor plot in Fig. 10. The Fig. 11 shows the pseudocolor plot of *frequency–wave-number* spectrum,

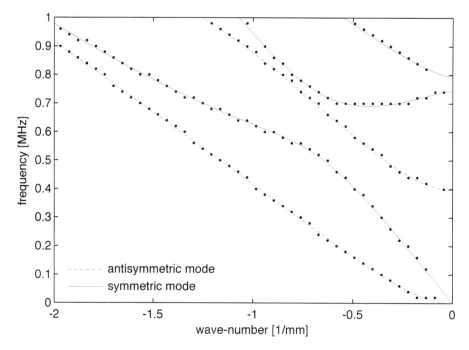

Fig. 12. Estimation of dispersion curves by the frequency method

which was computed by 2D FT in MATLAB. The dispersion curves obtained analytically for given testing plate are plotted by solid lines (symmetric mode) and dashed lines (antisymmetric mode). Negative values of wave-number correspond to the waves propagating from center of the plate. The experimental data do not contain reflected waves than positive values of wave-number are not considered. Note that there are many both symmetric and antisymmetric excited modes in wide range of frequencies. It is caused by very narrow laser pulse. The most significant are the first symmetric and antisymmetric modes.

5. Determination of dispersion curves

In this section dispersion curves in a thick plate were determined by means of FEM task in frequency domain and following 1D FT. The geometry of the plate, its elastic constants and mesh parameters were the same as in section 3. The plate was excited at the left edge by constant constrain. Because of more accurate determination of particular modes, the task was divided into two parts: identification of symmetric modes and identification of antisymmetric ones. The symmetric modes were excited by displacement in x direction; the antisymmetric modes were excited by displacement in y direction. The value of displacement was $1\,\mu$m for both cases. FEM task was performed for frequencies from $20\,$kHz to $1\,$MHz with step $20\,$kHz.

The dots in Fig. 12 represent the peak values of *frequency–wave-number* spectrum, which was computed by the spatial FT of the frequency spectra in particular places on the surface of the plate. The dispersion curves obtained analytically are plotted for verification by solid lines (symmetric mode) and dashed lines (antisymmetric mode).

Note that the dispersion curves obtained by means of FEM are in good agreement with the dispersion curves computed analytically. The accuracy of the dispersion curves determination is sufficient for an ultrasonic methodology using guided waves to examine structural components. The accuracy is possible to increase by decreasing of the frequency step.

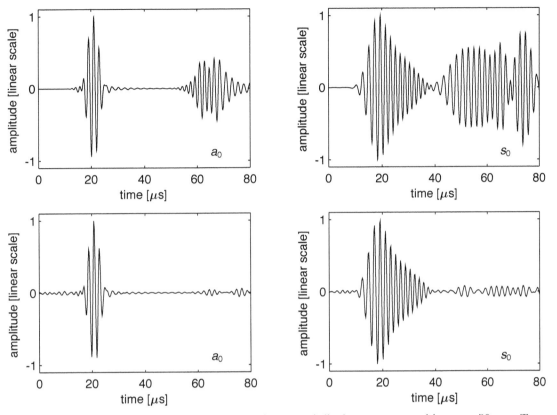

Fig. 13. Temporal waveform corresponding to the normal displacement at position $x = 50$ mm. Temporal method (upper plots) and frequency method (lower plots)

6. Conclusion

This paper reports on methods for determination of Lamb wave dispersion curves. Both algorithm working in the frequency domain and algorithm working in the more conventional temporal domain is possible to use for the estimate of dispersion curves also for other geometries than only thick plate.

Since a small number of frequencies are sufficient to achieve a correct representation of a wide variety of temporal excitations, the frequency method considerably speeds up the computation by avoiding the temporal FT and by decreasing the number of calculation steps.

Number of degrees of freedom for time method is 16 354 and for frequency method is 19 074. Number of elements for time method is 1 920 and for frequency method is 2 240. Higher values for frequency method are caused by presence of absorbing part. Despite of this fact the computation by frequency method is approximately 17 times faster then the computation by temporal method. The computations have also less requirements on the storage space of computer. Working in the frequency domain leads to FE calculations shorter than 90 seconds on a 2.67 GHz, 8-processor (Xeon Pentium), HP Workstation with 32 GB RAM.

The next advantage of this method is that the viscoelastic behavior could be taken into account by considering complex moduli as input data to the constitutive relations, since these are expressed in the frequency domain.

The frequency method can be also used for calculation of the temporal waveform of displacements. Temporal waveforms can be then reconstructed by applying the inverse FT to the set of harmonic displacements predicted for the given frequencies by FEM software. Fig. 13

shows the signal corresponding to the normal displacement predicted at position $x = 50\,\text{mm}$. The case when the input at $x = 0$ was designed to excite only a_0 is depicted in the left plots and the case when the input at $x = 0$ was designed to excite only s_0 is shown in the right plots. Upper plots are results of transient analysis in FEM and lower ones are results of frequency analysis in FEM and following by the inverse FT. Note that the reflected waves are soften due to the absorbing part of the plate.

Acknowledgements

The work has been supported by the Institute Research Plan AV0Z20760514 and by the grants GA CR No 101/09/1630. Authors thank Michal Landa from the Institute of Thermomechanics AS CR for the experimental data.

References

[1] Alleyne, D. N., Cawley, P., A two-dimensional Fourier transform method for the measurement of propagateing multimode signals, Journal of the Acoustical Society of America 89 (3) (1991) 1 159–1 168.

[2] Alleyne, D. N., The nondestructive testing of plates using ultrasonic Lamb waves, Ph.D. thesis, Imperial College of Science, Technology and Medicine, London, 1991.

[3] Castaings, M., Bacon, C., Hosten, B., Finite element predictions for the dynamic response of two thermo–viscoelastic material structures, Journal of the Acoustical Society of America 115 (3) (2004) 1 125–1 133.

[4] Cheung, Y. K., Tham, L. G., Finite strip method, CRC Press, 1997.

[5] COMSOL, Inc., http://www.comsol.com.

[6] Delsanto, P. P., Chaskelis, H. H., Whitcombe, T., Mignogna, R. B., Connection machine simulation of boundary effects in ultrasonic NDE, In Nondestructive Characterization of Materials IV, Plenum Press, 1991.

[7] Graff, K. F., Wave motion in elastic solids, Dover, New York, 1991.

[8] Hesthaven, J., Gottlieb, S., Gottlieb, D., Spectral methods for time-dependent problems, Cambridge UK, 2007.

[9] Knopoff, L., A matrix method for elastic wave problems, Bulletin of the Seismological Society of America 54 (1) (1964) 431–438.

[10] Krautkrämer, J., Krautkrämer, H., Ultrasonic testing of materials, American Society for Testing and Materials, Springer, New York, 1990.

[11] Miklowitz, J., The theory of elastic waves and waveguides, North-Holland, Amsterdam, 1978.

[12] The MathWorks, Inc., http://www.mathworks.com.

[13] Yim, H., Sohn, Y., Numerical simulation and visualization of elastic waves using mass-spring lattice model, Transactions on Ultrasonics, Ferroelectrics and Frequency Control 47 (3) (2000) 549–558.

4

Comparison of various refined beam theories for the bending and free vibration analysis of thick beams

A. S. Sayyad[a,*]

[a]*Department of Civil Engineering, SRES's College of Engineering, Kopargaon-423601, M.S., India.*

Abstract

In this paper, unified shear deformation theory is used to analyze simply supported thick isotropic beams for the transverse displacement, axial bending stress, transverse shear stress and natural frequencies. This theory enables the selection of different in-plane displacement components to represent shear deformation effect. The numbers of unknowns are same as that of first order shear deformation theory. The governing differential equations and boundary conditions are obtained by using the principle of virtual work. The results of displacement, stresses, natural bending and thickness shear mode frequencies for simply supported thick isotropic beams are presented and discussed critically with those of exact solution and other higher order theories. The study shows that, while the transverse displacement and the axial stress are best predicted by the models 1 through 5 whereas models 1 and 2 are overpredicts the transverse shear stress. The model 4 predicts the exact dynamic shear correction factor ($\pi^2/12 = 0.822$) whereas model 1 overpredicts the same.

Keywords: Thick beam, shear deformation, principle of virtual work, bending analysis, transverse shear stress, free flexural vibration, natural frequencies, dynamic shear correction factor

1. Introduction

Beams are common structural elements in most structures and they are analyzed using classical or refined shear deformation theories to evaluate static and dynamic characteristics. Elementary theory of beam bending underestimates deflections and overestimates the natural frequencies since it disregards the transverse shear deformation effect. Timoshenko [24] was the first to include refined effects such as rotatory inertia and shear deformation in the beam theory. This theory is now widely referred to as Timoshenko beam theory or first order shear deformation theory. In this theory transverse shear strain distribution is assumed to be constant through the beam thickness and thus requires problem dependent shear correction factor. The accuracy of Timoshenko beam theory for transverse vibrations of simply supported beam in respect of the fundamental frequency is verified by Cowper [6,7] with a plane stress exact elasticity solution.

The limitations of elementary theory of beam and first order shear deformation theory led to the development of higher order shear deformation theories. Many higher order shear deformation theories are available in the literature for static and dynamic analysis of beams [2–5,11,15]. Levinson [17] has developed a new rectangular beam theory for the static and dynamic analysis of beam. Reddy [18] has developed well known third order shear deformation theory for the non-linear analysis of plates with moderate thickness. The trigonometric shear deformation theories are presented by Touratier [25], Vlasov and Leont'ev [26] and Stein [20] for thick

---

*Corresponding author. e-mail: attu_sayyad@yahoo.co.in.

beams. However, with these theories shear stress free boundary conditions are not satisfied at top and bottom surfaces of the beam. Ghugal and Shipmi [9] and Ghugal [10] has developed a trigonometric shear deformation theory which satisfies the shear stress free condition at top and bottom surfaces of the beam. Soldatos [22] has dveloped hyperbolic shear deformation theory for homogeneous monoclinic plates. Recently Ghugal and Sharma [8] employed hyperbolic shear deformation theory for the static and dynamic analysis of thick isotropic beams. A study of literature [1,12–14,19] indicates that the research work dealing with flexural analysis of thick beams using refined shear deformation theories is very scant and is still in infancy. Sayyad [21] has carried out comparison of various shear deformation theories for the free vibration analysis of thick isotropic beams.

In the present study, various shear deformation theories are used for the bending and free vibration analysis of simply supported thick isotropic beams.

2. Beam under consideration

Consider a beam made up of isotropic material as shown in Fig. 1. The beam can have any boundary and loading conditions. The beam under consideration occupies the region given by

$$0 \leq x \leq L, \qquad -b/2 \leq y \leq b/2, \qquad -h/2 \leq z \leq h/2, \tag{1}$$

where x, y, z are Cartesian co-ordinates, L is length, b is width and h is the total depth of the beam. The beam is subjected to transverse load of intensity $q(x)$ per unit length of the beam.

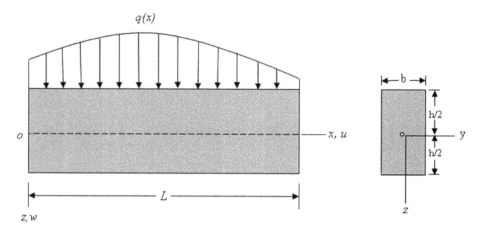

Fig. 1. Beam under bending in $x - z$ plane

2.1. Assumptions made in theoretical formulation

1. The in-plane displacement u in x direction consists of two parts:

    (a) A displacement component analogous to displacement in elementary beam theory of bending;

    (b) Displacement component due to shear deformation which is assumed to be parabolic, sinusoidal, hyperbolic and exponential in nature with respect to thickness coordinate.

2. The transverse displacement w in z direction is assumed to be a function of x coordinate.

3. One dimensional constitutive law is used.

4. The beam is subjected to lateral load only.

2.2. The displacement field

Based on the before mentioned assumptions, the displacement field of the present unified shear deformation theory is given as below

$$u(x, z, t) = -z\frac{\partial w}{\partial x} + f(z)\phi(x, t), \tag{2}$$

$$w(x, z, t) = w(x, t). \tag{3}$$

Here u and w are the axial and transverse displacements of the beam center line in x and z-directions respectively and t is the time. The ϕ represents the rotation of the cross-section of the beam at neutral axis which is an unknown function to be determined. The functions $f(z)$ assigned according to the shearing stress distribution through the thickness of the beam are given in Table 1.

Table 1. Functions $f(z)$ for different shear stress distribution

Model	Author	Function $f(z)$
Model 1	(Ambartsumian [2])	$f(z) = \left[\frac{z}{2}\left(\frac{h^2}{4} - \frac{z^2}{3}\right)\right]$
Model 2	(Kruszewski [15])	$f(z) = \left[\frac{5z}{4}\left(1 - \frac{4z^2}{3h^2}\right)\right]$
Model 3	(Reddy [18])	$f(z) = z\left[1 - \frac{4}{3}\left(\frac{z}{h}\right)^2\right]$
Model 4	(Touratier [25])	$f(z) = \frac{h}{\pi}\sin\frac{\pi z}{h}$
Model 5	(Soldatos [22])	$f(z) = \left[z\cosh\left(\frac{1}{2}\right) - h\sinh\left(\frac{z}{h}\right)\right]$
Model 6	(Karama et al. [14])	$f(z) = z\exp\left[-2\left(\frac{z}{h}\right)^2\right]$
Model 7	(Akavci [1])	$f(z) = \frac{3\pi}{2}\left[h\tanh\left(\frac{z}{h}\right) - z\sec^2 h\left(\frac{1}{2}\right)\right]$

2.3. Necessity of refined theories

The shear deformation effects are more pronounced in the thick beams than in the slender beams. These effects are neglected in elementary theory of beam (ETB) bending. In order to describe the correct bending behavior of thick beams including shear deformation effects and the associated cross sectional warping, shear deformation theories are required. This can be accomplished by selection of proper kinematics and constitutive models. The functions $f(z)$ is included in the displacement field of higher order theories to take into account effect of transverse shear deformation and to get the zero shear stress conditions at top and bottom surfaces of the beam.

2.4. Strain-displacement relationship

Normal strain and transverse shear strain for beam are given by

$$\varepsilon_x = \frac{\partial u}{\partial x} = -z\frac{\partial^2 w}{\partial x^2} + f(z)\frac{\partial \phi}{\partial x}, \tag{4}$$

$$\gamma_{zx} = \frac{\partial u}{\partial z} + \frac{\partial w}{\partial x} = f'(z)\phi. \tag{5}$$

2.5. Stress-Strain relationship

According to one dimensional constitutive law, the axial stress/normal bending stress and transverse shear stress are given by

$$\sigma_x = E\varepsilon_x = E\left[-z\frac{\partial^2 w}{\partial x^2} + f(z)\frac{\partial \phi}{\partial x}\right], \tag{6}$$

$$\tau_{zx} = G\gamma_{zx} = Gf'(z)\phi. \tag{7}$$

3. Governing equations and boundary conditions

Using Eqns. (4) through (7) and the principle of virtual work, variationally consistent governing differential equations and boundary conditions for the beam under consideration can be obtained. The principle of virtual work when applied to the beam leads to

$$\int_0^L \int_{-h/2}^{+h/2} (\sigma_x \delta\varepsilon_x + \tau_{zx}\delta\gamma_{zx})\, \mathrm{d}z\, \mathrm{d}x + \tag{8}$$

$$\rho \int_0^L \int_{z-h/2}^{+h/2} \left(\frac{\partial^2 u}{\partial t^2}\delta u + \frac{\partial^2 w}{\partial t^2}\delta w\right)\, \mathrm{d}z\, \mathrm{d}x - \int_0^L q\delta w\, \mathrm{d}x = 0,$$

where the symbol δ denotes the variational operator. Integrating the preceding equations by parts, and collecting the coefficients of δw and $\delta \phi$, the governing equations in terms of displacement variables are obtained as follows

$$A_0\frac{\partial^4 w}{\partial x^4} - B_0\frac{\partial^3 \phi}{\partial x^3} - \frac{\rho A_0}{E}\frac{\partial^4 w}{\partial x^2 \partial t^2} + \frac{\rho B_0}{E}\frac{\partial^3 \phi}{\partial x \partial t^2} + \rho h\frac{\partial^2 w}{\partial t^2} = q, \tag{9}$$

$$B_0\frac{\partial^3 w}{\partial x^3} - C_0\frac{\partial^2 \phi}{\partial x^2} + D_0\phi - \frac{\rho B_0}{E}\frac{\partial^3 w}{\partial x \partial t^2} - \frac{\rho C_0}{E}\frac{\partial^2 \phi}{\partial t^2} = 0 \tag{10}$$

and the associated boundary conditions obtained are of following form

$$-A_0\frac{\partial^3 w}{\partial x^3} + B_0\frac{\partial^2 \phi}{\partial x^2} + \frac{\rho A_0}{E}\frac{\partial^3 w}{\partial x \partial t^2} - \frac{\rho B_0}{E}\frac{\partial^2 \phi}{\partial t^2} = 0 \qquad \text{or } w \text{ is prescribed} \tag{11}$$

$$A_0\frac{\partial^2 w}{\partial x^2} - B_0\frac{\partial \phi}{\partial x} = 0 \qquad \text{or } \frac{\mathrm{d}w}{\mathrm{d}x} \text{ is prescribed} \tag{12}$$

$$-B_0\frac{\partial^2 w}{\partial x^2} + C_0\frac{\partial \phi}{\partial x} = 0 \qquad \text{or } \phi \text{ is prescribed} \tag{13}$$

where A_0, B_0, C_0 and D_0 are the stiffness coefficients given as follows

$$A_0 = E\int_{-h/2}^{+h/2} z^2\, \mathrm{d}z, \quad B_0 = E\int_{-h/2}^{+h/2} zf(z)\, \mathrm{d}z, \tag{14}$$

$$C_0 = E\int_{-h/2}^{+h/2} f^2(z)\, \mathrm{d}z, \quad D_0 = G\int_{-h/2}^{+h/2} [f'(z)]^2\, \mathrm{d}z.$$

3.1. Illustrative examples

In order to prove the efficacy of the present theories, the following numerical examples are considered. The following material properties for beam are used

$$E = 210 \,\text{GPa}, \qquad \mu = 0.3, \qquad G = \frac{E}{2(1+\mu)} \text{ and } \rho = 7\,800 \,\text{kg/m}^3, \qquad (15)$$

where E is the Young's modulus, ρ is the density, and μ is the Poisson's ratio of beam material.

Example 1: Bending analysis of beam

A simply supported uniform beam shown in Fig. 2 subjected to uniformly distributed load $q(x) = \sum_{m=1}^{m=\infty} q_m \sin\left(\frac{m\pi x}{L}\right)$ acting in the z-direction, where q_m is the coefficient of single Fourier expansion of load. The value of q_m for uniformly distributed load given as follows

$$q_m = \frac{4q_0}{m\pi}, \qquad m = 1, 3, 5, \ldots,$$

$$q_m = 0 \qquad , \qquad m = 2, 4, 6, \ldots, \qquad (16)$$

where q_0 is the intensity of uniformly distributed load.

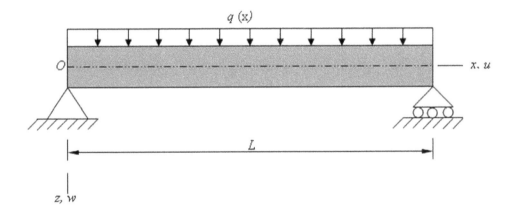

Fig. 2. Simply supported beam subjected to uniformly distributed load

The governing equations for bending analysis of beam (static flexure), discarding all the terms containing time derivatives become

$$A_0 \frac{d^4 w}{dx^4} - B_0 \frac{d^3 \phi}{dx^3} = q, \qquad (17)$$

$$B_0 \frac{d^3 w}{dx^3} - C_0 \frac{d^2 \phi}{dx^2} + D_0 \phi = 0. \qquad (18)$$

The following is the solution form assumed for $w(x)$ and $\phi(x)$ which satisfies the boundary conditions exactly

$$w(x) = \sum_{m=1}^{\infty} w_m \sin\frac{m\pi x}{L}, \qquad \phi(x) = \sum_{m=1}^{\infty} \phi_m \cos\frac{m\pi x}{L}, \qquad (19)$$

where w_m and ϕ_m are the unknown coefficients of the respective Fourier expansion and m is the positive integer. Substituting this form of solution and the load $q(x)$ into governing equations, yields the following two algebraic simultaneous equations

$$\left(A_0 \frac{m^4 \pi^4}{L^4}\right) w_m - \left(B_0 \frac{m^3 \pi^3}{L^3}\right) \phi_m = q_m, \tag{20}$$

$$-\left(B_0 \frac{m^3 \pi^3}{L^3}\right) w_m + \left(C_0 \frac{m^2 \pi^2}{L^2} + D_0\right) \phi_m = 0. \tag{21}$$

Solving Eqns. (20) and (21) simultenously to determine unknowns w_m and ϕ_m

$$w_m = \frac{q_m \left(C_0 \frac{m^2 \pi^2}{L^2} + D_0\right)}{\left(C_0 \frac{m^2 \pi^2}{L^2} + D_0\right) \left(A_0 \frac{m^4 \pi^4}{L^4}\right) - \left(B_0 \frac{m^3 \pi^3}{L^3}\right) \left(B_0 \frac{m^3 \pi^3}{L^3}\right)}, \tag{22}$$

$$\phi_m = \frac{q_m \left(B_0 \frac{m^3 \pi^3}{L^3}\right)}{\left(C_0 \frac{m^2 \pi^2}{L^2} + D_0\right) \left(A_0 \frac{m^4 \pi^4}{L^4}\right) - \left(B_0 \frac{m^3 \pi^3}{L^3}\right) \left(B_0 \frac{m^3 \pi^3}{L^3}\right)}. \tag{23}$$

Using Eqns. (22) and (23) substitute Eqn. (19) in the displacement field [Eqns. (2) and (3)] and stress-strain relationships [Eqns. (6) and (7)] to obtain expressions for axial displacement, transverse displacement, axial bending stress and transverse shear stress

Axial displacement: $\qquad u = \left[-z \frac{m\pi}{L} w_m + f(z)\phi_m\right] \cos \frac{m\pi x}{L}. \tag{24}$

Transverse displacement: $\qquad w = w_m \sin \frac{m\pi x}{L}. \tag{25}$

Axial bending stress: $\qquad \sigma_x = E\left[z \frac{m^2 \pi^2}{L^2} w_m - f(z)\frac{m\pi}{L}\phi_m\right] \sin \frac{m\pi x}{L}. \tag{26}$

Transverse shear stress: $\qquad \tau_{zx} = G f'(z)\phi_m \cos \frac{m\pi x}{L}. \tag{27}$

Example 2: Free flexural vibration of beam

The governing equations for free flexural vibration of simply supported beam can be obtained by setting the applied transverse load equal to zero in Eqns. (9) and (10). A solution to resulting governing equations, which satisfies the associated initial conditions, is of the form

$$w = w_m \sin \frac{m\pi x}{L} \sin \omega_m t, \tag{28}$$

$$\phi = \phi_m \cos \frac{m\pi x}{L} \sin \omega_m t, \tag{29}$$

where w_m and ϕ_m are the amplitudes of translation and rotation respectively, and ω_m is the natural frequency of the mth mode of vibration. Substitution of this solution form into the governing equations of free vibration of beam results in following algebraic equations

$$\left[\left(A_0 \frac{m^4 \pi^4}{L^4}\right) w_m - \left(B_0 \frac{m^3 \pi^3}{L^3}\right) \phi_m\right] - \omega^2 \left[\left(\frac{\rho A_0}{E} \frac{m^2 \pi^2}{L^2} + \rho h\right) w_m - \frac{\rho B_0}{E} \frac{m\pi}{L} \phi_m\right] = 0, \tag{30}$$

$$\left[-B_0 \frac{m^3 \pi^3}{L^3} w_m + \left(C_0 \frac{m^2 \pi^2}{L^2} + D_0\right) \phi_m\right] - \omega^2 \left[-\left(\frac{\rho B_0}{E} \frac{m\pi}{L}\right) w_m + \frac{\rho C_0}{E} \phi_m\right] = 0. \tag{31}$$

The Eqns. (30) and (31) can be written in the following matrix form

$$\left(\begin{bmatrix} K_{11} & K_{12} \\ K_{12} & K_{22} \end{bmatrix} - \omega^2 \begin{bmatrix} M_{11} & M_{12} \\ M_{12} & M_{22} \end{bmatrix} \right) \left\{ \begin{array}{c} w_m \\ \phi_m \end{array} \right\} = 0. \tag{32}$$

Above Eqn. (32) can be written in following more compact form

$$([K] - \omega_m^2[M])\{\Delta\} = 0, \tag{33}$$

where $\{\Delta\}$ denotes the vector, $\{\Delta\}^T = \{W_m, \phi_m\}$. The $[K]$ and $[M]$ are symmetric matrices. The elements of the coefficient matrix $[K]$ are given by

$$K_{11} = \left(A_0 \frac{m^4\pi^4}{L^4} \right), \quad K_{12} = K_{21} = -\left(B_0 \frac{m^3\pi^3}{L^3} \right), \quad K_{22} = \left(C_0 \frac{m^2\pi^2}{L^2} + D_0 \right). \tag{34}$$

The elements of the coefficient matrix $[M]$ are given by

$$M_{11} = \left(\frac{\rho A_0}{E} \frac{m^2\pi^2}{L^2} + \rho h \right), \quad M_{12} = M_{21} = -\frac{\rho B_0}{E} \frac{m\pi}{L}, \quad M_{22} = \frac{\rho C_0}{E}. \tag{35}$$

For nontrivial solution of Eqn. (33), $\{\Delta\} \neq 0$, the condition expressed by

$$([K] - \omega_m^2[M]) = 0, \tag{36}$$

yields the eigen-frequencies ω_m. From this solution natural frequencies of beam for various modes of vibration can be obtained.

4. Numerical results

The results for transverse displacement (w), axial bending stress (σ_x), transverse shear stress (τ_{zx}) and fundamental frequency ω_m are presented in the following non-dimensional form

$$\bar{w} = \frac{10Ebh^3 w}{q_0 L^4}, \quad \bar{\sigma}_x = \frac{b\sigma_x}{q_0}, \quad \bar{\tau}_{zx} = \frac{b\tau_{zx}}{q_0}, \quad \bar{\omega} = \omega_m \left(\frac{L^2}{h} \right) \sqrt{\frac{\rho}{E}}, \quad S = \frac{L}{h}, \tag{37}$$

where S is the aspect ratio.

The percentage error in results obtained by theories/models of various researchers with respect to the corresponding results obtained by theory of elasticity is calculated as follows

$$\text{error} = \left(\frac{\text{value by a particular model} - \text{value by exact elasticity solution}}{\text{value by exact elasticity solution}} \right) \times 100\,\%. \tag{38}$$

The results obtained for the above examples (static and dynamics) solved in this paper are presented in Tables 2 through 5.

5. Discussion of results

The results obtained from the present theories are compared with the elementary theory of beam (ETB), first order shear deformation theory (FSDT) of Timoshenko [24], higher order shear deformation theories of Heyliger and Reddy [12], Ghugal [10] and exact elasticity solutions given by Timoshenko and Goodier [23] and Cowper [7]. The value of dynamic shear correction factor is compared with its exact value given by Lamb [16]:

a) **Transverse Displacement** ($\bar{w}$): The comparison of maximum transverse displacement for the simply supported thick isotropic beams subjected to uniformly distributed load is presented in Table 2. The maximum transverse displacement predicted by models 5 and 6 is in excellent agreement with the exact solution for all the aspect ratios whereas the error in predicting transverse displacement by other models decreases with increase in aspect ratio. The FSDT overestimates the maximum transverse displacement whereas ETB underestimates the same for all the aspect ratios as compared to that of exact solution.

b) **Axial Bending Stress** ($\bar{\sigma}_x$): Table 2 shows the comparison of axial bending stress for the simply supported thick isotropic beams subjected to uniformly distributed load. Among all the models, model 6 overestimates the value of axial bending stress for all the aspect ratios as compared to that of exact solution whereas axial bending stress predicted by rest of the models is in excellent agreement with that of exact solution. The values of axial

Table 2. Comparison of transverse displacement $\bar{w}$ at $(x = L/2, z = 0)$, axial bending stress $\bar{\sigma}_x$ at $(x = L/2, z = \pm h/2)$ and transverse shear stress $\bar{\tau}_{zx}$ at $(x = 0, z = 0)$ for isotropic beam subjected to uniformly distributed load

S	Theory	$\bar{w}$	% Error	$\bar{\sigma}_x$	% Error	$\bar{\tau}_{zx}$	% Error
2	Model 1 [2]	2.357	−3.913	3.210	0.312	1.156	−22.93
	Model 2 [15]	2.515	2.527	3.261	1.906	1.333	−11.13
	Model 3 [18]	2.532	3.220	3.261	1.906	1.415	−5.667
	Model 4 [25]	2.529	3.098	3.278	2.437	1.451	−3.267
	Model 5 [22]	2.513	2.445	3.206	0.187	1.442	−3.866
	Model 6 [14]	2.510	2.323	3.322	3.812	1.430	−4.667
	Model 7 [1]	2.523	2.853	3.253	1.656	1.397	−6.866
	Timoshenko [FSDT] [24]	2.538	3.465	3.000	−6.250	0.984	−34.40
	Bernoulli-Euler [ETB]	1.563	−3.628	3.000	−6.250	—	—
	Timoshenko and Goodier [Exact] [23]	2.453	0.000	3.200	0.000	1.500	0.000
4	Model 1 [2]	1.762	−1.288	12.212	0.098	2.389	−20.36
	Model 2 [15]	1.805	1.120	12.262	0.508	2.836	−5.466
	Model 3 [18]	1.806	1.176	12.263	0.516	2.908	−3.066
	Model 4 [25]	1.805	1.120	12.280	0.655	2.993	−0.233
	Model 5 [22]	1.802	0.952	12.207	0.057	2.982	−0.600
	Model 6 [14]	1.801	0.896	12.324	1.016	2.957	−1.433
	Model 7 [1]	1.804	1.064	12.254	0.442	2.882	−3.933
	Timoshenko [FSDT] [24]	1.806	1.176	12.000	−1.639	1.969	−34.36
	Bernoulli-Euler [ETB]	1.563	−12.43	12.000	−1.639	—	—
	Timoshenko and Goodier [Exact] [23]	1.785	0.000	12.200	0.000	3.000	0.000
10	Model 1 [2]	1.595	−0.187	75.216	0.021	6.066	−19.12
	Model 2 [15]	1.602	0.250	75.266	0.087	7.328	−2.293
	Model 3 [18]	1.602	0.250	75.268	0.090	7.361	−1.853
	Model 4 [25]	1.601	0.187	75.284	0.111	7.591	1.213
	Model 5 [22]	1.601	0.187	75.211	0.014	7.576	1.013
	Model 6 [14]	1.601	0.187	75.330	0.172	7.513	0.173
	Model 7 [1]	1.601	0.187	75.259	0.078	7.312	−2.506
	Timoshenko [FSDT] [24]	1.602	0.250	75.000	−0.265	4.922	−34.37
	Bernoulli-Euler [ETB]	1.563	−2.190	75.000	−0.265	—	—
	Timoshenko and Goodier [Exact] [23]	1.598	0.000	75.200	0.000	7.500	0.000

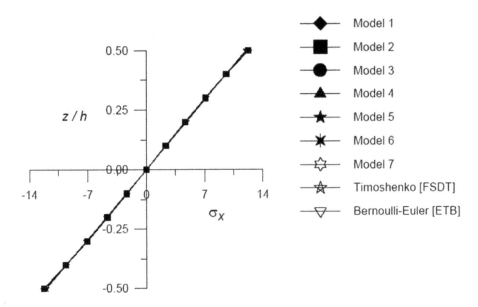

Fig. 3. Variation of axial bending stress ($\bar{\sigma}_x$) through thickness of beam subjected to uniformly distributed load for aspect ratio 4 at ($x = L/2, z$)

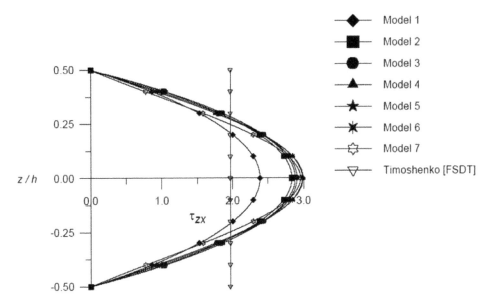

Fig. 4. Variation of transverse stress ($\bar{\tau}_{zx}$) through thickness of beam subjected to uniformly distributed load for aspect ratio 4 at ($x = 0, z$)

bending stress predicted by FSDT and ETB are identical for all the aspect ratios. The through thickness variation of axial bending stress is non-linear in nature as shown in Fig. 3.

c) **Transverse Shear Stress** ($\bar{\tau}_{zx}$)**:** The comparison of maximum transverse shear stress for the simply supported thick isotropic beams subjected to uniformly distributed load is presented in Table 2. The transverse shear stress is obtained using constitutive relation. Examination of Table 2 reveals that model 1 overestimates the value of transverse shear

Table 3. Comparison of non-dimensional fundamental ($m = 1$) flexural and thickness shear mode frequencies of the isotropic beam

Model	$S = 4$			$S = 10$		
	$\bar{\omega}_w$	% Error	$\bar{\omega}_\phi$	$\bar{\omega}_w$	% Error	$\bar{\omega}_\phi$
Model 1 [2]	2.625	0.884	37.237	2.808	0.143	217.439
Model 2 [15]	2.597	−0.192	33.704	2.802	−0.071	194.752
Model 3 [18]	2.596	−0.230	34.259	2.802	−0.071	198.109
Model 4 [25]	2.596	−0.230	34.238	2.802	−0.071	198.109
Model 5 [22]	2.596	−0.230	34.263	2.802	−0.071	198.258
Model 6 [14]	2.608	0.230	34.711	2.805	0.036	201.290
Model 7 [1]	2.598	−0.154	33.748	2.803	−0.036	195.055
Bernoulli-Euler [ETB]	2.779	6.802	–	2.838	1.212	–
Timoshenko [FSDT] [24]	2.624	0.845	34.320	2.808	0.143	198.616
Ghugal [10]	2.602	0.000	34.135	2.804	0.000	198.105
Heyliger and Reddy [12]	2.596	−0.230	34.250	2.802	−0.071	198.235
Cowper [7]	2.602	0.000	–	2.804	0.000	–

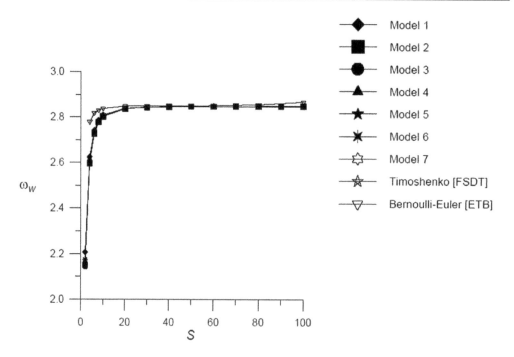

Fig. 5. Variation of fundamental bending frequency ($\bar{\omega}_w$) of beam with aspect ratio

stress whereas it is in excellent agreement when predicted by models 3 through 7 as compared to that of exact solution for all the aspect ratios. The transverse shear stress is overpredicted by models 1 and 2. Fig. 4 shows the through thickness variation of transverse shear stress for the thick isotropic beam subjected to uniformly distributed load for aspect ratio 4.

d) **Fundamental Flexural mode frequency** ($\bar{\omega}_w$): The comparison of lowest natural frequency in flexural mode is shown in Table 3. Observation of Table 2 shows that, Model 1

Table 4. Comparison of non-dimensional flexural frequency $(\bar{\omega}_w)$ of the isotropic beam for various modes of vibration

S	Model	Modes of vibration				
		$m = 1$	$m = 2$	$m = 3$	$m = 4$	$m = 5$
4	Model 1 [2]	2.625	8.823	16.491	24.713	33.165
	Model 2 [15]	2.597	8.598	15.957	23.923	32.304
	Model 3 [18]	2.596	8.569	15.793	23.435	31.240
	Model 4 [25]	2.596	8.573	15.811	23.483	31.339
	Model 5 [22]	2.596	8.569	15.791	23.429	31.228
	Model 6 [14]	2.608	8.691	16.202	24.357	32.935
	Model 7 [1]	2.598	8.612	16.004	24.027	32.493
	Cowper [7]	2.602	—	—	—	—
10	Model 1 [2]	2.808	10.791	22.903	37.999	55.142
	Model 2 [15]	2.802	10.711	22.582	37.228	53.740
	Model 3 [18]	2.802	10.709	22.566	37.164	53.557
	Model 4 [25]	2.802	10.710	22.570	37.175	53.583
	Model 5 [22]	2.802	10.709	22.566	37.163	53.554
	Model 6 [14]	2.805	10.742	22.708	37.537	54.317
	Model 7 [1]	2.803	10.715	22.598	37.271	53.827
	Cowper [7]	2.804	—	—	—	—

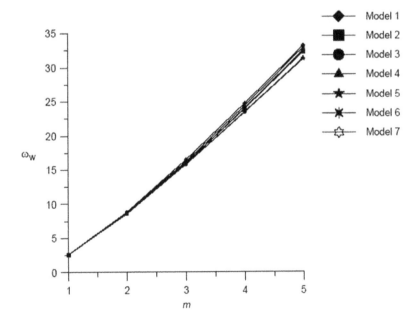

Fig. 6. Variation of fundamental bending frequency $(\bar{\omega}_w)$ of beam with various modes of vibration (m)

overestimates the lowest natural frequencies, in flexural mode by 0.884 % and 0.143 % for aspect ratios 4 and 10 respectively. The fundamental frequencies, in flexural mode predicted by models 2 through 6 is identical and in excellent agreement with that of exact solution Ghugal [10] yields the exact value of lowest natural frequencies, in flexural mode for aspect ratios 4 and 10. FSDT of Timoshenko overestimates the flexural mode

Table 5. Comparison of non-dimensional fundamental frequency of thickness shear mode ($\bar{\omega}_\phi$) of the isotropic beam for various modes of vibrations

S	Model	Modes of vibration				
		$m = 1$	$m = 2$	$m = 3$	$m = 4$	$m = 5$
4	Model 1 [2]	37.237	44.378	53.547	63.736	74.521
	Model 2 [15]	33.704	41.042	50.402	60.787	71.772
	Model 3 [18]	34.259	41.593	50.941	61.302	72.257
	Model 4 [25]	34.238	41.571	50.917	61.279	72.235
	Model 5 [22]	34.263	41.597	50.945	61.306	72.261
	Model 6 [14]	34.711	41.968	51.251	61.562	72.478
	Model 7 [1]	33.748	41.078	50.431	60.811	71.792
10	Model 1 [2]	217.439	226.391	240.105	257.416	277.363
	Model 2 [15]	194.752	204.080	218.272	236.080	256.514
	Model 3 [18]	198.235	207.555	221.739	239.539	259.959
	Model 4 [25]	198.109	207.425	221.606	239.401	259.819
	Model 5 [22]	198.258	207.578	221.763	239.563	259.984
	Model 6 [14]	201.290	210.468	224.467	242.071	262.302
	Model 7 [1]	195.055	204.368	218.539	236.327	256.740

frequency by 0.845 % and 0.143 % for aspect ratios 4 and 10 respectively whereas ETB overestimates the same by 6.802 % and 1.212 % due to neglect of shear deformation in the theory. The variation of lowest natural frequency in flexural mode with the aspect ratios is shown in Fig. 5. The comparison of flexural frequency for various modes of vibration (m) is shown in Table 4. The examination of Table 4 reveals that, the flexural frequencies obtained by various models are in excellent agreement with each other. The variation of flexural frequencies with various modes of vibration (m) is shown in Fig. 6.

e) **Fundamental frequency** ($\bar{\omega}_\phi$): Table 3 shows comparison of lowest natural frequency in thickness shear mode. Exact solution for the lowest natural frequency in thickness shear mode is not available in the literature. From the Table 3 it is observed that, thickness shear mode frequencies predicted by models 2 through 6 are in excellent agreement with each other whereas model 1 overestimates the same. Table 5 shows comparison of thickness shear mode frequencies for various modes of vibration and found in good agreement with each other. The solution for the circular frequency of thickness shear mode ($m = 0$) for thin rectangular beam is given by

$$\omega_\phi = \sqrt{\frac{K_{22}}{M_{22}}} = \sqrt{K_d \frac{GA}{\rho I}}, \tag{39}$$

where K_d is dynamic shear correction factor.

Table 6. Dynamic shear correction factors

Model	Model 1	Model 2	Model 3	Model 4	Model 5	Model 6	Model 7	Exact
K_d	0.995	0.794	0.824	0.822	0.824	0.850	0.797	0.822
% Error	21.046	−3.406	0.243	0.000	0.243	3.406	−3.041	0.000

Dynamic shear correction predicted by model 4 is same as the exact solution given by Lamb [16]. The corresponding values of shear factor for $m = 0$ according to models 3 and 5 is identical. The model 1 yields the higher value of dynamic shear correction factor whereas model 7 shows lower value for the same, Table 6.

6. Conclusions

From the study of comparison of various shear deformation theories for the bending and free vibration analysis of thick isotropic beams following conclusions are drawn.

1. The maximum transverse displacement predicted by all the models is in excellent agreement as compared to that of exact solution.

2. The axial bending stress predicted by the models 1 through 5 and 7 is in tune with exact solution whereas model 6 overestimates it for all the aspect ratios.

3. Through thickness variation of axial bending stress is non-linear in nature.

4. The maximum transverse shear stress predicted by models 3 through 7 is in excellent agreement as compared to that of exact solution whereas model 1 and 2 overestimates the value of transverse shear stress for all the aspect ratios.

5. Results of lowest natural frequencies for flexural mode predicted by models 3 through 5 are identical and in excellent agreement with that of exact solution. Model 1 overestimates the flexural mode frequency as compared to that of exact solution. Flexural mode frequencies predicted by models 2 and 7 are in tune with the exact solution.

6. The results of thickness shear mode frequencies are in excellent agreement with each other for all modes of vibration.

7. Model 4 yields the exact value of dynamic shear correction factor and it is in excellent agreement when predicted by models 3 and 5.

References

[1] Akavci, S. S., Buckling and free vibration analysis of symmetric and anti-symmetric laminated composite plates on an elastic foundation, Journal of Reinforced Plastics and Composites 26 (18) (2007) 1 907–1 919.

[2] Ambartsumyan, S. A., On the theory of bending plates, Izv Otd Tech Nauk AN SSSR 5 (1958) 69–77.

[3] Baluch, M. H., Azad, A. K., Khidir, M. A., Technical theory of beams with normal strain, Journal of Engineering Mechanics Proceedings ASCE 110 (1984) 1 233–1 237.

[4] Bhimaraddi, A., Chandrashekhara, K., Observations on higher-order beam theory, Journal of Aerospace Engineering Proceeding ASCE Technical Note 6 (1993) 408–413.

[5] Bickford, W. B., A consistent higher order beam theory. Development of Theoretical and Applied Mechanics, SECTAM 11 (1982) 137–150.

[6] Cowper, G. R., The shear coefficients in Timoshenko beam theory, ASME Journal of Applied Mechanics 33 (1966) 335–340.

[7] Cowper, G. R., On the accuracy of Timoshenko's beam theory, ASCE Journal of Engineering Mechanics Division 94 (6) (1968) 1 447–1 453.

[8] Ghugal, Y. M., Sharma, R., Hyperbolic shear deformation theory for flexure and vibration of thick isotropic beams, International Journal of Computational Methods 6 (4) (2009) 585–604.

[9] Ghugal, Y. M., Shimpi, R. P., A review of refined shear deformation theories for isotropic and anisotropic laminated beams, Journal of Reinforced Plastics and Composites 21 (2002) 775–813.

[10] Ghugal, Y. M., A simple higher order theory for beam with transverse shear and transverse normal effect, Department Report 4, Applied of Mechanics Department, Government College of Engineering, Aurangabad, India, 2006.

[11] Hildebrand, F. B., Reissner, E. C., Distribution of stress in built-in beam of narrow rectangular cross section, Journal of Applied Mechanics 64 (1942) 109–116.

[12] Heyliger, P. R., Reddy, J. N., A higher order beam finite element for bending and vibration problems, Journal of Sound and Vibration 126 (2) (1988) 309–326.

[13] Krishna Murty, A. V., Toward a consistent beam theory, AIAA Journal 22 (1984) 811–816.

[14] Karama, M., Afaq, K. S., Mistou, S., Mechanical behavior of laminated composite beam by new multi-layered laminated composite structures model with transverse shear stress continuity, International Journal of Solids and Structures, 40 (2003) 1525–1546.

[15] Kruszewski, E. T., Effect of transverse shear and rotatory inertia on the natural frequency of a uniform beam, NACATN (1909).

[16] Lamb, H., On waves in an elastic plates, Proceeding of Royal Society London Series A. 93 (1917) 114–128.

[17] Levinson, M., A new rectangular beam theory, Journal of Sound and Vibration 74 (1981) 81–87.

[18] Reddy, J. N., A general non-linear third order theory of plates with moderate thickness, International Journal of Non-linear Mechanics 25 (6) (1990) 677–686.

[19] Rehfield, L. W., Murthy, P. L. N., Toward a new engineering theory of bending: fundamentals, AIAA Journal 20 (1982) 693–699.

[20] Stein, M., Vibration of beams and plate strips with three-dimensional flexibility, Transaction ASME Journal of Applied Mechanics 56 (1) (1989) 228–231.

[21] Sayyad, A. S., Comparison of various shear deformation theories for the free vibration of thick isotropic beams, International Journal of Civil and Structural Engineering 2 (1) (2011) 85–97.

[22] Soldatos, K. P., A transverse shear deformation theory for homogeneous monoclinic plates, Acta Mechanica 94 (1992) 195–200.

[23] Timoshenko, S. P., Goodier, J. N., Theory of Elasticity, McGraw-Hill 3rd Int. ed. Singapore 1970.

[24] Timoshenko, S. P., On the correction for shear of the differential equation for transverse vibrations of prismatic bars, Philosophical Magazine 41 (6) (1921) 742–746.

[25] Touratier, M., An efficient standard plate theory, International Journal of Engineering Science 29 (8) (1991) 901–916.

[26] Vlasov, V. Z., Leont'ev, U. N., Beams, Plates and Shells on Elastic foundation, (Translated from Russian) Israel Program for Scientific Translation Ltd. Jerusalem 1996.

5

Possibility of identification of elastic properties in laminate beams with cross-ply laminae stacking sequences

M. Zajíček[a,*], J. Dupal[a]

[a] *Faculty of Applied Sciences, University of West Bohemia, Univerzitní 22, 306 14 Plzeň, Czech Republic*

Abstract

The goal of this work is to show the possibility of the identification of laminate beam specimens elastic properties with cross-ply laminae stacking sequences using prescribed eigenfrequencies. These frequencies are not determined experimentally in this paper but they are calculated numerically by means of the finite element (FE) software MSC.Marc. The composite material properties of the FE model based on Euler-Bernoulli theory have been subsequently tuned to correlate the determined frequencies in cross-ply laminate beams with the eigenfrequencies obtained by the software package. A real-coded genetic algorithm (GA) and a micro-genetic algorithm (μGA) are applied as the inverse technique for the identification problem. Because a small efficiency of the GAs in searching for Poisson's ratio values was found, this parameter and the in-plane shear modulus have been estimated by using the law of mixtures. Some numerical examples are given to illustrate the proposed technique.

Keywords: cross-ply laminate, beam, genetic algorithm, inverse technique, finite element model

1. Introduction

Composite materials are widely employed in modern industry. Analysis and design of structures manufactured from these materials depend directly upon accurate knowledge of their properties. Hence the property evaluation is one of the important goal of research.

Chu and Rokhlin [5] determined the elastic properties of composite from ultrasonic bulk wave velocity data. Balasubramaniam and Rao [2] carried out the reconstruction of material stiffness properties of unidirectional fiber-reinforced composites especially from incident ultrasonic bulk wave data. Computer-generated ultrasonic phase velocity data were used as the input to the GA that has been implemented for the parameters reconstruction. In the above two references, the Christoffel equation was applied to establish the relationship between material properties and bulk wave velocity. Complicated techniques were needed to measure the phase velocity of ultrasonic bulk waves and only single-ply anisotropic materials were considered in their works.

A number of researchers developed numerical-experimental methods in which experimental eigenfrequencies were used to identify elastic properties of composites. An indirect identification method for prediction the composite properties of plate specimens using measured eigenfrequencies is presented in [18]. The authors applied the Mindlin plate theory in combination with a FE model for the laminate analysis. Frederiksen [6] identified the elastic constants of thick orthotropic plates, whereas a mathematical model based on the higher-order shear deformation theory has been applied. This solution provides reliable estimations of the two transverse

---

*Corresponding author. e-mail: zajicek@kme.zcu.cz.

shear moduli. Ip et al. [10] also investigated eigenfrequencies in the orthotropic material. Furthermore, the mode shapes were measured on specimens with balanced symmetric lamination which were excited by an impact hammer. In parallel, an analytical model describing the modal responses of composite shells was developed using the Rayleigh-Ritz method. This model was subsequently tuned to correlate the theoretical frequencies with the measurements via Bayesian estimation. In the study [19], physical experiments were performed on the sample plates to measure the eigenfrequencies by a real-time television holography. The basic idea of the proposed approach corresponds to simple FE models which are determined only in the reference points of the experiment design. Therefore, a significant reduction against the conventional methods of minimization can be achieved in calculations of the cost function.

Liu et al. [15] developed the hybrid numerical method (HNM) which has been employed to calculate the transient waves in anisotropic laminated plates excited by impact loads. It combines the finite element method (FEM) with the method of Fourier transforms and it is described in [16]. The HNM and its modified version is then used as a forward solver in some identification problems, see e.g. [13,14]. The GA or μGA, alternatively combined with another method (for example with the nonlinear least squares method [13]), were usually adopted in these works as the inverse operator controlling the forward solver for material characterization using elastic waves. In the work [14], the dynamic displacement responses were obtained at only one receiving point of laminate surfaces. The robustness of procedure of the measurement noise effect has been investigated by adding Gauss noise to the input displacement response. Han et al. [9] utilized HNM to reconstruct the elastic constants of the cross-ply laminated axisymmetric cylinders subjected to an impact load. In this case, the laminated cylinder was divided into layered cylindrical elements in the thickness direction.

In addition, other techniques of material properties identification have been introduced in recent years. For instance, Genovese et al. [7] published a novel hybrid procedure for the mechanical characterization of orthotropic materials. This identification reverse problem has been solved by combining spectral interferometry and a combinatorial optimization technique, known as simulated annealing. Another numerical-experimental method for the identification of orthotropic materials is given in [12]. A biaxial tensile test was performed on a cruciform test specimen. The displacement field observed by a CCD camera and measured by a digital image correlation technique has been compared with a strain field which was computed by FEM. Newton-Raphson algorithm was used as an optimisation procedure. Kam and Liu [11] presented method for the determination of bending stiffness distribution of laminated shafts. The difference between predicted and measured deflections was minimized at any two points on the shaft using a quasi-Newton method. The view of material properties identification techniques is covered by Chen and Kam [4] who developed a two-level optimization method for material characterization by using two symmetric angle-ply beams with different fiber angles subjected to three-point bending. The best estimates of shear modulus and Poisson's ratio of the beam with fiber angles $45°$ are determined in the first-level optimization process. In the second level, the known shear modulus and Poisson's ratio are kept constant and Young's moduli of the second angle-ply beam with fiber angles different from $45°$ are identified.

In the present study, the possibility of the prediction of elastic properties in laminate beam specimens with different cross-ply laminae stacking sequences using prescribed eigenfrequencies is presented. The frequencies are determined by the FE software package in place of using the experimental method. These values were compared with the spectral analysis results of the FE models of beam. The GA and μGA are applied to manage the inverse problem.

2. FE formula in forward analysis

The FE model of beam based on Euler-Bernoulli theory is used for the calculation of the eigen-frequencies. In the (x_1, x_2, x_3) coordinate system, the displacement field is given by

$$u_1(x_1, x_3, t) = u(x_1, t) + x_3\, \psi(x_1, t)\,, \quad u_2(x_1, x_2, x_3, t) = 0\,, \quad u_3(x_1, t) = w(x_1, t)\,, \quad (1)$$

where $u(x_1, t)$ and $w(x_1, t)$ are the displacements due to extension and bending, respectively, and $\psi(x_1, t)$ denotes rotation about the x_2-axis. Besides, the displacement $u(x_1, t)$ can be re-written in the form

$$u(x_1, t) = u_c(x_1, t) - z_c\psi(x_1, t)\,, \qquad \text{where} \qquad z_c = B_{11}/A_{11}\,. \qquad (2)$$

The symbol $u_c(x_1, t)$ denotes the centroidal axis displacement. The stiffness parameters A_{11} and B_{11} are defined as

$$(A_{11}, B_{11}) = \sum_{k=1}^{n} Q_{11}^{k} \int_{h_{k-1}}^{h_k} b(x_3)(1, x_3)\mathrm{d}x_3\,, \qquad (3)$$

where

$$Q_{11}^{k} = \frac{E_k}{1 - \nu_{LT}\nu_{TL}} \quad \text{and} \quad E_k = \begin{cases} E_L & \text{for } \theta_k = 0\,, \\ E_T & \text{for } \theta_k = \pi/2\,. \end{cases} \qquad (4)$$

The longitudinal E_L and transverse E_T Young's modulus including the Poisson's ratios ν_{LT}, ν_{TL} represent the material properties of beam FE model that is consisted of n layers which are supposed to be orthotropic in the (L, T, T') directions, see Fig. 1. Each layer k is extended from lower face h_{k-1} to upper face h_k in the x_3 direction. The angle θ_k is orientated with respect to the x_1-axis and takes only values 0 or $\pi/2$. It is also depicted in Fig. 1 that the FE model is symmetric in the $x_1 - x_3$ plane. The beam cross-section is assumed to be uniform with a various shape having width $b(x_3)$ and the overall thickness h. The length of the FE is l_e.

The two-noded elements are used for the beam discretization. The linear and cubic polynomials are chosen as the displacement shape functions of the element, i.e.

$$u_c(x_1, t) = [1, x_1][a_0(t), a_1(t)]^T, \quad w(x_1, t) = [1, x_1, x_1^2, x_1^3][a_2(t), a_3(t), a_4(t), a_5(t)]^T. \qquad (5)$$

Consequently, the functions $u(x_1, t)$, $w(x_1, t)$ and $\psi(x_1, t)$ which describe deformations of a beam element can be expressed in terms of the nodal displacement components $\mathbf{q}_e(t) = [\mathbf{q}_1(t), \mathbf{q}_2(t)]^T$, where

$$\mathbf{q}_1(t) = [u(0, t), u(l_e, t)]^T \qquad \text{and} \qquad \mathbf{q}_2(t) = [w(0, t), w(l_e, t), \psi(0, t), \psi(l_e, t)]^T. \qquad (6)$$

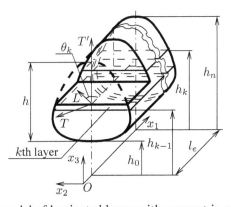

Fig. 1. FE model of laminated beam with symmetric cross-section

Because the derivation of the FE beam is based on Euler-Bernoulli theory, the relation $\psi = -\partial w/\partial x_1$ is valid. Then only one stress tensor component $\sigma_{11} = \sigma_1$ is generally nonzero. Therefore the constitutive equations of the kth layer reduce to

$$\sigma_1^k = Q_{11}^k \, \varepsilon_1 \,, \tag{7}$$

where ε_1 is a strain tensor component ε_{11} and the parameter Q_{11}^k is given by relation (4).

Using the principle of virtual work, the element governing equation of the free vibrations can be written as

$$\boldsymbol{M}_e \ddot{\boldsymbol{q}}_e(t) + \boldsymbol{K}_e \boldsymbol{q}_e(t) = \boldsymbol{0} \,, \tag{8}$$

where $\boldsymbol{M}_e$ and $\boldsymbol{K}_e$ are element mass matrix and element stiffness matrix, respectively. The detail form of these matrices is stated in [22] where the viscoelastic orthotropic beam element was derived according to Timoshenko theory. However, the FE beam based on Euler-Bernoulli theory can be easily obtained by omitting the transverse shear strain as it is mentioned in that paper. Assembling all the element matrices, the free vibration equation of the beam can be expressed in the form

$$\boldsymbol{M} \ddot{\boldsymbol{q}}(t) + \boldsymbol{K} \boldsymbol{q}(t) = \boldsymbol{0} \,, \tag{9}$$

where $\boldsymbol{M}$ is the total mass matrix, $\boldsymbol{K}$ is the total stiffness matrix and $\boldsymbol{q}$ is the global nodal displacement vector. When the periodic motion is considered, i.e. $\boldsymbol{q} = \boldsymbol{\nu} \, \mathrm{e}^{\mathrm{i}\Omega t}$, Eq. (9) can be rewritten as

$$(\boldsymbol{K} - \Omega^2 \boldsymbol{M})\boldsymbol{\nu} = \boldsymbol{0} \,, \tag{10}$$

whose solution leads to eigenvalue problem. The symbols $\boldsymbol{\nu}$ and Ω denote eigenvector and eigenfrequency, respectively. It could be found in [22] that the matrix $\boldsymbol{M}$ depends only on the material density and the beam geometry provided that Euler-Bernoulli theory has been considered. Therefore, the new eigenfrequencies can be obtained by solving Eq. (10) when the matrix $\boldsymbol{K}$ is updated with different elastic material constants. The eigenfrequencies calculated from Eq. (10) are then taken as an input for the inverse analysis.

3. Inverse problem formulation

The main aim of inverse methods is the determination of a selected set of unknown parameters in a numerical model. It is necessary to define the objective function which has to be minimized in the feasible domain of optimization parameters. In our case, this function is constructed using the sum of relative difference squares of the components of two vectors containing eigenfrequencies. Then the parametric optimization problem could be formulated as follows:

$$\widehat{\boldsymbol{p}} = \arg \left\{ f_o(\widehat{\boldsymbol{p}}) = \min_{\boldsymbol{p} \in \mathcal{D}} [f(\boldsymbol{p})] = \min_{\boldsymbol{p} \in \mathcal{D}} \left[\sum_{i=1}^n (1 - \Omega_i(\boldsymbol{p})/\Omega_i^{\mathrm{exp}})^2 \right] \right\} \,, \tag{11}$$

where $\boldsymbol{p} = [\alpha_1, \ldots, \alpha_s]^T \in \mathcal{D} \subseteq \boldsymbol{R}^s$ is a vector of unknown optimization parameters. The domain $\mathcal{D}$ is a convex set and is defined by constraints: $\alpha_k^l \leq \alpha_k \leq \alpha_k^u$ for $k = 1, 2, \ldots, s$. The vector $\boldsymbol{\Omega}^{\mathrm{exp}} = [\Omega_1^{\mathrm{exp}}, \ldots, \Omega_n^{\mathrm{exp}}]^T$ represents eigenfrequencies that are considered as known. Finally, $\boldsymbol{\Omega}(\boldsymbol{p}) \in \boldsymbol{R}^n$ is the vector of eigenfrequencies which are computed from Eq. (10) for admissible values of a vector $\boldsymbol{p}$. The GAs are then employed to search these unknown parameters α_k. The main procedure of the GAs for our problem is presented below.

4. Genetic algorithms

These are very effective algorithms searching for optimal or near-optimal solutions over the investigated finite domain. These methods have been developed by Goldberg [8] according to the idea of Darwin's Theory of Evolution. The GAs are suitable for finding the global optimum of optimization problems which have many local maxima and minima. They also posses the advantages of easy solving of the mixed problems with continuous and discrete variables, and without need of the objective function continuity. For these reasons, the GAs have been adopted to our optimization problem (11).

In the traditional GA, all variables of interest are encoded as binary digits which are known as genes. Collection of these genes further forms so-called chromosome. After manipulation of a binary-coded GA, the final binary numbers are then decoded as original real numbers. On the other hand, a real-coded GA has been also proposed in recent years, see e.g. [3]. The main discrepancy is that all genes in a chromosome are real numbers. It is more convenient to deal with most practical engineering applications because the changes from a real number to the binary digits may be the cause of a loss of the number precision. A real-coding also promotes the calculation efficiency because of straightforward using numbers in representation. Moreover, the various types of genetic operations can be simply adjusted or defined, as given in [17]. For these reasons the real-coded GAs are used in this paper to estimate the elastic properties in laminate beams with cross-ply laminae stacking sequences. In addition to the above, it should be mentioned that attempts to utilize the combination of binary and real genes to identify the unknown system can be found in literature. For example, a hybrid GA taking the advantage of binary and real digits including quantum computing is presented in [21].

In this work, the GA making use of the traditional genetic operators is applied as optimization technique. The algorithm starts with an initial population of N chromosomes randomly selected in the searching space $\mathcal{D}$. The selection, crossover and mutation operations are consequently performed to create next generation. The elitism operator is also adopted to replicate the best individual of current generation. This process is repeated until the convergence criterion or the maximum generation number N_g is achieved.

It has been pointed out in previous studies that a μGA is more robust algorithm for solving multi-parameter inverse problems than the traditional GA. According to Liu and Xi [16], the robustness of a uniform μGA lies in producing of a new genetic information due to the population restart process. Therefore, the μGA has been as well used besides the GA to search the unknown parameters. The structure of this algorithm is a similar to the above described GA but some differences can be found. The population is very small and usually includes only 5 or 6 individuals. Due to this fact, small discrepancies among individuals in the population are observed in a few generations and the convergence to some optimum occurs. At this point, a new population is randomly generated while keeping the best individual from the previously converged generation and the evolution process is thus restarted. The mutation operator is not applied for the population evaluation in the μGA because of new chromosomes keep flowing when the micro population is reborn.

Before the brief description of used real-coded genetic operations, some notations should be introduced. Let the vector of unknown parameters $\boldsymbol{p}_i = [\alpha_{1i}, \ldots, \alpha_{si}]^T$, see Eq. (11), means the ith chromosome in the gth generation of a population $P(g) = \{\boldsymbol{p}_1, \ldots, \boldsymbol{p}_{P_s}\}$ where $i = 1, \ldots, P_s$ while P_s is population size. Furthermore, the parameters p_c and p_m evaluate a probability of the performance of crossover and mutation operations, respectively.

4.1. Selection operation

The tournament selection has been chosen to generate offsprings since it is quite simple and suitable for checking whether a chromosome is reproduced or not according to its corresponding objective function. We randomly select 2 (may be up to $P_s - 1$) individuals from the current population and the best one minimum objective function value is added to the next population which is consequently subjected to other genetic operations. The process is repeated P_s times.

In addition, the roulette wheel selection mechanism is described in this part. This method is utilized below in the selection of suitable crossover type. The roulette wheel selection can be visualized by an imaginary wheel. Each parameter of observed set occupies an area that is related to its objective function value. When a spinning wheel stops, which is in practice represented by a randomly generated number from the range $\langle 0, 1 \rangle$, a fixed marker determines selected parameter. Such a selection mechanism needs more numerical computations but the probability of more frequent selection of one parameter can be easily increased, see the next subsection.

4.2. Crossover operation

As mentioned earlier, probability of crossover p_c is one of the parameters of genetic system. This probability gives us the expected number $p_c \times P_s$ of chromosomes which undergo the crossover operation. The process is started by generating a random number r from the range $\langle 0, 1 \rangle$ for each chromosome in the population that was subjected to the tournament selection. If $r < p_c$, the chromosome is selected for crossover. These parents are consequently mated randomly to make offsprings, i.e. new chromosomes. The result of this operation significantly depends on selected type of operator. In this work, the three various types of crossover operators are applied in evolutional algorithms. Note that the parameter $a \in (0, 1)$ contained in the following relations is generated randomly.

4.2.1. Simple crossover

Let parents $p_i = [\alpha_{1i}, \ldots, \alpha_{si}]^T$ and $p_j = [\alpha_{1j}, \ldots, \alpha_{sj}]^T$ are selected to be crossed after the kth position where $k \in \{1, 2 \ldots, s - 1\}$ is a random number. The offsprings $\tilde{p}_i$ and $\tilde{p}_j$ are then in the form

$$\tilde{p}_i = [\alpha_{1i}, \ldots, \alpha_{ki}, (1 - a)\, \alpha_{k+1,i} + a\, \alpha_{k+1,j}, \ldots, (1 - a)\, \alpha_{si} + a\, \alpha_{sj}]^T,$$
$$\tilde{p}_j = [\alpha_{1j}, \ldots, \alpha_{kj}, (1 - a)\, \alpha_{k+1,j} + a\, \alpha_{k+1,i}, \ldots, (1 - a)\, \alpha_{sj} + a\, \alpha_{si}]^T. \tag{12}$$

The results obtained from test cases in [17] showed that the system without simple crossover is less stable than the system without arithmetical crossover. In these tests, the traditional GA has been used in solving of optimization problems.

4.2.2. Arithmetical crossover

This operator is defined as a linear combination of two vectors. If parents p_i and p_j are crossed, the offsprings are given as follows:

$$\tilde{p}_i = a\, p_i + (1 - a)\, p_j, \qquad\qquad \tilde{p}_j = a\, p_j + (1 - a)\, p_i. \tag{13}$$

It follows from tests performed by Michalewicz [17] that the GA without arithmetical crossover has slower convergence.

4.2.3. Heuristic crossover

This operator is unique because it utilizes values of the objective function $f(p)$ defined in Eq. (11) to the determination of next search direction. Moreover, the heuristic crossover produces only a single offspring. Regarding to this fact, we generate the parameter a twice for the given parents p_i and p_j, and the following process of determination of a chromosome $\tilde{p}$ is also repeated two times,

$$\tilde{p} = \begin{cases} p_j + a\,(p_j - p_i) & \text{for} \quad f(p_j) \leq f(p_i), \\ p_i + a\,(p_i - p_j) & \text{otherwise.} \end{cases} \tag{14}$$

It seems that this operator can help us in searching for more accurate solution. It is in particular useful to fine local tuning when all chromosomes are already near each other in the population. Note that this operator may produce offspring outside the domain $\mathcal{D}$. In such a case another random value a is generated and another offspring is created. If after three attempts no new feasible solution is found, this crossover is replaced with the arithmetical crossover.

It is obvious that all three types of crossover operations are useful in the evolutional process and should be applied. But, some operators are better to use at the start of searching and some of them are more useful to use when the evolutional process finishes. Therefore we defined the probabilities of a selection for each crossover operation type in the following way:

$$p_1 = 1 - (p_2 + p_3), \qquad p_1 \in (0,1) \qquad \text{for} \quad \text{simple crossover,} \tag{15}$$
$$p_2 = s_1 + s_2 s_3^{-c}, \qquad p_2 \in (0,1) \qquad \text{for} \quad \text{arithmetical crossover,} \tag{16}$$
$$p_3 = s_4 + s_5 s_6^{-c}, \qquad p_3 \in (0,1) \qquad \text{for} \quad \text{heuristic crossover,} \tag{17}$$

where $s_1, \ldots, s_6$ are chosen real parameters and $c \geq 0$. Let S_i denotes the standard deviation of numbers in the ith row of a population $P(g)$ and $\overline{\alpha_i}$ is the arithmetic mean of the same numbers. Then the parameter c is defined as

$$c = \max \left\{ |S_1/\overline{\alpha_1}|, |S_2/\overline{\alpha_2}|, \ldots, |S_s/\overline{\alpha_s}| \right\}. \tag{18}$$

If the probabilities are calculated according to Eqs. (15)–(17), the selection process based on the roulette wheel is performed and the selected type of crossover is applied in the algorithm.

Note that the parameter c which means the absolute value of the variation coefficient is also used in the restart process. When the condition

$$c < \varepsilon \qquad \text{for} \quad \varepsilon > 0 \tag{19}$$

is satisfied, a new population is always generated in the evolutional process.

4.3. Mutation operation

This operator is performed on a gene-by-gene. The probability of mutation p_m provides the expected number $p_m \times P_s \times s$ of mutated genes. Every gene should have an equal chance to undergo mutation. Hence, we generate a random number r from the range $\langle 0, 1 \rangle$ for each gene in the whole population. If $r < p_m$, the gene is replaced. The main purpose of mutation is to keep the variety of population in the evolutional process and allows possible movement away from a local optimum in the search for a better result. On the other hand, a frequent mutation can be the reason of a low convergence of the whole algorithm. Therefore parameter p_m is usually set less than 0.05. These are the reasons why we decided to use only a uniform mutation in our algorithm. In this case, a selected gene is replaced by a random number from the admissible domain $\mathcal{D}$.

5. In-plane shear modulus estimation

It is clear from Eqs. (4) and (7) that the FE model based on Euler-Bernoulli theory described in the section 2 is not able to identify the in-plane shear modulus G_{LT} of a cross-ply laminate. But if some fibres and matrix material properties of longitudinal composite layers are known, the modulus G_{LT} could be estimated by means of the law of mixtures [1] as presented below.

Let us consider that the values of Poisson's ratios and density of fibres and matrix are known. These parameters are usually constant for the specified material type and could be commonly found on the web site of manufacturers. Besides this, we assume that the composite material density ρ can be calculated with the aid of weight and volume of the specimen. Then the fibre volume fraction V_f can be obtained from the simplified relations expressed by

$$\rho = \rho_f V_f + \rho_m V_m \qquad \text{and} \qquad V_m = 1 - V_f, \tag{20}$$

where ρ_f and ρ_m are fibres and matrix density, respectively. The parameter V_m denotes matrix volume fraction. Further, we expect that the parametric optimization problem has been performed and the Young's moduli E_L and E_T were identified. If we suppose the validity of the law of mixtures for the moduli E_L and E_T, i.e.

$$E_L = E_f V_f + E_m V_m \qquad \text{and} \qquad \frac{1}{E_T} = \frac{V_f}{E_{Tf}} + \frac{V_m}{E_m}, \tag{21}$$

and if we define the coefficients

$$r_{mf} = \frac{E_m}{E_f} \in (0,1), \qquad r_{ff} = \frac{E_f}{E_{Tf}} \geq 1 \qquad \text{and} \qquad r_{LT} = \frac{E_L}{E_T} \geq 1, \tag{22}$$

where E_m is matrix Young's modulus and E_f and E_{Tf} are longitudinal and transverse fibres Young's moduli, respectively, we can obtain the quadratic equation of variable r_{mf} in the form

$$(V_f V_m r_{ff}) r_{mf}^2 + (V_f^2 r_{ff} + V_m^2 - r_{LT}) r_{mf} + V_f V_m = 0. \tag{23}$$

Solution of this equation for the "reasonable" input coefficients r_{ff} and r_{LT} leads to finding an admissible root r_{mf} from feasible domain defined in (22)$_1$. Consequently, the relation

$$\frac{1}{G_{LT}} = \frac{V_f}{G_{LTf}} + \frac{V_m}{G_m} \tag{24}$$

is used to determine the in-plane shear modulus G_{LT}. The symbol G_{LTf} denotes the in-plane shear modulus of fibers and G_m is the shear modulus of matrix. It is concurrently assumed that these moduli could be determined for isotropic matrix material and generally orthotropic fibres material using the relations

$$G_m = \frac{E_m}{2(1+\nu_m)} \qquad \text{and} \qquad G_{LTf} = \frac{E_f}{2(1+\nu_f)r_{Gf}}, \tag{25}$$

where $r_{Gf} = 1$ for isotropic and $r_{Gf} > 1$ for orthotropic fibre material. The symbols ν_f and ν_m mean the major Poisson's ratio of fibres and the Poisson's ratio of matrix, respectively. By combining Eqs. (21)$_1$, (22)$_1$ and (25) with Eq. (24), we obtain

$$G_{LT} = \frac{E_L r_{mf}}{2(V_f + V_m r_{mf})[V_f(1+\nu_f)r_{Gf} r_{mf} + V_m(1+\nu_m)]}. \tag{26}$$

6. Numerical examples and discussion

Identification of elastic properties is demonstrated on examples of single-end clamped beams. These beams are assumed to be made of cross-ply laminates with various stacking sequences of layers as shown in Table 1. The geometric properties are chosen as follows: length and width of beams are 240 mm and 8 mm, respectively, each laminate has the same material properties of all layers which have the uniform thickness 0.5 mm. The mechanical properties of two investigated orthotropic composite materials are taken from [20] and are introduced in Table 2.

Table 1. Stacking sequences of used cross-ply laminates

Laminate model ID	VAR1	VAR2	VAR3
A	$[0_2/90_4/0_4]$	$[0_2/90_2/0]_\mathrm{S}$	$[0/90]_5$
B	$[90_2/0_4/90_4]$	$[90_2/0_2/90]_\mathrm{S}$	$[90/0]_5$

Table 2. Mechanical properties of used unidirectional laminae

Composite material type	T300/BSL914C epoxy	E-Glass/MY750/HY917/DY063 epoxy
Material model ID	MAT1	MAT2
Fibre volume fraction, V_f [$-$]	0.6	0.6
Longitudinal modulus, E_L [GPa]	138	45.6
Transverse modulus, E_T [GPa]	11	16.2
Shear modulus, G_{LT} [GPa]	5.5	5.83
Major Poisson's ratio, ν_{LT} [$-$]	0.28	0.278
Transverse Poisson's ratio, $\nu_{TT'}$ [$-$]	0.4	0.4

The vector Ω^exp of frequencies is not stated experimentally but is calculated numerically with the help of software MSC.Marc. This approach was chosen to verify the quality of the material identification process when the FE model with beam elements described in this work has been employed. At first, we analyzed the influence of the number of used FEs on values of first four flexural frequencies. The mesh of 32×2 regular four-node isoparametric shell elements with linear approximations of displacements was created using the software of MSC.Mentat. On the contrary, the beam element model contained only 8 elements. It was observed that the increase of element number in both model has a very small influence on the change of flexural eigenfrequencies values. As shown in Table 3 for laminate model VAR1, the eigenfrequency errors are about 5 % in comparison to both numerical model results. We came to similar conclusions in all other calculated cases.

Table 3. First four flexural frequencies in cycles per time, laminate model VAR1

FE model	ID	MAT1				MAT2			
		1	2	3	4	1	2	3	4
Shell elements (MSC.Marc)	A	124.2	770.4	2124	4073	63.54	397.5	1111	2171
	B	51.01	319.5	894.7	1754	42.72	267.8	750.5	1474
Beam elements	A	124.9	782.2	2190	4292	64.66	405.1	1134	2223
	B	53.64	336.0	940.5	1844	45.94	287.7	805.5	1579

The identification process of elastic Young's moduli and the major Poisson's ratio was performed according to optimization problem (11). However, it could happen that the same eigenfrequencies $\Omega(p)$ can be obtained from different values of input material parameters $p = [E_L, E_T, \nu_{LT}]^T$ because the beam is made from orthotropic layers of a various orientation. Therefore, the objective function $f(p)$ in Eq. (11) is constructed as a weighted sum of functions

$$f(p) = \xi_1 f_A(p) + \xi_2 f_B(p), \qquad \xi_1 = 1, \quad \xi_2 = 1, \qquad (27)$$

while the same material model ID is assumed. The indexes A and B mean various layer sequences stated in the columns of Table 1. This modification of objective function should lead to the unique solution of elastic parameters. Only first four flexural eigenfrequencies for every variant A and B are accepted in the whole optimization process. This limit is set for the reason of ability to reliably detect their maximum values in experimental way in the future.

The GA and the μGA were used to solve the optimization problem (11) where the objective function was constructed according to (27). The tournament selection and elitism were applied in all presented simulations. The restart process with parameter $\varepsilon = 0.001$ was also active in all calculations. The crossover and the mutation probabilities were set to be equal to 0.95 and 0.05, respectively, in the case of GA. When the μGA was employed, the crossover operation was used with the same probability but the mutation operation was omitted. The parameters in Eqs. (16) and (17) which have influence on the selection of the crossover operation type were chosen as follows: $s_1 = 0.3$, $s_2 = 0.1$, $s_4 = 0.2$, $s_5 = 0.3$ and $s_3 = s_6 = 1 \cdot 10^6$. The maximum number of generated chromosomes during the process was invariable and was set equal to 3000. Furthermore, every optimization problem was independently repeated 50 times and the obtained results were evaluated statistically.

Table 4. Statistical evaluation of E_L [GPa] and E_T [GPa], laminate model VAR1

$P_s \cdot N_g$	MAT1				MAT2			
	E_L^a	S_{E_L}	E_T^a	S_{E_T}	E_L^a	S_{E_L}	E_T^a	S_{E_T}
100·30	131.3	0.415	9.609	0.012	44.52	0.548	13.70	0.008
60·50	131.3	0.519	9.612	0.029	44.49	0.594	13.71	0.026
30·100	130.7	2.691	9.634	0.149	44.21	1.247	13.73	0.129
6·500	131.3	0.612	9.605	0.014	44.22	0.817	13.71	0.028

The identification process was started for the following estimation intervals of unknown material parameters: $E_L \in \langle 10, 1000 \rangle$ [GPa], $E_T \in \langle 1, 100 \rangle$ [GPa], $\nu_{LT} \in \langle 0.1, 0.4 \rangle$ [$-$]. It can be seen from Table 4 for the case of laminate model VAR1 that the average values of longitudinal modulus E_L^a and transverse modulus E_T^a give very similar results for various population size P_s. The results in the last row of Table 4 were obtained by using the μGA whereas the others by using the GA. The average estimations of the longitudinal modulus for MAT1 and MAT2 were different about 5% from expected values in Table 2. But the errors about 15% (difference of 2.5 GPa between real and calculated values) were detected for the transverse moduli. Besides, it can be concluded from Table 4 that the robustness of the GA and the μGA in regard to choice of initial Young's moduli values is very good because the standard deviations S_{E_L} and S_{E_T} are small in comparison to average values of E_L^a and E_T^a. Much worse results were obtained in the Poisson's ratio calculation. As shown in Table 5, determined average Poisson's ratios ν_{LT}^a are not near to expected values of this parameter, even though the best Poisson's ratios ν_{LT}^b are rather close to the real values given in Table 2. It is obvious that the identification problem has

Table 5. Statistical evaluation of ν_{LT} [$-$], laminate model VAR1

$P_s \cdot N_g$	MAT1			MAT2		
	ν_{LT}^b	ν_{LT}^a	S_ν	ν_{LT}^b	ν_{LT}^a	S_ν
100·30	0.281	0.241	0.068	0.286	0.239	0.078
60·50	0.281	0.245	0.087	0.277	0.242	0.080
30·100	0.282	0.263	0.118	0.276	0.252	0.106
6·500	0.266	0.233	0.105	0.276	0.270	0.101

a small sensitivity in relation to this parameter, which can be shown by means of the standard deviation S_ν in Table 5, because the ratios of S_ν to ν_{LT}^a give relatively large values against to coefficients of variation S_{E_L}/E_L^a and S_{E_T}/E_T^a. However, the value of the Poisson's ratio can be usually found in a local range for specified material type and it can be estimated reliable using the law of mixtures

$$\nu_{LT} = \nu_f V_f + \nu_m V_m .$$ (28)

In view of the fact that similar results were also obtained in the case of the laminate model VAR2 and VAR3, the Poisson's ratio ν_{LT} has been next calculated according to Eq. (28) and was removed from a vector of unknown parameters p. The value of ν_{LT} is then equal to 0.26 for both material models MAT1 and MAT2 while the Poisson's ratios of fibres and matrix have been adopted from [20], see Tables 6 and 7. Differences of computed and real (Table 2) values are less than 7 %.

Table 6. Mechanical properties of used fibres

Fibre type	T300	Silenka E-Glass 1200tex
Longitudinal modulus, E_f [GPa]	230	74
Transverse modulus, E_{Tf} [GPa]	15	74
In-plane shear modulus, G_{LTf} [GPa]	15	30.8
Transverse shear modulus, $G_{TT'f}$ [GPa]	7.0	30.8
Major Poisson's ratio, ν_f [$-$]	0.2	0.2

Table 7. Mechanical properties of used matrices

Matrix type	BSL914C epoxy	MY750/HY917/DY063 epoxy
Modulus, E_m [GPa]	4.0	3.35
Shear modulus, G_m [GPa]	1.48	1.24
Poisson's ratio, ν_m [$-$]	0.35	0.35

The identification process has been repeated again to generate a new vector of unknown material parameters $p = [E_L, E_T]^T$. The Poisson's ratio ν_{LT} was assumed constant during every computation and equals to 0.26 for all laminate and material models. The longitudinal and transverse moduli obtained using the μGA and the GA for various numbers of P_s and N_g gave very similar results. Therefore, only the results of the μGA are presented in the following text. The ranges of investigated parameters were chosen as above, i.e. $E_L \in \langle 10, 1000 \rangle$ [GPa], $E_T \in \langle 1, 100 \rangle$ [GPa].

As it is obvious from Tables 8 and 9 for VAR1 and VAR2, the value of the Poisson's ratio ν_{LT} has a neglected influence on computation of Young's moduli E_L and E_T in comparison to results in Table 4. The low values of the standard deviations S_{E_L} and S_{E_T} mean that our optimization algorithm is robust for searching parameters. However, these conclusions are not

Table 8. Statistical evaluation of longitudinal modulus E_L [GPa], $\nu_{LT} = 0.26$ [−]

Model	MAT1			MAT2		
	VAR1	VAR2	VAR3	VAR1	VAR2	VAR3
E_L^b	131.7	131.9	137.5	44.53	44.80	45.52
E_L^a	131.3	131.6	95.29	44.46	44.57	27.97
S_{E_L}	0.109	0.148	35.40	0.033	0.079	15.19

Table 9. Statistical evaluation of transverse modulus E_T [GPa], $\nu_{LT} = 0.26$ [−]

Model	MAT1			MAT2		
	VAR1	VAR2	VAR3	VAR1	VAR2	VAR3
E_T^b	9.619	10.43	11.00	13.72	13.88	15.65
E_T^a	9.606	10.30	38.64	13.71	13.77	26.83
S_{E_T}	0.006	0.037	27.03	0.005	0.029	11.73

valid in the case of VAR3 because incorrect results were obtained for average Young's moduli. In addition, large values of standard deviation were computed. The reason of errors lies in the unsuitable assemblage of composite layers, see Table 1. The laminate model VAR3 can be taken into account as quasi-isotropic on a macroscopic scale in both variants A and B. Therefore, the vector of computed eigenfrequencies is the same in variant A and B. Due to this fact, there is not only one solution of the optimization problem in the domain $\mathcal{D}$. This conclusion follows from the very low objective function value at the end of the optimization process when the number of generated chromosomes is equal to $3\,000$. This occurrence was observed in all considered cases.

It was discussed in the section 5 that the presented mathematical model of the beam is not able to identify the in-plane shear modulus G_{LT}. Therefore, the method of this modulus estimation has been proposed. The calculations of G_{LT} were performed for the average Young's moduli given in Tables 8 and 9, and for the Poisson's ratios given in Tables 6 and 7. The coefficients r_{ff} and r_{Gf} were set according to elastic moduli values of fibers in Table 6. While E-glass fibers have isotropic properties and due to the both coefficients were set equal to 1, T300 carbon fibres are orthotropic and the coefficients were determined as follows: $r_{ff} = 230/15 \doteq 15.3$, see Eq. (22)$_2$; $G_{LTf} \equiv E_{Tf} \implies r_{Gf} = r_{ff}/[2\,(1 + \nu_f)] = 15.3/[2\,(1 + 0.2)] \doteq 6.39$, see Eqs. (22)$_2$ and (25)$_2$. The volume fraction $V_f = 0.6$ (Table 2) was assumed constant in all calculated problems. Note that the mechanical properties of a certain material class, as Young's moduli or the Poisson's ratio, are usually almost invariable. This fact can be also supported for the carbon fibres (namely for fibre types AS4 and T300) by experimental data given in [20]. Therefore, we can suggest the set of coefficients $r_{ff} = 15$ and $r_{Gf} = 6.2 \div 6.4$ when the specific material properties of the carbon fibres are unknown.

The results of G_{LT} calculations for VAR1 and VAR2 are stated in Table 10. The variant VAR3 was not considered. When we compare the values of shear modulus from Tables 2 and 10 it is evident that the best computed value was found in the case of the material model MAT1 and VAR2 where the relative error was less than $4\,\%$. In the rest cases, the relative difference between the known and calculated values is about $12\,\%$, which is result comparable to results obtained for the transverse moduli. If we compute the shear modulus G_{LT} only with real data from Tables 2, 6 and 7, we obtain even more better results. These moduli are as follows: $G_{LT} = 5.69$ [GPa] for MAT1 (error $3.4\,\%$) and $G_{LT} = 6.09$ [GPa] for MAT2 (error $4.5\,\%$).

Table 10. Calculated values of average shear modulus G_{LT} [GPa], $V_f = 0.6$ [−]

Model	MAT1		MAT2	
	VAR1	VAR2	VAR1	VAR2
r_{LT}	13.67	12.78	3.243	3.237
r_{mf}	0.030	0.034	0.088	0.089
G_{LT}	4.804	5.286	5.145	5.168

7. Conclusion

It has been shown in this paper that the robustness of genetic algorithms is very good for the determination of Young's moduli values. This method is particularly effective for a low time-consuming computation of chromosomes in a generation. Acceptable values of the modulus E_L were calculated but worse results were obtained for moduli E_T and G_{LT}. These results are partially influenced by the selection of used mathematical model of a beam because of the differences between eigenfrequencies obtained from shell and beam elements. The calculation of the shear modulus G_{LT} was then directly dependent on the values of E_L and E_T. The computation of the Poisson's ratio ν_{LT} gave high standard deviation values when inverse procedure was used for identification. However, a small influence of ν_{LT} value on values of Young's moduli was found. In practice the value of ν_{LT} is quite close in vicinity of 0.3 and can be stayed close to this value or can be computed by using the law of mixtures when the volume fraction of fibers and the Poisson's ratios of the fibers and the matrix are known.

The proposed methodology has some disadvantages. The measurement of eigenfrequencies has to be performed on two independent specimens with different layer sequences. Due to this fact, the beam specimen made of a quasi-isotropic laminate is not suitable to use for the identification of material properties. In addition, we have to determine the volume fraction of fibers and we have to know some material properties of fibers and matrix when we want to apply the law of mixtures in the calculation of ν_{LT} and G_{LT}. It requires the knowledge of fibers and matrix material properties. On the other hand, the proposed methodology brings some advantages. The method enables to utilize the simple specimens. Mechanical properties are directly calculated for a final laminate. Only a few eigenfrequencies given by the simple test are needed to be computed values of Young's moduli E_L and E_T. The required measuring aparatures are not as expensive as typical static testing machines.

In future, the real measuring of eigenfrequencies is prepared and different mathematical model of the beam which gives higher accuracy of calculated frequencies should be used.

Acknowledgements

This work has been supported by the European project NTIS-New Technologies for Information Society No.: CZ.1.05/1.1.00/02.0090.

References

[1] Altenbach, H., Altenbach, J., Kissing, W., Mechanics of composite structural elements, Springer, Berlin, 2004.

[2] Balasubramaniam, K., Rao, N. S., Inversion of composite material elastic constants from ultrasonic bulk wave phase velocity data using genetic algorithms, Composites 29 B (1998) 171–180.

[3] Chang, W. D., Nonlinear system identification and control using a real-coded genetic algorithm, Applied Mathematical Modelling 31 (2007) 541–550.

[4] Chen, C. M., Kam, T. Y., A two-level optimization procedure for material characterization of composites using two symmetric angle-ply beams, International Journal of Mechanical Sciences 49 (2007) 1 113–1 121.

[5] Chu, Y. C., Rokhlin, S. I., Stability of determination of composite moduli from velocity data in planes of symmetry for weak and strong anisotropies, Journal of the Acoustical Society of America 95 (1) (1994) 213–225.

[6] Frederiksen, P. S., Application of an improved model for the identification of material parameters, Mechanics of Composite Materials and Structures 4 (4) (1997) 297–316.

[7] Genovese, K., Lamberti, L., Pappalettere, C., A new hybrid technique for in-plane characterization of orthotropic materials, Experimental Mechanics 44 (6) (2004) 584–592.

[8] Goldberg, D. E., Genetic algorithms in search, optimization and machine learning, Addison-Wesley, Boston, 1989.

[9] Han, X., Liu, G. R., Li, G. Y., Transient response in cross-ply laminated cylinders and its application to reconstruction of elastic constants, Computers, Materials and Continua 1 (1) (2004) 39–49.

[10] Ip, K. H., Tse, P. CH., Lai, T. CH., Material characterization for orthotropic shells using modal analysis and Rayleigh-Ritz models, Composites 29 B (1998) 397–409.

[11] Kam, T. Y., Liu, C. K., Stiffness identification of laminated composite shafts, International Journal of Mechanical Sciences 40 (9) (1998) 927–936.

[12] Lecompte, D., Smits, A., Sol, H., Vantomme, J., Van Hemelrijck, D., Mixed numerical-experimental technique for orthotropic parameter identification using biaxial tensile tests on cruciform specimen, International Journal of Solids and Structures 44 (2007) 1 643–1 656.

[13] Liu, G. R., Han, X., Lam, K. Y., A combined genetic algorithm and nonlinear least squares method for material characterization using elastic waves, Computer Methods in Applied Mechanics and Engineering 191 (2002) 1 909–1 921.

[14] Liu, G. R., Ma, W. B., Han, X., An inverse procedure for determination of material constants of composite laminates using elastic waves, Computer Methods in Applied Mechanics and Engineering 191 (2002) 3 543–3 554.

[15] Liu, G. R., Tani, J., Ohyoshi, T., Watanabe, K., Transient waves in anisotropic laminated plates, Journal of Vibration and Acoustics 113 (1) (1991) 230–239.

[16] Liu, G. R., Xi, Z. C., Elastic waves in anisotropic laminates, CRC Press LLC, Boca Raton, 2002.

[17] Michalewicz, Z., Genetic algorithms + data structures = evolution programs, Springer, Berlin, 1995.

[18] Mota Soares, C. M., de Freitas, M. M., Araújo, A. L., Pedersen, P., Identification of material properties of composite plate specimens, Composite Structures 25 (2) (1993) 277–285.

[19] Rikards, R., Chate, A., Steinchen, W., Kessler, A., Bledzki, A. K., Method for identification of elastic properties of laminates based on experiment design, Composites 30 B (1999) 279–289.

[20] Soden, P. D., Hinton, M. J., Kaddour, A. S., Lamina properties lay-up configurations and loading conditions for range of fibre-reinforced composite laminates, Composites Science and Technology 58 (1998) 1 011–1 022.

[21] Wang, L., Tang, F., Wu, H., Hybrid genetic algorithm based on quantum computing for numerical optimization and parameter estimation, Appl. Math. Comput. 171 (2005) 1 141–1 156.

[22] Zajíček, M., Adámek, V., Dupal, J., Finite element for non-stationary problems of viscoelastic orthotropic beams, Applied and Computational Mechanics 5 (1) (2011) 89–100.

6

Biomechanical study of the bone tissue with dental implants interaction

P. Marcián[a,*], L. Borák[a], Z. Florian[a], S. Bartáková[b], O. Konečný[c], P. Navrátil[a]

[a]Institute of Solid Mechanics, Mechatronics and Biomechanics, Faculty of Mechanical Engineering, Brno University of Technology, Technická 2896/2, 616 69 Brno, Czech Republic

[b]Stomatological Clinic, Faculty of Medicine, Masaryk University, Komenského náměstí 2, 662 43 Brno, Czech Republic

[c]Faculty of Eletrical Engeenering and Communication, Brno University of Technology, Technická 3058/10, 616 00 Brno, Czech Republic

Abstract

The article deals with the stress-strain analysis of human mandible in the physiological state and after the dental implant application. The evaluation is focused on assessing of the cancellous bone tissue modeling-level. Three cancellous bone model-types are assessed: Non-trabecular model with homogenous isotropic material, non-trabecular model with inhomogeneous material obtained from computer tomography data using CT Data Analysis software, and trabecular model built from mandible section image. Computational modeling was chosen as the most suitable solution method and the solution on two-dimensional level was carried out. The results show that strain is more preferable value than stress in case of evaluation of mechanical response in cancellous bone. The non-trabecular model with CT-obtained material model is not acceptable for stress-strain analysis of the cancellous bone for singularities occurring on interfaces of regions with different values of modulus of elasticity.

Keywords: dental implant, trabecular bone, finite elements method, stress and strain analysis, modeling level

1. Introduction

Since 1960's when Brånemark and Linkow defined the direction of modern dental implantology, this medical field developed into a large extent and today it provides very effective solution to many dental problems: Starting with restoring the natural function of masticatory system and ending with dental aesthetic improvements. In 2003, there were more than 220 different types of implants manufactured by about 80 producers [12]. New materials such as technically clean titanium, nanostructural titanium or titanium alloys etc. as well as deepening understanding of the implant-living tissue interaction are emphasized nowadays. Despite the reliability of dental implants which is nowadays around 95 % [18,27], it would be mistake to underestimate influences of other factors on successful treatment. Therefore, it is necessary to continue in efforts to improve quality of dental implants.

Application of a dental implant which is essentially a technical object is not only a medical procedure but an interdisciplinary process. It is therefore necessary to solve the dental implantology problems together with technical specialists, i.a. biomechanicians whose task is to assess the mechanical interaction of the system components, including living tissues as well as technical materials. In this respect, a very effective tool is computational modeling. Implant failures depend on many various factors as, for instance, mechanical ones. These factors are

---

*Corresponding author. e-mail: ymarci00@stud.fme.vutbr.cz.

expressed by mechanical quantities stress and strain. In biomechanics computational models basically consists of models of geometry, material, loading and boundary conditions.

2. Material and methods

The paper deals with the interaction of a tooth or a dental implant with bone tissue regarding various modeling levels of the bone model. As the most suitable solution method was chosen the finite element method (FEM) which is used very often in such biomechanical problems like this. Thanks to latest technologies it is possible to build models on a quite high modeling-level. An example of building of such model is given in this article as well. In our case ANSYS 11.0 (Ansys Inc., Canonsburg, PA, USA) was used.

Prior to calculations, suitable geometry model is needed. The model of trabecular model of cancellous bone is obtained from the image of the mandible section. Therefore, 2-D geometry level is used throughout this study. This approach is sufficient enough for the comparison of the three proposed material models of mandibular bone. 3-D level of the same case is also possible but the model of trabecular structure would require data obtained directly from micro-CT [21].

In this study, the solution is divided into three parts:

1) An assessment of modeling level of bone tissue material.

2) Stress-strain analysis of mandible in the physiological state.

3) Stress-strain analysis of mandible with an applied screw implant.

2.1. An assessment of modeling level of bone tissue material

There are two types of bone tissue: Hard outer cortical bone and inner trabecular structure, so-called cancellous bone. All bones are inhomogeneous anisotropic biomaterials. Both these types should be incorporated into the model.

In our study, three general modeling-levels of cancellous bone material are assessed: Non-trabecular model (the lowest level), trabecular model (considering the trabecular architecture of the real bone) and CT-obtained model (non-trabecular but inhomogeneous material). Furthermore, these general models are divided into "sub-levels" according to the bone quality as will be explained below. Cortical bone is modeled in all cases as homogenous material with no special architecture.

The shape of bone was obtained based on the image of mandibular cross section [15] and also by using the computer tomography. Using CAD software SolidWorks and Rhinoceros, 2-D geometry model was created, specifically the outer shape corresponding to the 1st premolar region as well as the trabecular architecture of cancellous bone (see Fig. 1).

Linear isotropic material model is used for all parts of the system, i.e. two material characteristics are needed, specifically Young's modulus E [MPa] and Poisson's Ratio μ [–].

At the lowest level, bone is modeled as non-trabecular homogenous material with apparent material characteristics of cancellous bone (no cortical bone is incorporated, see Fig. 1a). Sensitivity study is performed to obtain proper value of the Young's modulus which is changed ranging from 150 to 750 MPa. The changing value refer to varying bone quality, specifically the lower Young's modulus value means lower bone quality and therefore lower bone density (e.g. in osteoporotic bones [29]). As for the interaction of an implant with bone, the main disadvantage of bone with lower density is lower bone support of applied implant.

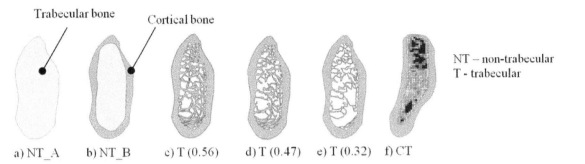

Fig. 1. Model of geometry and analyzed material modeling levels overview: a) non-trabecular model (only cancellous bone) – NT_A; b) non-trabecular model (cancellous as well as cortical bone) – NT_B; c) trabecular bone of good quality (BA/TA – 0.56); d) trabecular bone of normal quality (BA/TA – 0.47); e) trabecular bone of low quality (BA/TA – 0.32); f) CT-obtained model

Next modeling-level used in the study and which is very often used by other researchers ([4,7,19]) is non-trabecular homogenous model with incorporated cortical layer (see Fig. 1b). In this case, Young's modulus of cortical bone is of 13 700 MPa [10,31] and similar sensitivity study as in previous modeling-level is performed in case of cancellous bone, i.e. Young's modulus ranged from 150 to 750 MPa.

Contrary to the previous models, model shown in Fig. 1c includes trabecular structure of real cancellous bone. At the micro-level trabeculae has the same material properties as cortical bone [25,32] and therefore the Young's modulus of such cancellous bone model is of 13 700 MPa. In such model, lower bone quality can be modeled by using less dense structure and with thinner trabeculae (see Fig. 1d,e). The bone quality is here identified by a special characteristics often used in these cases, i.e. by so-called bone area fraction BA/TA which is defined as the area of bone tissue per total area (in 2-D analysis) [33].

The last modeling-level used in the study is non-trabecular inhomogeneous model. This model takes into account the distribution of bone density which is in the model characterized by varying value of the Young's modulus. This material property values can be obtained directly from CT images in the following way.

Prior to dental implant application, densitometry investigation is often performed by using CT [28,34]. From CT images, CT numbers (which signify pixel intensity of the image) can be read. CT numbers can be converted to Hounsfield units (HU) [34] by which individual tissues are identified. There are observed correlation between HU and bone density or Young's modulus of that (mainly cancellous bone) tissue [16,23,24]. This correlation is the basement for material model. For purpose of its creation, new software CT Data Analysis [35] was developed at Institute of Solid Mechanics, Mechatronics and Biomechanics, Brno University of Technology. This software allows the user to load CT images and after defining a desired region to export CT numbers (HU units) matrix. This matrix is then converted into the Young's moduli matrix which can be exported into Ansys program as CT-obtained material model. It should be noted that two correlations can be used:

1) Linear correlation: The reference value is the CT number of cortical bone which has known Young's modulus of 13 700 MPa. According to the known formulas, Young's moduli of other tissues (cancellous bone in various positions within the region) can be calculated (see Fig. 2a)

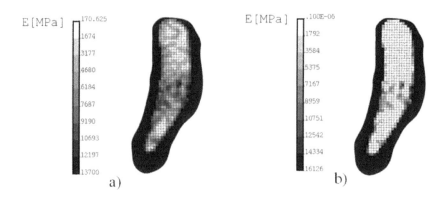

Fig. 2. CT-obtained material model: a) linear correlation used – CTlin, b) nonlinear correlation used – CTnonlin

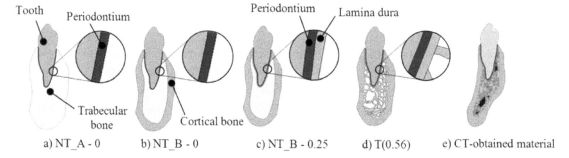

Fig. 3. Model of geometry: a) case 1A, b) case 1B, c) case 1B, d) case 1B, e) case 1B

2) Nonlinear correlation: Relation between bone density and Young's modulus is given by formulae (1) [23,24] (See Fig. 2b and note that the bone shape is different to that in Fig. 1 since it is obtained from different mandible)

$$\rho = 1.205 \cdot HU + 139, \qquad E = 2.349 \cdot \rho^{2.15}. \qquad (1)$$

Material characteristics used for all models with the exception of non-trabecular inhomogeneous model of cancellous bone are presented in Table 1.

Table 1. Material characteristics: Young's modulus and Poisson's ratio

	E [MPa]	μ [−]	Reference
Cortical bone	13 700	0.3	[10,31]
Cancellous bone	150–750	0.3	[8,9]
Dentin	17 600	0.25	[17]
PDL	10	0.45	[14,30]
Titanium	110 000	0.33	[36]

2.2. Stress-strain analysis of mandible in the physiological state

For the analysis of the physiological state geometry model is modified in alveolar region so as 2-D tooth model (1st premolar) including periodontal layer could be embodied. The layer is of 0.25 mm in thickness. In case of NT_A model (see Fig. 1a) the tooth with periodontium (PDL) is embodied directly into cancellous bone (see Fig. 3a). In case of NT_B model (see Fig. 1b)

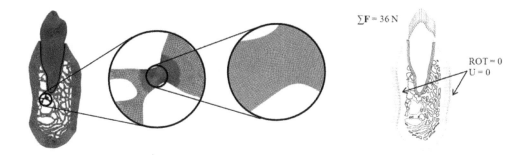

Fig. 4. FE mesh and the boundary condition

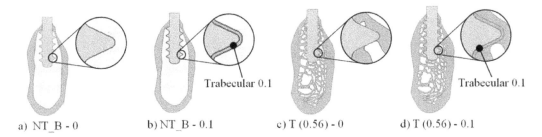

a) NT_B - 0   b) NT_B - 0.1   c) T (0.56) - 0   d) T (0.56) - 0.1

Fig. 5. Implant-Bone interface: Cases overview

two other variants are modeled: Model without Lamina dura layer (see Fig. 3b) and model with Lamina dura layer of 0.25 mm in thickness (see Fig. 3c). In case of all trabecular cancellous bone models ($T(0.56), T(0.47)$ and $T(0.32)$ – see Fig. 3d) as well as in case of non-trabecular inhomogeneous model (see Fig. 3e) Lamina dura of the same thickness is included as well.

All models are discretized with 8-node quadratic element PLANE 182 with global size of 0.02 mm. Since periodontium, which is largely deformable soft tissue, is part of the model, the finite strain theory was selected for the analysis, i.e. the finite strain tensor is enlarged by higher order derivatives which are negligible in infinitesimal strain theory.

In order to carry out the FEM solution, it is inevitable to have boundary conditions. In our case, the model is in all cases constrained in the middle of buccal as well as lingual cortical bone boundary (see Fig. 4). The model is loaded by the force applied in coronoapical direction on the tooth crown; specifically the force of 36 N is applied. This value is equivalent to the loading force of 200 N applied in similar 3-D model [11].

2.3. Stress-strain analysis of mandible with an applied screw implant

The same cases as in the previous section are modified so as the screw implant could be inserted into the presented models. Two specific cases are furthermore analyzed. Either there is no occurrence of special bony structure similar to lamina dura (see Fig. 5a,c) or there is. If the structure occurs, the layer of this special bone is of 0.1 mm in thickness [1,13,22]. In case of $T(0.47), T(0.32)$ and CTmat only cases with the special bony structure are analyzed.

Implants are osseointegrated [1,5]. Three stages of the osseointegration are analyzed:

A) implant is osseointegrated by its whole surface (see Fig. 6a),

B) implant neck region is not osseointegrated with cortical bone (see Fig. 6b),

C) implant is not osseointagrated at all (unacceptable situation, see Fig. 6c).

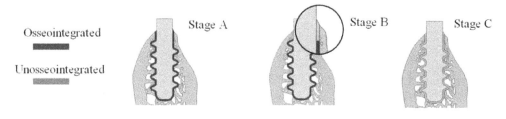

Fig. 6. Stages of osseointegration

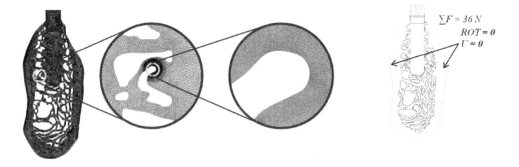

Fig. 7. FE mesh and the boundary conditions (model with the implant)

The osseointegration is modeled in ANSYS by rigid connection of osseointegrated parts, i.e. implant and bone have common FE nodes. In the un-osseointegrated regions (stages B and C), the implant bone connection is modeled by using contact elements CONTA172 and TARGE169 (pure penalty algorithm, standard contact, friction coefficient $f = 0$, normal contact stiffnes $FKN = 1$).

Model discretisation is similar as in the previous section. The same are also the boundary condition as well as the loading (see Fig. 7). Implant is made of technically pure titanium and it is modeled using linear isotropic model (see Table 1).

A total number of 35 computational models are created. Intel Core 4 Duo inside 2GHz 8GB RAM processor, PCG iterative solver and Full Newton-Raphson procedure was used. Unless specified otherwise all solver options were left default.

3. Results

3.1. Stress-strain analysis of mandible in the physiological state

Strain intensity and stress intensity are analyzed specifically their maximum values which occur in various locations of the system. Besides, maximum 1st and 3rd principal stresses (S1, S3) are analyzed as well. Representative strain intensity distribution for NT_B-0.25 and $T(0.56)$ cases is shown in Fig. 8. Results of all cases are shown in bar graphs in Fig. 9 and 10. These graphs show significant difference in stress and strain results.

By comparing the maximum stresses in cancellous bone, it is obvious that non-trabecular model gives significantly lower values that trabecular model. In case of non-trabecular model, the significant effect of lamina dura layer is also observed (NT_B-0.25). This layer is a part of load-bearing structure consisting mainly of cortical bone. Thus, this makes the cancellous bone less stressed than in case without the layer (NT_B-0). The highest stresses are found in trabecular cancellous bone models, especially with the trabeculae of low quality ($T(0.32)$).

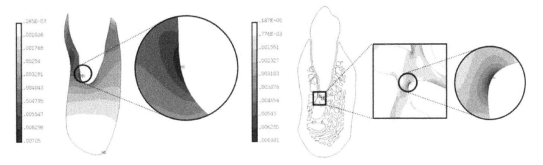

Fig. 8. Strain intensity in cancellous bone with the tooth. Comparison of cases NT_B-0.25, $T(0.56)$

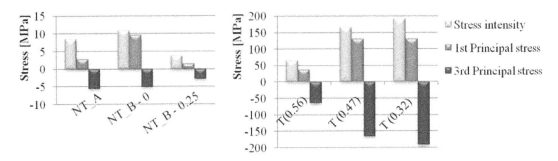

Fig. 9. Maximum stresses

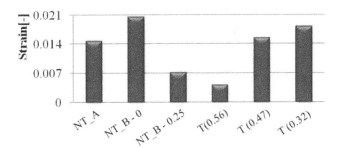

Fig. 10. Maximum strain intensities

On the other hand, by comparing the maximum strains in cancellous bone, the highest values are found in non-trabecular cancellous bone models without lamina dura (NT_A and NT_B-0) and, on the contrary, the lowest values are in trabecular cancellous bone models with trabeculae of low quality ($T(0.32)$) (see Fig. 10). From a quantitative point of view, the best accordance can be observed between the non-trabecular model including lamina dura (NT_B-25) and the trabecular model $T(0.56)$. Even the maximum values position is similar in these two cases (see Fig. 8). Therefore, it is obvious from the presented graphs that there is significant difference between stress and strain assessment. Stresses can be of tens to hundreds MPa in trabecular cancellous bone models, increasing with lowering the quality of trabeculae, whereas in non-trabecular cancellous bone stresses are much more lower, decreasing with lowering quality of bone (lowering its Young's modulus).

Neither stresses nor strains can be analyzed in models with CT-obtained material properties because of singularities occurring at interfaces of too many regions of different Young's modulus.

3.2. *Stress-strain analysis of mandible with an applied screw implant*

Results of the three analyzed stages of osseointegration are shown in Fig. 11 through Fig. 13. The maximum values of stress and strain intensities occurred always in different locations. It is found that not-surprisingly the best case is the fully osseointegrated implant. Two representative modeling-levels are presented. Firstly, model without the special bone layer around the implant $(T(0.56) - 0)$, and secondly, model with the special bone layer $(T(0.56) - 0.1)$. It is found that the layer reduces stresses and strains in cancellous bone and it also changes their distribution. Fig. 13 shows 1st principal stress distribution in models with lower quality of trabeculae $(T(0.45) - 0.1$ and $T(0.32) - 0.1)$.

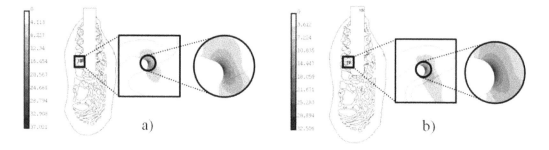

Fig. 11. 1st Principal stress. Stage A: a) $T(0.56) - 0$, b) $T(0.56) - 0.1$

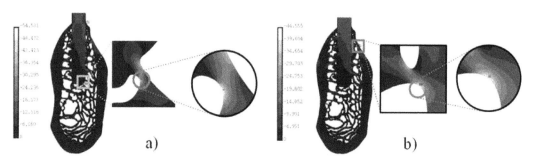

Fig. 12. 1st Principal stress. Stage B: a) $T(0.56) - 0$, b) $T(0.56) - 0.1$

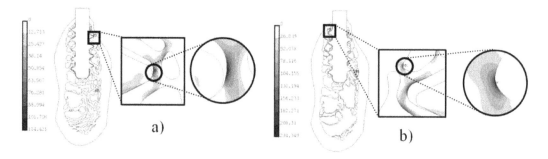

Fig. 13. 1st Principal stress. Stage C: a) $T(0.47) - 0.1$, b) $T(0.32) - 0.1$

As in physiological state analysis, results are presented in bar graphs again (see Fig. 14 and Fig. 15). The series of three bars (each for one stage of osseointegration) are presented for each material used in the model. Only stress and strain intensities are presented. From the graphs, significant difference between stress and strain results is obvious in all analyzed cases. The

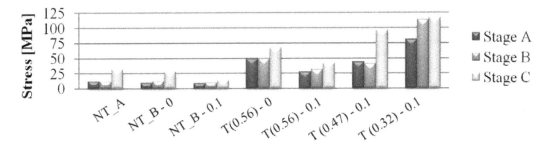

Fig. 14. Maximum stress intensities in cancellous bone

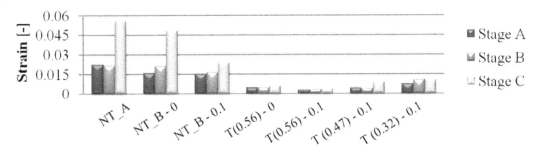

Fig. 15. Maximum strain intensities in cancellous bone

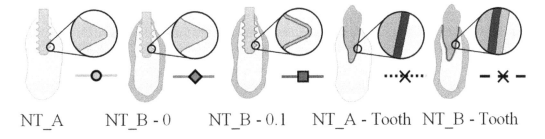

Fig. 16. Non-trabecular material models: Analyzed cases overview

stress intensity is of the highest value in non-trabecular cancellous bone model ($T(0.32) - 0.1$), whereas the strain values are, on the contrary, of the highest value in NT_A model (see Fig. 15). As stated in the previous section, neither stress nor strain analysis are applicable in model with CT-obtained material properties.

3.3. Stress-strain analysis of non-trabecular cancellous bone model

Results presented in the previous section are related to the trabecular model whose lower density (quality) is modeled by assuming thinner trabeculae. However, in most publications considering osteoporotic bone tissue non-trabecular cancellous bone model is prevailing and lower bone quality is modeled with decreasing apparent Young's modulus (see Table 1). Results obtained from such model are presented in this section.

Maximum values of stress intensity and strain intensity were analyzed again. Two states were compared: The physiological state on the one hand, and after implantation on the other (see Fig. 16). The maximum values of stresses change their position as Young's modulus of cancellous bone changes. On the contrary, the position of maximum strain values does not

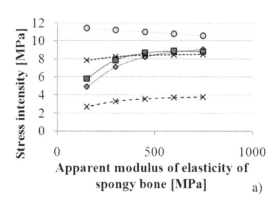
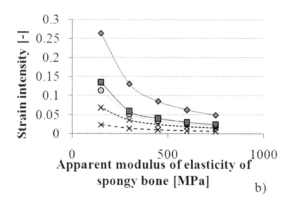

Fig. 17. Maximum stress and strain intensities. Non-trabecular models comparison (for graph legend see Fig. 16)

change (no strain redistribution occurs). As in the previous section, three osseointegration stages are analyzed. Fig. 17 shows results of fully osseointegrated stage. Each graph shows also the comparison with physiological state in two variants (dashed line).

4. Discussion

The loading of real cancellous bone of lower density (i.e. with lower BA/TA value) causes increasing stresses and strains in this tissue. It depends on the trabeculae distribution within the bone, their number and thickness. The lower number and thickness, the higher stresses occur (it is similar to loading of a beam in mechanical engineering). By comparing stresses in Fig. 9 and strains in Fig. 10 it is clear that a significant difference occurs in various material models of cancellous bone. This comparison is not possible in case of CT-obtained material model since no acceptable stresses and strains are calculated due to above mentioned singularities.

As for the models with the implant, a significant difference in maximum stresses and strains occurs in case of various material models. Much higher stresses occur in trabecular model of cancellous bone than in non-trabecular one. Quite the opposite result gives the strain evaluation; the highest values are obtained from non-trabecular model of cancellous bone while the lowest are obtained from trabecular one. Therefore, one should be aware of these differences between both cases (trabecular/non-trabecular). It should be seriously considered which of the two models is suitable for the analysis that one would like to perform. Our study shows that non-trabecular model is no longer suitable for analysis of interaction bone-tooth/implant. Model at higher modeling-level (i.e. trabecular model) should be used instead. To these days, cancellous bone is usually modeled as non-trabecular mass with no complicated structure of trabeculae [3,6,26]. An apparent Young's modulus obtained from experiment or from CT is prescribed to this structure. Stresses obtained from such models are inconsistent with reality because with decreasing bone quality stresses should increase, not decrease as shown by non-trabecular model results presented in this study. This inconsistency does not appear in strain evaluation. With decreasing bone quality strains increase (see Fig. 17a). The reason why maximum stresses in cancellous bone decrease as strains increase can be explained by the presence of cortical bone which plays significant role of the load bearing structure. This is obvious from Fig. 17a which shows that stresses in the model without the cortical bone (NT_A) are almost independent on Young's modulus change while stresses in the model with cortical bone (NT_B)

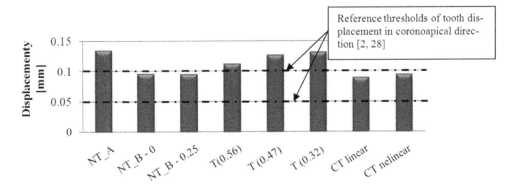

Fig. 18. Coronoapical tooth displacements

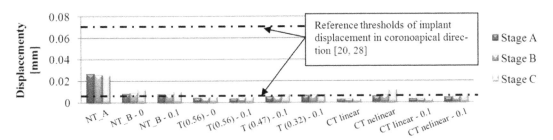

Fig. 19. Coronoapical implant displacements

decrease with the changing Young's modulus. This is because the load is carried via the cortical bone and the cancellous bone load is reduced. The study shows that strains are more appropriate mechanical quantity for the evaluation of mechanical interaction.

Verification of model quality, and thus quality of calculation results, is quite difficult in biomechanics. In case of dental biomechanics there is a possibility to utilize known values of tooth movement and implant movement. Specifically, we can compare calculated tooth and implant displacements in the coronoapical direction (i.e. in direction of the applied force) with that of obtained from literature. The displacement, in case of a tooth embedded in a bone of normal quality (see model $T(0.56)$) and loaded by force of 200 N, is of 0.092 2 mm. This value falls within the interval 0.05–0.1 mm which is valid for the same loading [2,28]. Displacement results for all models are shown in Fig. 18. Maximum calculated value of implant displacement ranges from 0.004 to 0.027 mm (see Fig. 19). These results are also in agreement with values obtained from literature (0.005–0.073 mm; [20,28]).

5. Conclusion

This biomechanical study is focused on assessment of the modeling-level of cancellous bone and its interaction with tooth or dental implant. 39 computational models are created and analyzed. Maximum values of stress and strains are evaluated and for better clarity shown in several graphs. Following conclusion can be stated:

1. It is more suitable to evaluate strains than stresses in an assessment of mechanical response of cancellous bone on its loading.

2. CT-obtained material model of cancellous bone is not acceptable for stress-strain analysis because of singularities occurring at interfaces of too many regions with different Young's modulus. However, this model can be used for stress evaluation in implants.

3. Trabecular model of cancellous bone is more suitable for analysis of interaction of bone with the embodied tooth/implant. Further improvement of modeling-level should be researched. One possible direction of this research could be using micro CT, which can be helpful for creating 3-D trabecular model of cancellous bone.

Acknowledgements

This work was supported by grant specific research FSI-J-11-3/942 and FSI-S-11-12/1225.

References

[1] Albrektsson, T., Brånemark, P. I., Hannsson, H. A., Lindström, J., Osseointegrated titanium implants. Requirements for ensuring a long-lasting direct bone anchorage in man, Acta Orthopaedica Scandinavica, 52 (1981) 155–170.

[2] Amarsaikhan, B., Miura, H., Okada, D., Masuda, T., Ishihara, H., Shinki, T., Kanno, T., Influence of environmental factors on tooth displacement, Journal of Medical and Dental Sciences, 49 (2002) 19–26.

[3] Bonnet, A. S., Postaire, M., Lipinski, P., Biomechanical study of mandible bone supporting a four-implant retained bridge: finite element analysis of the influence of bone anisotropy and foodstuff position, Medical Engineering & Physics, 31 (2009) 806–815.

[4] Borák, L., Marcián, P., Florian, Z., Bartáková, S., Biomechanical study of disk implants, Engineering Mechanics, 17 (1) (2010) 1–12.

[5] Branemark, P. I., Hansson, B. O., Adell, R., Osseointegrated implants in the treatment of the edentulous jaw. Experience from a 10-year period, Scandinavian journal of plastic and reconstructive surgery and hand surgery, 16 (1977) 1–132.

[6] Daniel, L., Qing, L., Wei, L., Michael, S., Dental implant induced bone remodelling and associated algorithms, Journal of the Mechanical Behaviour of Biomedical Material, 2 (5) (2009) 410–432.

[7] Cattaneo, P. M., Dalstra, M., Melsen, B., The finite element method: a tool to study orthodontic tooth movement, Journal of Dental Research, 84 (5) (2005) 428–433.

[8] Clason, Ch., Hinz, A. M., Schieferstein, H., A method for material parameter determinative for the human mandible based on simulation and experiment, Computer Methods in Biomechanics and Biomedical Engineering, 7 (5) (2004) 265–276.

[9] Futterling, S., Klein, R., Strasser, W., Weber, H., Automated finite element modelling of human mandible with dental implants, 6-th International Conference in Central Europe on Computer Graphics and Visualization, 1998.

[10] Gei, M., Genna, F., Bigoni, D., An interface model for the periodontal ligament, Journal of Biomechanical Engineering, 124 (5) (2002) 536–546.

[11] Hajian, M. R., In-vivo-Bite Force: Comparison between conventional and implant-supported dentures, University of Würzburg, Ph.D. thesis, 2004 (in German).

[12] Jokstad, A., Braegger, U., Brunski, J. B., Carr, A. B., Naert, I., Wennerberg, A., Quality of dental implants, International Dental Journal, 53 (2003) 409–443.

[13] Jokstad, A., Osseointegration and dental implants, John Wiley & Sons, Inc., 2008. ISBN 978-0-813-81341-7.

[14] Kawarizadeh, A., Bourauel, C., Jager, A., Experimental and numeric determinative of initial tooth mobility and material properties of the periodontal ligament in rat molar specimen, European Journal of Orthodontics, 25 (2003) 569–578.

[15] Kingsmill, V. J., Post-extraction remodelling of the adult mandible, Critical Reviews in Oral Biology & Medicine, 10 (3) (1999) 384–404.

[16] Kohles, S., Bowers, J., Vailas, A., Ultrasonic wave velocity measurement in small polymeric and cortical bone specimens, Journal of Biomechanical Engineering, 119 (3) (1997) 232–238.

[17] Korioth, T. W. P., Hannam, A. G., Deformation of the human mandible during simulated tooth clenching, Journal of Dental Research, 73 (1) (1994) 56–66.

[18] Levin, L., Dealing with dental implant failures, Journal of Applied Oral Science, 16 (3) (2008) 171–175.

[19] Marcián, P., Florian, Z., Borák, L., Krpalek, D., Valášek, J., Biomechanical Study of Disk Implants, Part II. Engineering Mechanics, 17 (2) (2010) 111–121.

[20] Mish, C., Contemporary Implant Dentistry, Mosby Elsevier, Indiana University, 2008. ISBN 0323043739.

[21] Moon, H. S., Won, Y. Y., Kim, K. D., Ruprecht, A., Kim, H. J., Kook, H. K., Chung, M. K., The three-dimensional microstructure of the trabecular bone in the mandible, Surgical and Radiologic Anatomy, 26 (2004) 466–473.

[22] Natali, A. N., Dental biomechanics, Taylor & Francis Books, London, 2003. ISBN 0-415-30666-3.

[23] O'Mahony, A. M., Williams, J. L., Spencer, P., Anisotropic elasticity of cortical and cancellous bone in the posterior mandible increases peri-implant stress and strain under oblique loading, Clinical Oral Implants Research, 12 (6) (2001) 648–657.

[24] Rho, J. Y., Hobatho, M. C., Ashman, B. R., Relations of mechanical properties to density and CT numbers in human bone, Medical Engineering & Physics, 17 (5) (1995) 347–355.

[25] Rho, J. Y., Tsui, T. Y., Pharr, G. M., Elastic properties of osteon and trabecular bone measured by nanoindentation, Journal of Biomechanics, 31 (1) (1998) 21–21.

[26] Samira, F., Sinan, M., Load transfer along the bone-dental implant interface, Journal of Biomechanics, 43 (2010) 1 761–1 770.

[27] Simsek, B., Simsek, S., Evaluation of success rates of immediate and delayed implants after tooth extraction, Chinese Medical Journal, 116 (8) (2003) 1 216–1 219.

[28] Šímůnek, A., Dental Implantology, Nucleus, 2008 (in Czech). ISBN 978-80-87009-30-7.

[29] Stephan, G., Radiology of Osteoporosis, 2nd Revised Edition, 2008 Springer-Verlag Berlin Heidelberg. ISBN 978-3-540-25888-9.

[30] Takahashi, N., Kitagami, T., Komori, T., Behaviour of teeth under various loading condition with finite element method, Journal of Oral Rehabilitation, 7 (1980) 453–461.

[31] Tanaka, E., Tanne, K., Sakuda, M., A free-dimensional finite element model of the mandible including TMJ and its application to stress analysis in the TMJ during clenching, Medical Engineering & Physics, 16 (1994) 316–322.

[32] Turner, Ch. H., The elastic properties of trabecular and cortical bone tissues are simile: Results from two microscopic measurement techniques, Journal of Biomechanics, 32 (4) (1999) 437–441.

[33] Ulm, C., Tepper, G., Blahout, R., Rausch, F. X., Hienz, S., Matejka, M., Characteristic features of trabecular bone in edentulous mandibles, Clinical Oral Implants, 20 (2009) 594–600.

[34] Valášek, J., Marcián, P., Krpalek, D., Borák, L., Florian, Z., Konečný, O., Material properties of bone tissue obtained from CT for biomechanics purposes, In MENDEL 2010. Mendel Journal series. (2010) 483–490.

[35] CT Data Analysis — Software. URL <www.biomechanika.fme.vutbr.cz>.

[36] MatWeb — Database of material properties. URL <http://www.matweb.com>.

7

Bending and free vibration analysis of thick isotropic plates by using exponential shear deformation theory

A. S. Sayyad[a,*], Y. M. Ghugal[b]

[a]Department of Civil Engineering, SRES's College of Engineering Kopargaon-423601, M.S., India
[b]Department of Applied Mechanics, Government Engineering College, Karad, Satara-415124, M.S., India

Abstract

This paper presents a variationally consistent an exponential shear deformation theory for the bi-directional bending and free vibration analysis of thick plates. The theory presented herein is built upon the classical plate theory. In this displacement-based, refined shear deformation theory, an exponential functions are used in terms of thickness co-ordinate to include the effect of transverse shear deformation and rotary inertia. The number of unknown displacement variables in the proposed theory are same as that in first order shear deformation theory. The transverse shear stress can be obtained directly from the constitutive relations satisfying the shear stress free surface conditions on the top and bottom surfaces of the plate, hence the theory does not require shear correction factor. Governing equations and boundary conditions of the theory are obtained using the dynamic version of principle of virtual work. The simply supported thick isotropic square and rectangular plates are considered for the detailed numerical studies. Results of displacements, stresses and frequencies are compared with those of other refined theories and exact theory to show the efficiency of proposed theory. Results obtained by using proposed theory are found to be agree well with the exact elasticity results. The objective of the paper is to investigate the bending and dynamic response of thick isotropic square and rectangular plates using an exponential shear deformation theory.

Keywords: shear deformation, isotropic plates, shear correction factor, static flexure, transverse shear stresses, free vibration

1. Introduction

The wide spread use of shear flexible materials has stimulated interest in the accurate prediction of structural behavior of thick plates. Thick beams and plates, either isotropic or anisotropic, basically form two and three dimensional problems of elasticity theory. Reduction of these problems to the corresponding one and two dimensional approximate problems for their analysis has always been the main objective of research workers. The shear deformation effects are more pronounced in the thick plates when subjected to transverse loads than in the thin plates under similar loading. These effects are neglected in classical plate theory. In order to describe the correct bending behavior of thick plates including shear deformation effects and the associated cross sectional warping, shear deformation theories are required. This can be accomplished by selection of proper kinematics and constitutive models. These theories can be classified into two major classes on the basis of assumed fields: Stress based theories and displacement based theories. In stress based theories, the stresses are treated as primary variables. In displacement based theories, displacements are treated as primary variables.

Kirchhoff [5,6] developed the well-known classical plate theory (CPT). It is based on the Kirchhoff hypothesis that straight lines normal to the undeformed midplane remain straight and

---

*Corresponding author. e-mail: attu_sayyad@yahoo.co.in.

normal to the deformed midplane. In accordance with the kinematic assumptions made in the CPT all the transverse shear and transverse normal strains are zero. The CPT is widely used for static bending, vibrations and stability of thin plates in the area of solid structural mechanics. Since the transverse shear deformation is neglected in CPT, it cannot be applied to thick plates wherein shear deformation effects are more significant. Thus, its suitability is limited to only thin plates. First-order shear deformation theory (FSDT) can be considered as improvement over the CPT. It is based on the hypothesis that the normal to the undeformed midplane remain straight but not necessarily normal to the midplane after deformation. This is known as FSDT because the thicknesswise displacement field for the inplane displacement is linear or of the first order. Reissner [13, 14] has developed a stress based FSDT which incorporates the effect of shear and Mindlin [9] employed displacement based approach. In Mindlin's theory, transverse shear stress is assumed to be constant through the thickness of the plate, but this assumption violates the shear stress free surface conditions on the top and bottom surfaces of the plate. Mindlin's theory satisfies constitutive relations for transverse shear stresses and shear strains by using shear correction factor. The limitations of CPT and FSDTs forced the development of higher order shear deformation theories (HSDTs) to avoid the use of shear correction factors, to include correct cross sectional warping and to get the realistic variation of the transverse shear strains and stresses through the thickness of plate. The higher order theory is developed by Reddy [12] to get the parabolic shear stress distribution through the thickness of plate and to satisfy the shear stress free surface conditions on the top and bottom surfaces of the plate to avoid the need of shear correction factors. Comprehensive reviews of refined theories have been given by Noor and Burton [10] and Vasil'ev [17], whereas Liew et al. [8] surveyed plate theories particularly applied to thick plate vibration problems. A recent review papers are presented by Ghugal and Shimpi [1] and Kreja [7]. The effect of transverse shear and transverse normal strain on the static flexure of thick isotropic plates using trigonometric shear deformation theory is studied by Ghugal and Sayyad [2]. Shimpi and Patel [15] developed two variable refined plate theory for the static flexure and free vibration analysis of isotropic plates; however, theory of these authors yields the frequencies identical to those of Mindlin's theory. Ghugal and Pawar [3] have developed hyperbolic shear deformation theory for the bending, buckling and free vibration analysis of thick shear flexible plates. Karama et al. [4] has proposed exponential shear deformation theory for the multilayered beam structures. Exact elasticity solution for bidirectional bending of plates is provided by Pagano [11], whereas Srinivas et. al. [16] provided an exact analysis for Vibration of simply supported homogeneous thick rectangular plates.

In this paper a displacement based an exponential shear deformation theory (ESDT) is used for the bi-directional bending and free vibration analysis of thick isotropic square and rectangular plates which includes effect of transverse shear deformation and rotary inertia. The displacement field of the theory contains three variables as in the FSDT of plate. The theory is shown to be simple and more effective for the bending and free vibration analysis of isotropic plates.

2. Theoretical formulation

2.1. Isotropic plate under consideration

Consider a plate made up of isotropic material as shown in Fig. 1. The plate occupies a region given by Eq. (1):

$$0 \le x \le a, \qquad 0 \le y \le b, \qquad -h/2 \le z \le h/2. \tag{1}$$

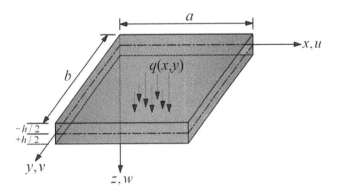

Fig. 1. Plate geometry and co-ordinate system

2.2. Assumptions made in the proposed theory

1. The displacement components u and v are the inplane displacements in x and y — directions respectively and w is the transverse displacement in z-direction. These displacements are small in comparison with the plate thickness.

2. The in-plane displacement u in x-direction and v in y-direction each consist of two parts:

    (a) a displacement component analogous to displacement in classical plate theory of bending;

    (b) displacement component due to shear deformation which is assumed to be exponential in nature with respect to thickness coordinate.

3. The transverse displacement w in z-direction is assumed to be a function of x and y coordinates.

4. The plate is subjected to transverse load only.

2.3. The proposed plate theory

Based upon the before mentioned assumptions, the displacement field of the proposed plate theory is given as below:

$$u(x, y, z, t) = -z\frac{\partial w(x, y, t)}{\partial x} + f(z)\phi(x, y, t),$$

$$v(x, y, z, t) = -z\frac{\partial w(x, y, t)}{\partial y} + f(z)\psi(x, y, t), \qquad (2)$$

$$w(x, y, z, t) = w(x, y, t),$$

where $f(z) = z\exp\left[-2\left(\frac{z}{h}\right)^2\right]$.

　　Here u, v and w are the displacements in the x, y and z-directions respectively. The exponential function in terms of thickness coordinate $[f(z)]$ in both the inplane displacements u and v is associated with the transverse shear stress distribution through the thickness of plate. The functions ϕ and ψ are the unknown functions associated with the shear slopes.

2.4. Superiority of the present theory

The present theory is a displacement-based refined theory, and refined shear deformation theories are known to be successful techniques for improving the accuracy of displacement and stresses. The kinematics of the present theory is much richer than those of the higher order shear deformation theories available in the literature, because if the exponential term is expanded in power series, the kinematics of higher order theories are implicitly taken into account to good deal of extent. Exponential function has all even and odd powers in its expansion unlike sine function which have only odd powers.

2.5. Strain-displacement relationships

Normal strains (ε_x and ε_y) and shear strains ($\gamma_{xy}, \gamma_{yz}, \gamma_{zx}$) are obtained within the framework of linear theory of elasticity using the displacement field given by Eq. (2).

$$
\begin{aligned}
\varepsilon_x &= \frac{\partial u}{\partial x} = -z\frac{\partial^2 w}{\partial x^2} + f(z)\frac{\partial \phi}{\partial x}, \\
\varepsilon_y &= \frac{\partial v}{\partial y} = -z\frac{\partial^2 w}{\partial y^2} + f(z)\frac{\partial \psi}{\partial y}, \\
\gamma_{xy} &= \frac{\partial u}{\partial y} + \frac{\partial v}{\partial x} = -2z\frac{\partial^2 w}{\partial x \partial y} + f(z)\left(\frac{\partial \phi}{\partial y} + \frac{\partial \psi}{\partial x}\right), \\
\gamma_{zx} &= \frac{\partial u}{\partial z} + \frac{\partial w}{\partial x} = \frac{\mathrm{d}f(z)}{\mathrm{d}z}\phi, \\
\gamma_{yz} &= \frac{\partial v}{\partial z} + \frac{\partial w}{\partial y} = \frac{\mathrm{d}f(z)}{\mathrm{d}z}\psi.
\end{aligned}
\tag{3}
$$

2.6. Stress-strain relationships

For a plate of constant thickness, composed of isotropic material, the effect of transverse normal stress σ_z on the gross response of the plate is assumed to be negligible in comparison with inplane stresses σ_x and σ_y. Therefore, for a linearly elastic material, stresses σ_x and σ_y are related to normal strains ε_x and ε_y and shear stresses τ_{xy}, τ_{yz} and τ_{zx} are related to shear strains γ_{xy}, γ_{yz} and γ_{zx} by the following constitutive relations:

$$
\begin{aligned}
\sigma_x &= \frac{E}{1-\mu^2}(\varepsilon_x + \mu\varepsilon_y) = \frac{E}{1-\mu^2}\left[-z\frac{\partial^2 w}{\partial x^2} + f(z)\frac{\partial \phi}{\partial x}\right] + \frac{\mu E}{1-\mu^2}\left[-z\frac{\partial^2 w}{\partial y^2} + f(z)\frac{\partial \psi}{\partial y}\right], \\
\sigma_y &= \frac{E}{1-\mu^2}(\varepsilon_y + \mu\varepsilon_x) = \frac{\mu E}{1-\mu^2}\left[-z\frac{\partial^2 w}{\partial x^2} + f(z)\frac{\partial \phi}{\partial x}\right] + \frac{E}{1-\mu^2}\left[-z\frac{\partial^2 w}{\partial y^2} + f(z)\frac{\partial \psi}{\partial y}\right], \\
\tau_{xy} &= G\gamma_{xy} = \frac{E}{2(1+\mu)}\left[-2z\frac{\partial^2 w}{\partial x \partial y} + f(z)\left(\frac{\partial \phi}{\partial y} + \frac{\partial \psi}{\partial x}\right)\right], \\
\tau_{zx} &= G\gamma_{zx} = \frac{E}{2(1+\mu)}\frac{\mathrm{d}f(z)}{\mathrm{d}z}\phi, \qquad \tau_{yz} = G\gamma_{yz} = \frac{E}{2(1+\mu)}\frac{\mathrm{d}f(z)}{\mathrm{d}z}\psi.
\end{aligned}
\tag{4}
$$

3. Governing equations and boundary conditions

Using the Eqs. (2)–(4) and the principle of virtual work, variationally consistent governing differential equations and associated boundary conditions for the plate under consideration can be obtained. The dynamic version of principle of virtual work when applied to the plate leads to

$$
\int_{z=-h/2}^{z=h/2} \int_{y=0}^{y=b} \int_{x=0}^{x=a} \left[\sigma_x \delta\varepsilon_x + \sigma_y \delta\varepsilon_y + \tau_{yz}\delta\gamma_{yz} + \tau_{zx}\delta\gamma_{zx} + \tau_{xy}\delta\gamma_{xy} \right] dx\,dy\,dz - \tag{5}
$$

$$
\int_{y=0}^{y=b} \int_{x=0}^{x=a} q(x,y)\,\delta w\,dx\,dy + \rho \int_{z=-h/2}^{z=h/2} \int_{y=0}^{y=b} \int_{x=0}^{x=a} \left[\frac{\partial^2 u}{\partial t^2}\delta u + \frac{\partial^2 v}{\partial t^2}\delta v + \frac{\partial^2 w}{\partial t^2}\delta w \right] dx\,dy\,dz = 0,
$$

where symbol δ denotes the variational operator. Employing Green's theorem in Eq. (5) successively, we obtain the coupled Euler-Lagrange equations, which are the governing equations and the associated boundary conditions of the plate. The governing differential equations in-terms of stress resultants are as follows:

$$
\frac{\partial^2 M_x}{\partial x^2} + 2\frac{\partial^2 M_{xy}}{\partial x \partial y} + \frac{\partial^2 M_y}{\partial y^2} + q = I_1\frac{\partial^2 w}{\partial t^2} - I_2\left(\frac{\partial^4 w}{\partial x^2 \partial t^2} + \frac{\partial^4 w}{\partial y^2 \partial t^2}\right) + I_3\left(\frac{\partial^3 \phi}{\partial x \partial t^2} + \frac{\partial^3 \psi}{\partial y \partial t^2}\right),
$$

$$
\frac{\partial N_{sx}}{\partial x} + \frac{\partial N_{sxy}}{\partial y} - N_{Tcx} = -I_3\frac{\partial^3 w}{\partial x \partial t^2} + I_4\frac{\partial^2 \phi}{\partial t^2}, \tag{6}
$$

$$
\frac{\partial N_{sy}}{\partial y} + \frac{\partial N_{sxy}}{\partial x} - N_{Tcy} = -I_3\frac{\partial^3 w}{\partial y \partial t^2} + I_4\frac{\partial^2 \psi}{\partial t^2}.
$$

The boundary conditions at $x = 0$ and $x = a$ obtained are of the following form:

$$
\begin{array}{llll}
\text{either} & V_x = 0 & \text{or } w \text{ is prescribed,} & \\
\text{either} & M_x = 0 & \text{or } \frac{\partial w}{\partial x} \text{ is prescribed,} & \\
\text{either} & N_{sx} = 0 & \text{or } \phi \text{ is prescribed,} & (7) \\
\text{either} & N_{sxy} = 0 & \text{or } \psi \text{ is prescribed.} &
\end{array}
$$

The boundary conditions at $y = 0$ and $y = b$ obtained are of the following form:

$$
\begin{array}{llll}
\text{either} & V_y = 0 & \text{or } w \text{ is prescribed,} & \\
\text{either} & M_y = 0 & \text{or } \frac{\partial w}{\partial y} \text{ is prescribed,} & \\
\text{either} & N_{sxy} = 0 & \text{or } \phi \text{ is prescribed,} & (8) \\
\text{either} & N_{sy} = 0 & \text{or } \psi \text{ is prescribed.} &
\end{array}
$$

Reaction at the corners of the plate is of the following form:

$$
\text{either} \quad M_{xy} = 0 \qquad \text{or } w \text{ is prescribed.} \tag{9}
$$

The stress resultants in the governing equations [Eq. (6)] and boundary conditions [Eqs. (7)–(9)] are given as:

$$
(M_x, M_y, M_{xy}) = \int_{-h/2}^{h/2} (\sigma_x, \sigma_y, \tau_{xy}) z\,dz, \quad (N_{sx}, N_{sy}, N_{sxy}) = \int_{-h/2}^{h/2} (\sigma_x, \sigma_y, \tau_{xy}) f(z)\,dz,
$$

$$
(N_{Tcx}, N_{Tcy}) = \int_{-h/2}^{h/2} (\tau_{zx}, \tau_{yz}) \frac{df(z)}{dz}\,dz, \tag{10}
$$

$$
V_x = \frac{\partial M_x}{\partial x} + 2\frac{\partial M_{xy}}{\partial y}, \quad V_y = \frac{\partial M_y}{\partial y} + 2\frac{\partial M_{xy}}{\partial x}.
$$

The governing differential equations in-terms of unknown displacement variables used in the displacement field (w, ϕ and ψ) obtained are as follows:

$$D_1 \left(\frac{\partial^4 w}{\partial x^4} + 2\frac{\partial^4 w}{\partial x^2 \partial y^2} + \frac{\partial^4 w}{\partial y^4} \right) - D_2 \left(\frac{\partial^3 \phi}{\partial x^3} + \frac{\partial^3 \phi}{\partial x \partial y^2} + \frac{\partial^3 \psi}{\partial y^3} + \frac{\partial^3 \psi}{\partial x^2 \partial y} \right) +$$

$$I_1 \frac{\partial^2 w}{\partial t^2} - I_2 \left(\frac{\partial^4 w}{\partial x^2 \partial t^2} + \frac{\partial^4 w}{\partial y^2 \partial t^2} \right) + I_3 \left(\frac{\partial^3 \phi}{\partial x \partial t^2} + \frac{\partial^3 \psi}{\partial y \partial t^2} \right) = q,$$

$$D_2 \left(\frac{\partial^3 w}{\partial x^3} + \frac{\partial^3 w}{\partial x \partial y^2} \right) - D_3 \left(\frac{\partial^2 \phi}{\partial x^2} + \frac{(1-\mu)}{2} \frac{\partial^2 \phi}{\partial y^2} \right) + \qquad (11)$$

$$D_4 \phi - D_3 \frac{(1+\mu)}{2} \frac{\partial^2 \psi}{\partial x \partial y} - I_3 \frac{\partial^3 w}{\partial x \partial t^2} + I_4 \frac{\partial^2 \phi}{\partial t^2} = 0,$$

$$D_2 \left(\frac{\partial^3 w}{\partial y^3} + \frac{\partial^3 w}{\partial x^2 \partial y} \right) - D_3 \left(\frac{(1-\mu)}{2} \frac{\partial^2 \psi}{\partial x^2} + \frac{\partial^2 \psi}{\partial y^2} \right) +$$

$$D_4 \psi - D_3 \frac{(1+\mu)}{2} \frac{\partial^2 \phi}{\partial x \partial y} - I_3 \frac{\partial^3 w}{\partial y \partial t^2} + I_4 \frac{\partial^2 \psi}{\partial t^2} = 0.$$

The associated consistent boundary conditions in-terms of unknown displacement variables obtained along the edges $x = 0$ and $x = a$ are as below:

either $\qquad D_1 \left[\frac{\partial^3 w}{\partial x^3} + (2-\mu)\frac{\partial^3 w}{\partial x \partial y^2} \right] -$

$D_2 \left[\frac{\partial^2 \phi}{\partial x^2} + (1-\mu)\frac{\partial^2 \phi}{\partial y^2} + \frac{\partial^2 \psi}{\partial x \partial y} \right] - I_2 \frac{\partial^3 w}{\partial x \partial t^2} - I_3 \frac{\partial^3 \phi}{\partial t^2} = 0 \qquad$ or w is prescribed,

either $\qquad D_1 \left(\frac{\partial^2 w}{\partial x^2} + \mu\frac{\partial^2 w}{\partial y^2} \right) - D_2 \left(\frac{\partial \phi}{\partial x} + \mu\frac{\partial \psi}{\partial y} \right) = 0 \qquad$ or $\frac{\partial w}{\partial x}$ is prescribed, $\quad (12)$

either $\qquad D_2 \left(\frac{\partial^2 w}{\partial x^2} + \mu\frac{\partial^2 w}{\partial y^2} \right) - 2D_3 \left(\frac{\partial \phi}{\partial x} + \mu\frac{\partial \psi}{\partial y} \right) = 0 \qquad$ or ϕ is prescribed,

either $\qquad D_3 \left(\frac{\partial \psi}{\partial x} + \frac{\partial \phi}{\partial y} \right) - D_2 \frac{\partial^2 w}{\partial x \partial y} = 0 \qquad$ or ψ is prescribed.

The associated consistent boundary conditions in-terms of unknown displacement variables obtained along the edges $y = 0$ and $y = b$ are as below:

either $\qquad D_1 \left[\frac{\partial^3 w}{\partial y^3} + (2-\mu)\frac{\partial^3 w}{\partial x^2 \partial y} \right] -$

$D_2 \left[\frac{\partial^2 \psi}{\partial y^2} + (1-\mu)\frac{\partial^2 \psi}{\partial x^2} + \frac{\partial^2 \phi}{\partial x \partial y} \right] - I_2 \frac{\partial^3 w}{\partial y \partial t^2} - I_3 \frac{\partial^3 \psi}{\partial t^2} = 0 \qquad$ or w is prescribed,

either $\qquad D_1 \left(\mu\frac{\partial^2 w}{\partial x^2} + \frac{\partial^2 w}{\partial y^2} \right) - D_2 \left(\mu\frac{\partial \phi}{\partial x} + \frac{\partial \psi}{\partial y} \right) = 0 \qquad$ or $\frac{\partial w}{\partial y}$ is prescribed, $\quad (13)$

either $\qquad D_3 \left(\frac{\partial \psi}{\partial x} + \frac{\partial \phi}{\partial y} \right) - D_2 \frac{\partial^2 w}{\partial x \partial y} = 0 \qquad$ or ϕ is prescribed,

either $\qquad D_2 \left(\mu\frac{\partial^2 w}{\partial x^2} + \frac{\partial^2 w}{\partial y^2} \right) - 2D_3 \left(\mu\frac{\partial \phi}{\partial x} + \frac{\partial \psi}{\partial y} \right) = 0 \qquad$ or ψ is prescribed.

The boundary condition in-terms of unknown displacement variables (w, ϕ and ψ) obtained along the corners of plate is:

$$\text{either}\quad 2D_1\frac{\partial^2 w}{\partial x \partial y} - D_2\left(\frac{\partial \phi}{\partial y} + \frac{\partial \psi}{\partial x}\right) = 0 \quad\text{or } w \text{ is prescribed,}\tag{14}$$

where constants D_i and I_i appeared in governing equations and boundary conditions are as follows:

$$D_1 = \frac{Eh^3}{12(1-\mu^2)},\qquad D_2 = \frac{A_0 E}{(1-\mu^2)},\qquad D_3 = \frac{B_0 E}{(1-\mu^2)},\qquad D_4 = \frac{C_0 E}{2(1+\mu)},\tag{15}$$

$$I_1 = \rho h,\qquad I_2 = \frac{\rho h^3}{12},\qquad I_3 = \rho A_0,\qquad I_4 = \rho B_0\tag{16}$$

and

$$A_0 = \int_{-h/2}^{h/2} z f(z)\,\mathrm{d}z,\qquad B_0 = \int_{-h/2}^{h/2} f^2(z)\,\mathrm{d}z,\qquad C_0 = \int_{-h/2}^{h/2}\left[\frac{\mathrm{d}f(z)}{\mathrm{d}z}\right]^2\,\mathrm{d}z.\tag{17}$$

4. Illustrative examples

Example 1: Bending analysis of isotropic plates subjected to uniformly distributed load

A simply supported isotropic square plates occupying the region given by the Eq. (1) is considered. The plate is subjected to uniformly distributed transverse load, $q(x,y)$ on surface $z = -h/2$ acting in the downward z-direction as given below:

$$q(x,y) = \sum_{m=1}^{\infty}\sum_{n=1}^{\infty} q_{mn} \sin\left(\frac{m\pi x}{a}\right)\sin\left(\frac{n\pi y}{b}\right),\tag{18}$$

where q_{mn} are the coefficients of Fourier expansion of load, which are given by

$$\begin{array}{ll} q_{mn} = \frac{16q_0}{mn\pi^2} & \text{for } m = 1,3,5,\ldots, \text{ and } n = 1,3,5,\ldots, \\ q_{mn} = 0 & \text{for } m = 2,4,6,\ldots, \text{ and } n = 2,4,6,\ldots. \end{array}\tag{19}$$

The governing differential equations and the associated boundary conditions for static flexure of square plate under consideration can be obtained directly from Eqs. (6)–(9). The following are the boundary conditions of the simply supported isotropic plate.

$$w = \psi = M_x = N_{sx} = 0 \quad\text{at } x = 0 \text{ and } x = a,\tag{20}$$

$$w = \phi = M_y = N_{sy} = 0 \quad\text{at } y = 0 \text{ and } y = b.\tag{21}$$

Example 2: Bending analysis of isotropic plates subjected to sinusoidal load progress

A simply supported square plates is subjected to sinusoidal load progress in both x and y directions, on surface $z = -h/2$, acting in the downward z direction. The load is expressed as:

$$q(x,y) = q_0 \sin\left(\frac{\pi x}{a}\right)\sin\left(\frac{\pi y}{b}\right),\tag{22}$$

where q_0 is the magnitude of the sinusoidal loading at the centre.

Example 3: Bending analysis of isotropic plate subjected to linearly varying load

A simply supported square plate is subjected to linearly varying transverse load $(q_0 x/a)$. The intensity of load is zero at the edge $x = 0$ and maximum (q_0) at the edge $x = a$. The magnitude of coefficient of Fourier expansion of load in the Eq. (18) is given by $q_{mn} = -(8q_0/mn\pi^2)\cos(m\pi)$.

4.1. The closed-form solution

The governing equations for bending analysis of plate (static flexure), discarding all the terms containing time derivatives becomes:

$$D_1 \left(\frac{\partial^4 w}{\partial x^4} + 2\frac{\partial^4 w}{\partial x^2 \partial y^2} + \frac{\partial^4 w}{\partial y^4} \right) - D_2 \left(\frac{\partial^3 \phi}{\partial x^3} + \frac{\partial^3 \phi}{\partial x \partial y^2} + \frac{\partial^3 \psi}{\partial y^3} + \frac{\partial^3 \psi}{\partial x^2 \partial y} \right) = q,$$

$$D_2 \left(\frac{\partial^3 w}{\partial x^3} + \frac{\partial^3 w}{\partial x \partial y^2} \right) - D_3 \left(\frac{\partial^2 \phi}{\partial x^2} + \frac{(1-\mu)}{2}\frac{\partial^2 \phi}{\partial y^2} \right) + D_4 \phi - D_3 \frac{(1+\mu)}{2}\frac{\partial^2 \psi}{\partial x \partial y} = 0, \quad (23)$$

$$D_2 \left(\frac{\partial^3 w}{\partial y^3} + \frac{\partial^3 w}{\partial x^2 \partial y} \right) - D_3 \left(\frac{(1-\mu)}{2}\frac{\partial^2 \psi}{\partial x^2} + \frac{\partial^2 \psi}{\partial y^2} \right) + D_4 \psi - D_3 \frac{(1+\mu)}{2}\frac{\partial^2 \phi}{\partial x \partial y} = 0.$$

The following is the solution form for $w(x, y), \phi(x, y)$, and $\psi(x, y)$ satisfying the boundary conditions perfectly for a plate with all the edges simply supported:

$$w(x, y) = \sum_{m=1}^{\infty} \sum_{n=1}^{\infty} w_{mn} \sin\left(\frac{m\pi x}{a}\right) \sin\left(\frac{n\pi y}{b}\right),$$

$$\phi(x, y) = \sum_{m=1}^{\infty} \sum_{n=1}^{\infty} \phi_{mn} \cos\left(\frac{m\pi x}{a}\right) \sin\left(\frac{n\pi y}{b}\right), \quad (24)$$

$$\psi(x, y) = \sum_{m=1}^{\infty} \sum_{n=1}^{\infty} \psi_{mn} \sin\left(\frac{m\pi x}{a}\right) \cos\left(\frac{n\pi y}{b}\right),$$

where w_{mn}, ϕ_{mn} and ψ_{mn} are unknown coefficients, which can be easily evaluated after substitution of Eq. (24) in the set of three governing differential Eq. (23) resulting in following three simultaneous equations, in case of sinusoidal load $m = 1$ and $n = 1$,

$$K_{11}w_{mn} + K_{12}\phi_{mn} + K_{13}\psi_{mn} = q_{mn},$$
$$K_{12}w_{mn} + K_{22}\phi_{mn} + K_{23}\psi_{mn} = 0, \quad (25)$$
$$K_{13}w_{mn} + K_{23}\phi_{mn} + K_{33}\psi_{mn} = 0,$$

where

$$K_{11} = D_1 \pi^4 \left(\frac{m^4}{a^4} + \frac{n^4}{b^4} + 2\frac{m^2 n^2}{a^2 b^2} \right), \quad K_{12} = -D_2 \pi^3 \left(\frac{m^3}{a^3} + \frac{mn^2}{ab^2} \right),$$

$$K_{13} = -D_2 \pi^3 \left(\frac{n^3}{b^3} + \frac{m^2 n}{a^2 b} \right), \qquad K_{22} = D_3 \pi^2 \left(\frac{(1-\mu)}{2}\frac{n^2}{b^2} + \frac{m^2}{a^2} \right) + D_4, \quad (26)$$

$$K_{23} = D_3 \frac{(1+\mu)}{2}\frac{mn\pi^2}{ab}, \qquad K_{33} = D_3 \pi^2 \left(\frac{(1-\mu)}{2}\frac{m^2}{a^2} + \frac{n^2}{b^2} \right) + D_4.$$

Having obtained the values of w_{mn}, ϕ_{mn} and ψ_{mn} from above set of Eqs. (25) and (24), one can then calculate all the displacement and stress components within the plate using displacement field given by Eq. (2) and stress strain relationships given by Eq. (4).

4.2. Computation of displacements and inplane stresses

Substituting the final solution for $w(x, y), \phi(x, y)$ and $\psi(x, y)$ in the displacement field, the final displacements (u, v and w) can obtained and using strain-displacement relations, final strains ($\varepsilon_x, \varepsilon_y, \gamma_{xy}, \gamma_{yz}$ and γ_{zx}) can be obtained. Finally, the inplane stresses (σ_x, σ_y and τ_{xy}) could be obtained by using stress-strain relations (constitutive relations) as given by the Eq. (5). Non-dimensional displacements are represented as $\bar{u}, \bar{v}$ and $\bar{w}$, whereas non-dimensional inplane stresses are represented as $\bar{\sigma}_x, \bar{\sigma}_y$ and $\bar{\tau}_{xy}$.

4.3. Computation of transverse shear stresses

The transverse shear stresses τ_{zx} and τ_{yz} can be obtained either by using the constitutive relations [Eq. (4)] or by integrating equilibrium equations with respect to the thickness coordinate. Equilibrium equations of three-dimensional elasticity, ignoring body forces, can be used to obtain transverse shear stresses. These equations are:

$$\frac{\partial \sigma_x}{\partial x} + \frac{\partial \tau_{xy}}{\partial y} + \frac{\partial \tau_{zx}}{\partial z} = 0 \quad \text{and} \quad \frac{\partial \tau_{xy}}{\partial x} + \frac{\partial \sigma_y}{\partial y} + \frac{\partial \tau_{yz}}{\partial z} = 0. \tag{27}$$

Integrating Eq. (27) both w.r.t the thickness coordinate z and imposing the following boundary conditions at top and bottom surfaces of the plate

$$[\tau_{zx}]_{z=\pm h/2} = 0, \qquad [\tau_{yz}]_{z=\pm h/2} = 0, \tag{28}$$

expressions for τ_{zx} and τ_{yz} can be obtained satisfying the requirements of zero shear stress conditions on the top and bottom surfaces of the plate. Non-dimensional transverse shear stresses are represented as $\bar{\tau}_{zx}$ and $\bar{\tau}_{yz}$. Further it may be noted that τ_{zx} and $\bar{\tau}_{zx}$ obtained by constitutive relations are indicated by τ_{zx}^{CR} and $\bar{\tau}_{zx}^{CR}$ and when they are obtained by using equilibrium equations, are indicated by τ_{zx}^{EE} and $\bar{\tau}_{zx}^{EE}$. In case of isotropic plate $u = v$, $\sigma_x = \sigma_y$ and $\tau_{zx} = \tau_{yz}$.

Example 4: Free vibration analysis of isotropic plates

The following is the solution form of $w(x, y, t)$, $\phi(x, y, t)$, and $\psi(x, y, t)$ for free vibration analysis satisfying the boundary conditions (time dependent), perfectly for a plate with all the edges simply supported:

$$w = \sum_{m=1}^{\infty} \sum_{n=1}^{\infty} w_{mn} \sin\left(\frac{m\pi x}{a}\right) \sin\left(\frac{n\pi y}{b}\right) \sin \omega_{mn} t,$$

$$\phi = \sum_{m=1}^{\infty} \sum_{n=1}^{\infty} \phi_{mn} \cos\left(\frac{m\pi x}{a}\right) \sin\left(\frac{n\pi y}{b}\right) \sin \omega_{mn} t, \tag{29}$$

$$\psi = \sum_{m=1}^{\infty} \sum_{n=1}^{\infty} \psi_{mn} \sin\left(\frac{m\pi x}{a}\right) \cos\left(\frac{n\pi y}{b}\right) \sin \omega_{mn} t,$$

where w_{mn} is the amplitude of translation and ϕ_{mn} and ψ_{mn} are the amplitudes of rotations. ω_{mn} is the natural frequency. The governing equations for free vibration of simply supported square and rectangular plate can be obtained by setting the applied transverse load $q(x, y)$ equal to zero in the set of Eq. (11). Substitution of solution form [Eq. (29)] into the governing equations of free vibration [Eqs. (11)] of plate results in following three simultaneous equations:

$$D_1 \left(\frac{m^4 \pi^4}{a^4} + 2\frac{m^2 n^2 \pi^4}{a^2 b^2} + \frac{n^4 \pi^4}{b^4}\right) w_{mn} -$$

$$D_2 \left(\frac{m^3 \pi^3}{a^3} + \frac{mn^2 \pi^3}{ab^2}\right) \phi_{mn} - D_2 \left(\frac{n^3 \pi^3}{b^3} + \frac{m^2 n\pi^3}{a^2 b}\right) \psi_{mn} - \tag{30}$$

$$\omega^2 \left(I_1 + I_2 \frac{m^2 \pi^2}{a^2} + I_2 \frac{n^2 \pi^2}{b^2}\right) w_{mn} + \omega_{mn}^2 I_3 \frac{m\pi}{a} \phi_{mn} + \omega_{mn}^2 I_3 \frac{n\pi}{b} \psi_{mn} = 0,$$

$$-D_2 \left(\frac{m^3\pi^3}{a^3} + \frac{mn^2\pi^3}{ab^2}\right) w_{mn} + \left(D_3\frac{m^2\pi^2}{a^2} + D_3\frac{(1-\mu)}{2}\frac{n^2\pi^2}{b^2} + D_4\right)\phi_{mn} + \tag{31}$$

$$D_3\frac{(1+\mu)}{2}\frac{mn\pi^2}{ab}\psi_{mn} + \omega_{mn}^2 I_3\frac{m\pi}{a}w_{mn} - \omega_{mn}^2 I_4\phi_{mn} = 0,$$

$$-D_2 \left(\frac{n^3\pi^3}{b^3} + \frac{m^2n\pi^3}{a^2b}\right) w_{mn} + D_3\frac{(1+\mu)}{2}\frac{mn\pi^2}{ab}\phi_{mn} + \tag{32}$$

$$\left(D_3\frac{n^2\pi^2}{b^2} + D_3\frac{(1-\mu)}{2}\frac{m^2\pi^2}{a^2} + D\right)\psi_{mn} + \omega_{mn}^2 I_3\frac{n\pi}{b}w_{mn} - \omega_{mn}^2 I_4\psi_{mn} = 0.$$

Eqs. (30)–(32) are written in following matrix form:

$$\left([K] - \omega_{mn}^2[M]\right)\{\Delta\} = 0, \tag{33}$$

where $[K]$ is stiffness matix, $[M]$ is mass matix and $\{\Delta\}$ is amplitude vector. The elements of stiffness matrix are given in Eq. (26). Elements of mass matrix and amplitude vector are given below:

$$M_{11} = \left(I_1 + I_2\frac{m^2\pi^2}{a^2} + I_2\frac{n^2\pi^2}{b^2}\right), \quad M_{12} = -I_3\frac{m\pi}{a}, \quad M_{13} = -I_3\frac{n\pi}{b}, \tag{34}$$

$$M_{22} = I_4, \quad M_{23} = 0, \quad M_{33} = I_4, \quad M_{21} = M_{12}, \quad M_{31} = M_{13},$$

$$\{\Delta\}^T = \{w_{mn} \ \phi_{mn} \ \psi_{mn}\}. \tag{35}$$

Following material properties of isotropic plates are used:

$$E = 210\,\text{GPa}, \quad \mu = 0.3, \quad G = \frac{E}{2(1+\mu)} \quad \text{and} \quad \rho = 7\,800\,\text{kg/m}^3, \tag{36}$$

where E is the Young's modulus, G is the shear modulus, μ is the Poisson's ratio and ρ is density of the material.

5. Numerical results and discussion

5.1. Numerical results

Results obtained for displacements, stresses and natural frequencies will now be compared and discussed with the corresponding results of higher order shear deformation theory (HSDT) of Reddy [12], trigonometric shear deformation theory (TSDT) of Ghugal and Sayyad [2], hyperbolic shear deformation theory (HPSDT) of Ghugal and Pawar [3], classical plate theory (CPT) of Kirchhoff [5,6], first order shear deformation theory (FSDT) of Mindlin [9], the exact elasticity solution for bidirectional bending of plate Pagano [11] and exact elasticity solution for free vibrational analysis of plate Srinivas et. al. [16]. The numerical results are presented in the following non-dimensional form for the purpose of presenting the results in this paper.

$$\bar{u} = \frac{uE_2}{qhS^3}, \ \bar{w} = \frac{100Ew}{qhS^4}, \ (\bar{\sigma}_x, \bar{\tau}_{xy}) = \frac{(\sigma_x, \tau_{xy})}{qS^2}, \ (\bar{\tau}_{zx}) = \frac{(\tau_{zx})}{qS}, \ \bar{\omega}_{mn} = \omega_{mn}h\sqrt{\rho/G}, \tag{37}$$

where $S(a/h) = $ Aspect Ratio. The percentage error in result of a particular theory with respect to the result of exact elasticity solution is calculated as follows:

$$\% \text{ error} = \frac{\text{value by a particular theory-valueby exact elasticity solution}}{\text{value by exact elasticity solution}} \times 100. \tag{38}$$

Table 1. Comparison of non-dimensional inplane displacement ($\bar{u}$) at ($x = 0$, $y = b/2$, $z = \pm h/2$), transverse displacement ($\bar{w}$) at ($x = a/2$, $y = b/2$, $z = 0$), inplane normal stress ($\bar{\sigma}_x$) at ($x = a/2$, $y = b/2$, $z = \pm h/2$), inplane shear stress ($\bar{\tau}_{xy}$) at ($x = 0$, $y = 0$, $z = \pm h/2$) and transverse shear stress $\bar{\tau}_{zx}$ at ($x = 0$, $y = b/2$, $z = 0$) in isotropic square plate subjected to uniformly distributed load

S	Theory	Model	$\bar{u}$	$\bar{w}$	$\bar{\sigma}_x$	$\bar{\tau}_{xy}$	$\bar{\tau}_{zx}^{CR}$	$\bar{\tau}_{zx}^{EE}$
4	Present	ESDT	0.079	5.816	0.300	0.223	0.481	0.472
	Reddy [12]	HSDT	0.079	5.869	0.299	0.218	0.482	0.452
	Ghugal and Sayyad [2]	TSDT	0.074	5.680	0.318	0.208	0.483	0.420
	Ghugal and Pawar [3]	HPSDT	0.079	5.858	0.297	0.185	0.477	0.451
	Mindlin [9]	FSDT	0.074	5.633	0.287	0.195	0.330	0.495
	Kirchhoff [5,6]	CPT	0.074	4.436	0.287	0.195	–	0.495
	Pagano [11]	Elasticity	0.072	5.694	0.307	–	0.460	–
10	Present	ESDT	0.075	4.658	0.289	0.204	0.494	0.490
	Reddy [12]	HSDT	0.075	4.666	0.289	0.203	0.492	0.486
	Ghugal and Sayyad [2]	TSDT	0.073	4.625	0.307	0.195	0.504	0.481
	Ghugal and Pawar [3]	HPSDT	0.074	4.665	0.289	0.193	0.489	0.486
	Mindlin [9]	FSDT	0.074	4.670	0.287	0.195	0.330	0.495
	Kirchhoff [5,6]	CPT	0.074	4.436	0.287	0.195	–	0.495
	Pagano [11]	Elasticity	0.073	4.639	0.289	–	0.487	–

Table 2. Comparison of non-dimensional inplane displacement ($\bar{u}$) at ($x = 0$, $y = b/2$, $z = \pm h/2$), transverse displacement ($\bar{w}$) at ($x = a/2$, $y = b/2$, $z = 0$), inplane normal stress ($\bar{\sigma}_x$) at ($x = a/2$, $y = b/2$, $z = \pm h/2$), inplane shear stress ($\bar{\tau}_{xy}$) at ($x = 0$, $y = 0$, $z = \pm h/2$) and transverse shear stress $\bar{\tau}_{zx}$ at ($x = 0$, $y = b/2$, $z = 0$) in isotropic square plate subjected to sinusoidal load

S	Theory	Model	$\bar{u}$	$\bar{w}$	$\bar{\sigma}_x$	$\bar{\tau}_{xy}$	$\bar{\tau}_{zx}^{CR}$	$\bar{\tau}_{zx}^{EE}$
4	Present	ESDT	0.046	3.748	0.213	0.114	0.238	0.236
	Reddy [12]	HSDT	0.046	3.787	0.209	0.112	0.237	0.226
	Ghugal and Sayyad [2]	TSDT	0.044	3.653	0.226	0.133	0.244	0.232
	Ghugal and Pawar [3]	HPSDT	0.047	3.779	0.209	0.112	0.236	0.235
	Mindlin [9]	FSDT	0.044	3.626	0.197	0.106	0.159	0.239
	Kirchhoff [5,6]	CPT	0.044	2.803	0.197	0.106	–	0.238
	Pagano [11]	Elasticity	0.049	3.662	0.217	–	0.236	–
10	Present	ESDT	0.044	2.954	0.200	0.108	0.239	0.238
	Reddy [12]	HSDT	0.044	2.961	0.199	0.107	0.238	0.229
	Ghugal and Sayyad [2]	TSDT	0.044	2.933	0.212	0.110	0.245	0.235
	Ghugal and Pawar [3]	HPSDT	0.044	2.959	0.199	0.107	0.237	0.238
	Mindlin [9]	FSDT	0.044	2.934	0.197	0.106	0.169	0.239
	Kirchhoff [5,6]	CPT	0.044	2.802	0.197	0.106	–	0.238
	Pagano [11]	Elasticity	0.044	2.942	0.200	–	0.238	–

Table 3. Comparison of inplane displacement ($\bar{u}$) at ($x = 0$, $y = b/2$, $z = \pm h/2$), transverse displacement ($\bar{w}$) at ($x = a/2$, $y = b/2$, $z = 0$), inplane normal stress ($\bar{\sigma}_x$) at ($x = a/2$, $y = b/2$, $z = \pm h/2$), inplane shear stress ($\bar{\tau}_{xy}$) at ($x = 0$, $y = 0$, $z = \pm h/2$) and transverse shear stress $\bar{\tau}_{zx}$ at ($x = 0$, $y = b/2$, $z = 0$) in isotropic square plate subjected to linearly varying load

S	Theory	Model	$\bar{u}$	$\bar{w}$	$\bar{\sigma}_x$	$\bar{\tau}_{xy}$	$\bar{\tau}_{zx}^{CR}$	$\bar{\tau}_{zx}^{EE}$
4	Present	ESDT	0.0396	2.908	0.150	0.111	0.240	0.236
	Reddy [12]	HSDT	0.0395	2.935	0.150	0.109	0.241	0.226
	Ghugal and Sayyad [2]	TSDT	0.0370	2.840	0.159	0.104	0.241	0.210
	Ghugal and Pawar [3]	HPSDT	0.0395	2.929	0.148	0.092	0.239	0.225
	Mindlin [9]	FSDT	0.0370	2.817	0.144	0.097	0.165	0.247
	Kirchhoff [5,6]	CPT	0.0370	2.218	0.144	0.097	–	0.247
	Pagano [11]	Elasticity	0.0360	2.847	0.153	–	0.230	–
10	Present	ESDT	0.0375	2.329	0.144	0.102	0.247	0.245
	Reddy [12]	HSDT	0.0375	2.333	0.144	0.101	0.246	0.243
	Ghugal and Sayyad [2]	TSDT	0.0365	2.313	0.153	0.097	0.252	0.241
	Ghugal and Pawar [3]	HPSDT	0.0370	2.332	0.144	0.096	0.245	0.243
	Mindlin [9]	FSDT	0.0370	2.335	0.143	0.097	0.165	0.248
	Kirchhoff [5,6]	CPT	0.0370	2.213	0.143	0.097	–	0.248
	Pagano [11]	Elasticity	0.0365	2.320	0.144	–	0.244	–

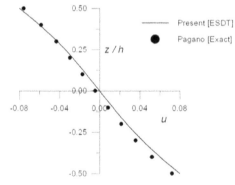

Fig. 2. Through thickness variation of inplane displacement of isotropic plate subjected to uniformly distributed load for aspect ratio 4

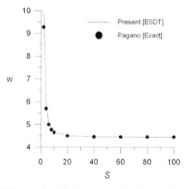

Fig. 3. Through thickness variation of transverse displacement of isotropic plate subjected to uniformly distributed load for aspect ratio 4

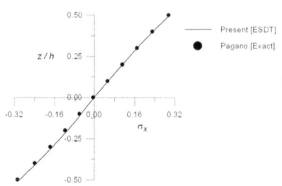

Fig. 4. Through thickness variation of inplane normal stress of isotropic plate subjected to uniformly distributed load for aspect ratio 4

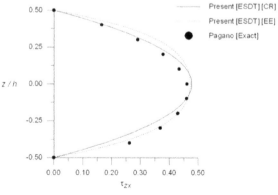

Fig. 5. Through thickness variation of transverse shear stress of isotropic plate subjected to uniformly distributed load for aspect ratio 4

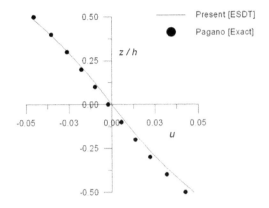

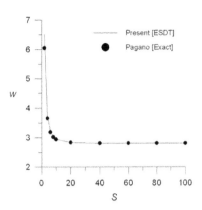

Fig. 6. Through thickness variation of inplane displacement of isotropic plate subjected to single sine load for aspect ratio 4

Fig. 7. Through thickness variation of transverse displacement of isotropic plate subjected to single sine load for aspect ratio 4

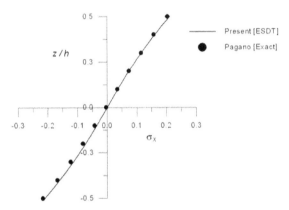

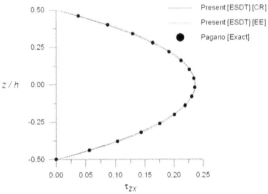

Fig. 8. Through thickness variation of inplane normal stress of isotropic plate subjected to single sine load for aspect ratio 4

Fig. 9. Through thickness variation of transverse shear stress of isotropic plate subjected to single sine load for aspect ratio 4

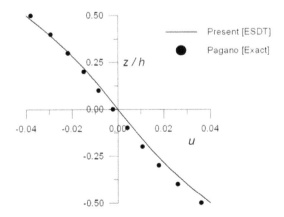

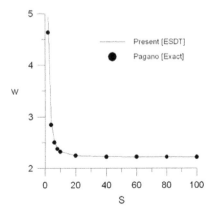

Fig. 10. Through thickness variation of inplane displacement of isotropic plate subjected to linearly varying load for aspect ratio 4

Fig. 11. Through thickness variation of transverse displacement of isotropic plate subjected to linearly varying load for aspect ratio 4

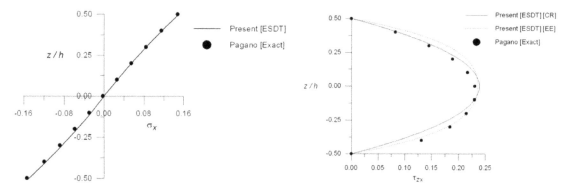

Fig. 12. Through thickness variation of inplane normal stress of isotropic plate subjected to linearly varying load for aspect ratio 4

Fig. 13. Through thickness variation of transverse shear stress of isotropic plate subjected to linearly varying load for aspect ratio 4

5.2. Discussion of results

Example 1: Table 1 shows the comparison of maximum displacements and stresses for the isotropic plate subjected to uniformly distributed load. The present theory and other higher order theories overestimate the results of inplane displacement as compared to those of exact solution. Through thickness variation of inplane displacement for isotropic plate subjected to uniformly distributed load is shown in Fig. 2. The value of maximum transverse displacement by present theory (ESDT), HPSDT and HSDT overestimate it by 2.142%, 2.880% and 3.073% for aspect ratio 4 and 0.409%, 0.560% and 0.582% for aspect ratio 10 respectively. TSDT gives the results which are in close agreement with exact value as compared to the theories of Kirchhoff and Mindlin for both aspect ratios. Variation of maximum transverse displacement with aspect ratio is shown in Fig. 3. Maximum values of inplane normal stress obtained by Present theory and HSDT are in close agreement with exact solution for aspect ratio 4, whereas yields exact value of it for aspect ratio 10. However CPT and FSDT underestimate the result by 6.51% for aspect ratio 4.

Its through thickness distribution is shown in Fig. 4. As exact elasticity solutions for inplane shear stress are not available, the results are compared with the other higher order theories, and corresponding values of FSDT and CPT. Present theory is in close agreement with the available solution in the literature. The transverse shear stress can be obtained directly by constitutive relations and equilibrium equations. The examination of Table 1 also reveals that the present theory overestimates the transverse shear stress by 4.656% than the exact elasticity solution when obtained using constitutive relation and underestimates the same by 2.608% when obtained using equilibrium equation for aspect ratio 4 (see Fig. 5). For aspect ratio 10, the results of transverse shear stresses, obtained by constitutive relations and equilibrium equation are in close agreement with the elasticity solution.

Example 2: The displacements and stresses for isotropic plate subjected to sinusoidal load are presented in Tables 2. The result of inplane displacement predicted by present theory and HSDT is identical for the aspect ratio 4 (overestimated by 6.976%), whereas HPSDT overestimates it by 9.302%. Inplane displacement predicted by TSDT is in excellent agreement for both the aspect ratios. Its variation through thickness of the plate is shown in Fig. 6.

Table 4. Comparison of natural bending mode frequencies ($\bar{\omega}_w$) and thickness shear mode frequencies ($\bar{\omega}_\phi$ and $\bar{\omega}_\psi$) of simply supported isotropic square plates ($S = 10$)

a/b	$\bar{\omega}$	(m,n)	Exact [16]	Present [ESDT]	Ghugal and Sayyad [2]	Reddy [12]	Mindlin [9]	CPT [5,6]
1.0	$\bar{\omega}_w$	$(1,1)$	0.0932	0.0931	0.0933	0.0931	0.0930	0.0955
		$(1,2)$	0.2226	0.2223	0.2231	0.2219	0.2219	0.2360
		$(1,3)$	0.4171	0.4163	0.4184	0.4150	0.4149	0.4629
		$(2,2)$	0.3421	0.3415	0.3431	0.3406	0.3406	0.3732
		$(2,3)$	0.5239	0.5228	0.5258	0.5208	0.5206	0.5951
		$(2,4)$	0.7511	0.7499	0.7542	0.7453	0.7446	0.8926
		$(3,3)$	0.6889	0.6874	0.6917	0.6839	0.6834	0.8090
		$(4,4)$	1.0889	1.0872	1.0945	1.0785	1.0764	1.3716
	$\bar{\omega}_\phi$	$(1,1)$	3.1729	3.1626	3.1729	3.1749	3.1730	–
		$(1,2)$	3.2192	3.2091	3.2191	3.2212	3.2193	–
		$(1,3)$	3.2949	3.2851	3.2949	3.2969	3.2951	–
		$(2,2)$	3.2648	3.2549	3.2648	3.2668	3.2650	–
		$(2,3)$	3.3396	3.3299	3.3396	3.3415	3.3397	–
		$(2,4)$	3.4414	3.4320	3.4414	3.4433	3.4416	–
		$(3,3)$	3.4126	3.4031	3.4126	3.4145	3.4128	–
		$(4,4)$	3.6094	3.6004	3.6094	3.6112	3.6096	–
	$\bar{\omega}_\psi$	$(1,1)$	3.2465	3.2428	3.2469	3.2555	3.2538	–
		$(1,2)$	3.3933	3.3994	3.3940	3.4125	3.4112	–
		$(1,3)$	3.6160	3.6381	3.6178	3.6517	3.6510	–
		$(2,2)$	3.5298	3.5455	3.5312	3.5589	3.5580	–
		$(2,3)$	3.7393	3.7709	3.7414	3.7848	3.7842	–
		$(2,4)$	4.0037	4.0576	4.0082	4.0720	4.0720	–
		$(3,3)$	3.9310	3.9786	3.9351	3.9928	3.9926	–
		$(4,4)$	4.4013	4.4944	4.4102	4.5092	4.5098	–

Table 5. Comparison of natural bending mode frequencies ($\bar{\omega}_w$) of simply supported isotropic rectangular plates ($S = 10$)

a/b	$\bar{\omega}$	(m,n)	Exact [16]	Present [ESDT]	Ghugal and Sayyad [2]	Reddy [12]	Mindlin [9]	CPT [5,6]
$\sqrt{2}$	$\bar{\omega}_w$	$(1,1)$	0.0704	0.0704	0.0705	0.0704	0.0703	0.0718
		$(1,2)$	0.1376	0.1376	0.1393	0.1374	0.1373	0.1427
		$(1,3)$	0.2431	0.2433	0.2438	0.2426	0.2424	0.2591
		$(1,4)$	0.3800	0.3803	0.3811	0.3789	0.3782	0.4182
		$(2,1)$	0.2018	0.2017	0.2023	0.2041	0.2012	0.2128
		$(2,2)$	0.2634	0.2639	0.2642	0.2628	0.2625	0.2821
		$(2,3)$	0.3612	0.3639	0.3623	0.3601	0.3595	0.3958
		$(2,4)$	0.4890	0.4928	0.4906	0.4874	0.4861	0.5513
		$(3,1)$	0.3987	0.3985	0.3999	0.3975	0.3967	0.4406
		$(3,2)$	0.4535	0.4552	0.4550	0.4520	0.4509	0.5073
		$(3,3)$	0.5411	0.5465	0.5431	0.5392	0.5375	0.6168

The maximum central deflection for single sine load obtained by the present theory over-estimates the value by 2.348% than the exact. TSDT yields the value very closed to the exact value, whereas HPSDT overestimates it by 3.194%. HSDT is in error by 3.413%. The FSDT underestimates the value of maximum transverse deflection by 0.983%, whereas CPT underestimates the same by 23.45% as compared to exact value due to the neglect of shear deformation for aspect ratio 4. For aspect ratio 10, the value of maximum transverse displacement by present theory overestimate it by 0.407%, HPSDT overestimate it by 0.577%, whereas TSDT underestimates it by 0.306%, and HSDT and FSDT overestimate it by 0.645% and 0.271% respectively. For the same aspect ratio, CPT underestimates the value by 4.758%. Fig. 7 shows the variation of transverse displacement with aspect ratio of the plate. The present theory underestimates inplane normal stress by 1.483%, HPSDT underestimates it by 3.686%. TSDT overestimates it by 4.147%, whereas HSDT and FSDT underestimate it by 3.686%, and 9.216% respectively for aspect ratio 4. It is observed that the values of present theory and other theories are in close agreement with those of exact solution for aspect ratio 10. The through thickness distribution of this stress is shown in Fig. 8. The transverse shear stress obtained by constitutive relations are much closed to those of elasticity solution for aspect ratio 4 and 10. The present theory predicts exact value of transverse shear stress for aspect ratios 4 and 10 using equations of equilibrium. HSDT and TSDT underestimate the transverse shear stress for both the aspect ratios. HPSDT underestimate the transverse shear stress for aspect ratio 4 and yield exact value of it for aspect ratio 10. The variation of this stress through the thickness is presented in Fig. 9.

Example 3: The numerical results of displacements and stresses of simply supported square plate subjected to linearly varying load are presented in Table 3 and found in excellent agreement with exact solution. Through thickness variations of displacements and stresses are shown in Figs. 10–13.

Example 4: Table 4 shows comparison of non-dimensional bending mode frequencies and thickness shear mode frequencies for simply supported isotropic square plates. The non-dimensional frequency corresponding to bending mode is denoted by $\bar{\omega}_w$. From the examination of Table 4, it is observed that the present theory (ESDT) yields excellent values of bending frequencies for all modes of vibration as compared to those of exact results. The value of bending frequencies for fundamental mode predicted by ESDT and HSDT are identical. TSDT overestimates the bending frequency for fundamental mode. HSDT underestimates the bending frequencies for higher modes. FSDT yields the lower values of bending frequency for all modes of vibration compared to those of other higher order and exact results, whereas CPT yields the higher values for this frequency. The non-dimensional frequency corresponding to thickness-shear modes are denoted by $\bar{\omega}_\phi$ and $\bar{\omega}_\psi$. The proposed ESDT shows excellent results for thickness shear mode frequency $\bar{\omega}_\phi$ for higher modes. HSDT yields the higher values of this frequency compared to those of present and exact theories. Results for thickness shear mode frequency ($\bar{\omega}_\phi$) obtained by FSDT is not satisfactory for higher modes. Thickness-shear mode frequencies ($\bar{\omega}_\psi$) for square plate predicted by ESDT shows good accuracy of results, whereas HSDT and FSDT overestimates the same. Comparison of non-dimensional bending mode frequency ($\bar{\omega}_w$) of simply supported isotropic rectangular plate is presented in Table 5. For rectangular plate, ESDT and HSDT show exact value for the bending frequency when $m = 1, n = 1$. FSDT underestimates the bending frequencies for rectangular plate, whereas CPT overestimates the same for fundamental mode.

6. Conclusions

From the study of bending and free vibration analysis of thick isotropic plates by using exponential shear deformation theory (ESDT), following conclusions are drawn:

1. The results of displacements and stresses obtained by present theory for the all loading cases are in excellent agreement with those of exact solution.

2. The results of displacements and stresses when plate subjected to linearly varying load are exactly half of those when plate subjected to uniformly distributed load

3. The frequencies obtained by the present theory for bending and thickness shear modes of vibration for all modes of vibration are in excellent agreement with the exact values of frequencies for the simply supported square plate.

4. The frequencies of bending and thickness shear modes of vibration according to present theory are in good agreement with those of higher order shear deformation theory for simply supported rectangular plate. This validates the efficacy and credibility of the proposed theory.

References

[1] Ghugal, Y. M., Shimpi, R. P., A review of refined shear deformation theories for isotropic and anisotropic laminated plates, Journal of Reinforced Plastics and Composites 21 (2002) 775–813.

[2] Ghugal, Y. M., Sayyad, A. S., Free vibration of thick orthotropic plates using trigonometric shear deformation theory, Latin American Journal of Solids and Structures 8 (2010) 229–243.

[3] Ghugal, Y. M., Pawar, M. D., Buckling and vibration of plates by hyperbolic shear deformation theory, Journal of Aerospace Engineering and Technology 1(1) (2011) 1–12.

[4] Karama, M., Afaq, K. S., Mistou, S., Mechanical behavior of laminated composite beam by new multi-layered laminated composite structures model with transverse shear stress continuity, International Journal of Solids and Structures 40 (2003) 1 525–1 546.

[5] Kirchhoff, G. R., Über das Gleichgewicht und die Bewegung einer elastischen Scheibe, Journal für die reine und angewandte Mathematik 40 (1850) 51–88. (in German)

[6] Kirchhoff, G. R., Über die Schwingungen einer kriesformigen elastischen Scheibe, Annalen der Physik und Chemie 81 (1850) 258–264. (in German)

[7] Kreja, I., A literature review on computational models for laminated composite and sandwich panels, Central European Journal of Engineering 1(1) (2011) 59–80.

[8] Liew, K. M., Xiang, Y., Kitipornchai, S., Research on thick plate vibration, Journal of Sound and Vibration 180 (1995) 163–176.

[9] Mindlin, R. D., Influence of rotatory inertia and shear on flexural motions of isotropic, elastic plates, ASME Journal of Applied Mechanics 18 (1951) 31–38.

[10] Noor, A. K., Burton, W. S., Assessment of shear deformation theories for multilayered composite plates, Applied Mechanics Reviews 42 (1989) 1–13.

[11] Pagano, N. J., Exact solutions for bidirectional composites and sandwich plates, Journal of Composite Materials 4 (1970) 20–34.

[12] Reddy, J. N., A simple higher order theory for laminated composite plates, ASME Journal of Applied Mechanics 51 (1984) 745–752.

[13] Reissner, E., The effect of transverse shear deformation on the bending of elastic plates, ASME Journal of Applied Mechanics 12 (1945) 69–77.

[14] Reissner, E., On the theory of bending of elastic plates, Journal of Mathematics and Physics 23 (1944) 184–191.

[15] Shimpi, R. P., Patel, H. G., A two variable refined plate theory for orthotropic plate analysis, International Journal of Solids and Structures 43 (2006) 6 783–6 799.

[16] Srinivas, S., Joga Rao, C. V., Rao, A. K., An exact analysis for vibration of simply supported homogeneous and laminated thick rectangular plates, Journal of sound and vibration 12(2) (1970) 187–199.

[17] Vasil'ev, V. V., The theory of thin plates, Mechanics of Solids, 27 (1992) 22–42.

8

Reduction of lateral forces between the railway vehicle and the track in small-radius curves by means of active elements

T. Michálek[a,*], J. Zelenka[a]

[a]University of Pardubice, Jan Perner Transport Faculty, Detached Branch Česká Třebová, Slovanská 452, 560 02 Česká Třebová, Czech Republic

Abstract

This paper deals with a possibility of reduction of guiding forces magnitude in small-radius curves by means of active elements. These guiding forces characterize the lateral force interaction between the rail vehicle and the track and influence the wear of wheels and rails in curves. Their magnitudes are assessed in the framework of vehicle authorization process. However, in case of new railway vehicles with axleload of approximately 20 t and more it is problematic to meet the condition of maximum value of the quasistatic guiding force which acts on the outer wheel of the 1st wheelset in small-radius curves. One of the possible ways how to reduce these forces is using the system of active yaw dampers. By means of computer simulations of guiding behaviour of a new electric locomotive, comparison of reached values of the quasistatic guiding forces in case of locomotive equipped with active yaw dampers and without them was performed. Influences of magnitude of force generated by the active yaw dampers, friction coefficient in wheel/rail contact and curve radius were analysed in this work, as well.

Keywords: active yaw dampers, quasistatic guiding forces, electric locomotive, simulations

1. Introduction

Under the term "guiding behaviour" we understand dynamical properties of railway vehicles during the passing through curves. The guiding behaviour is assessed in the framework of authorization process of new or modernised vehicles according to the European Standard EN 14363 [2] or the UIC Code No. 518 [6]. So-called quasistatic guiding force is one of the most important quantities in this respect. Guiding forces characterize the lateral force interaction between the vehicle and the track; if the rail vehicle passes through a curve these forces act on each wheel in wheel/rail contact and influence the wear of wheels and rails.

In case of modern locomotives with axleload of approximately 20 t and more (especially if they are designed for high speed operation or for freight haulage) it is problematic to meet the condition that value of the quasistatic guiding force acting on the outer wheel of the 1st wheelset in small-radius curves should not exceed the limit value $Y_{qst,lim} = 60$ kN [2]. In case of exceeding of this value the authorization process of the vehicle is complicated; due to higher degree of wear of wheels and rails, operational costs of such locomotive are higher, as well. Therefore, producers of railway vehicles search for technical solutions which allow reduction of the quasistatic guiding forces during the passing through curves. For a long time, bogie couplings have been used for these purposes in case of electric locomotives. However, an ability of the bogie couplings to reduce the guiding forces is limited. A more effective solution how to reduce the magnitudes of guiding forces is represented by active elements.

---

*Corresponding author. e-mail: tomas.michalek@student.upce.cz.

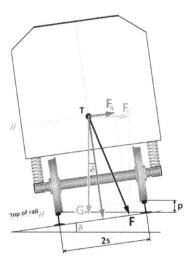

Fig. 1. Derivation of quantities F_q and a_q

Such a system of active elements, which serves for reduction of lateral forces in wheel/rail contact during the passing through curves (and wear of wheels and rails as well as operational costs, consequently), was developed by Liebherr and operationally tested on several Siemens' electric locomotives in 2006 [1]. In 2008 Czech producer of rail vehicles Škoda Transportation started testing of its new electric locomotive type 109E. Since 2009 an influence of hydraulic bogie coupling on the guiding behaviour of this locomotive has been observed, as well; see [9]. The aim of this paper is to assess the possible influence of system of active elements on guiding behaviour in small-radius curves if this device would be implemented into the running gear of the locomotive Škoda 109E.

2. Technical solutions for reduction of quasistatic guiding forces

If a standard four-axled locomotive with two bogies runs through a curve the maximum value of quasistatic guiding force usually acts on the outer wheel of the 1st wheelset and the second highest value of this force belongs to the outer wheel of the 3rd wheelset. Quasistatic guiding forces act also on all other wheels but their magnitudes are usually not so high. However, the sum of all these forces must always correspond to the Newton's law, i.e.:

$$\sum Y_{qst} = M_{loc} \cdot a_q = F_q, \tag{1}$$

where M_{loc} is total weight of locomotive and a_q is so-called unbalanced lateral acceleration. The product of M_{loc} and a_q is also named as unbalanced centrifugal force F_q.

Quantity a_q is often used in the branch of railway vehicle dynamics (see for example [6] or [4]) and its value can be derived on the base of Fig. 1. Unbalanced centrifugal force F_q — i.e. the component of resultant force F which is parallel with plane of the top of rail — is given as:

$$F_q = F_c \cdot \cos \delta - G \cdot \sin \delta, \tag{2}$$

where F_c is centrifugal force, G is force of gravity and δ is angle of the cant. This angle is given by superelevation p and tape line distance $2s$ (in case of standard-gauge rail vehicles, the tape line distance has a value of $2s = 1\,500$ mm) and its value is relatively small. Therefore, the force F_q can be approximately expressed as:

$$F_q = M_{loc} \cdot \frac{v^2}{R} - M_{loc} \cdot g \cdot \frac{p}{2s}, \tag{3}$$

where v is vehicle speed, R is curve radius and $g = 9.81$ m $\cdot$ s^{-2} is acceleration of gravity. It is evident that the unbalanced lateral acceleration a_q is given as:

$$a_q = \frac{v^2}{R} - g \cdot \frac{p}{2s}. \tag{4}$$

It is evident that under concrete conditions, which are defined by values of weight M_{loc} and unbalanced lateral acceleration a_q, the sum of all quasistatic guiding forces ΣY_{qst} must have an always constant value given by equation (1). A very important consequence of this equation is the fact that each technical device for reduction of quasistatic guiding forces can only redistribute these forces among the other wheels. It means that the maximum (and from the point of view of the current EN standard [2] also critical) values can be decreased; however, the sum of all these forces must stay unchanged.

2.1. Devices for implementation into the primary suspension

Practically all technical devices for reduction of quasistatic guiding forces work in such a way that they decrease the value of angle of attack of relevant wheelset or bogie. The angle of attack of the wheelset can be influenced by means of primary suspension and wheelset guiding. *Passive system of radial wheelset steering*, based on low stiffness of the wheelset guiding in longitudinal direction (possibly supplemented with cross-coupling of wheelsets), is usually not suitable for high speed vehicles because of problems with stability of run at higher speeds.

Nowadays, *active systems of radial wheelset steering* are being developed. These systems usually work in such a way that axleboxes of each wheelset are linked with bogie frame by means of actuators which allow steering of the wheelset into the radial position, i.e. decreasing of the angle of attack during the passing through curves. A technical device using this principle is being developed for example by Bombardier as an equipment of its "Flexx Tronic" bogie for double-decker EMUs — see [5,11]; it is named ARS (in German: "aktive Radsatzsteuerung"). The principle of active wheelset steering generally allows improving the stability of run in straight track (it means increasing of the critical speed of the vehicle), as well. However, these active systems are not used in technical practice in the present.

2.2. Devices for implementation into the secondary suspension

Technical solutions for reduction of quasistatic guiding forces, which belong into this category, influence angle of attack of the whole bogies. These systems can be designed passive as well as active. *Bogie coupling* is one of the widely used passive systems. The bogie coupling turns the bogies of the vehicle during the passing through curves about the vertical axis into opposite directions. The best state, which can be reached in this way, is equalisation of the quasistatic guiding forces acting on outer wheels of the 1st and 3rd wheelset. The bogie coupling (the force transmission between the bogies) can be realized mechanically as well as hydraulically. Mechanical bogie coupling is often applied in running gears of older electric locomotives. As it was mentioned, hydraulic bogie coupling was tested for example on the locomotive Škoda 109E; a more detailed description of this device as well as results of computer simulations are shown in [9].

Usage of active elements, especially *active yaw dampers*, is the second possibility how to reduce quasistatic guiding forces. During the passing through curve, this device works in such

Fig. 2. Active yaw damper (ADD) from Liebherr [11]

Fig. 3. Preparation for installation of ADD on the
locomotive Siemens Vectron

a way that both bogies of the vehicle are turned into the bogie direction; it means that angles
of attack of both bogies are decreasing. In straight track, these dampers work as classic yaw
dampers; it means that they damp yawing motion of bogies. System of active yaw dampers,
which is known as ADD (in German: "aktive Drehdämpfer"), was developed in co-operation
of Siemens and Liebherr and operationally tested on 6 Siemens' electric locomotives (types
ES64F4 and ES64U2); see for example [1, 11]. An advantage of this system is the fact that
actuator is integrated into the yaw damper (see Fig. 2) and its installation does not need any
modifications of the bogie or vehicle body frame. ADD seems to be a very effective way
how to reduce the magnitudes of quasistatic guiding forces in small-radius curves and Siemens
want to use this system on its new generation of locomotives "Vectron", as well. Preparation
for installation of ADD instead of classic yaw dampers (i.e. installation of the power supply
connectors) on the electric locomotive "Vectron" is shown in Fig. 3.

3. Simulations of guiding behaviour of an electric locomotive

For purposes of assessment of the active yaw dampers influence on guiding behaviour of an
electric locomotive series of computer simulations was performed. The simulations were carried out by means of multi-body program system "SJKV" which is being developed at the
Detached Branch of the Jan Perner Transport Faculty in Česká Třebová. Dynamical model of
the electric locomotive Škoda 109E was used and modified for these purposes. Magnitudes of
the quasistatic guiding forces reached with locomotive equipped with active yaw dampers were
compared with results of locomotive without them.

3.1. Program system "SJKV" for simulations of rail vehicle dynamics

The program system "SJKV" is original multi-body simulation (MBS) software for simulations
of running and guiding behaviour of rail vehicles which is being developed in the IDE Borland
Delphi. Its architecture is based on program units and allows creating of different modifications
for concrete rail vehicles. In comparison with commercial MBS software as for example SIM-
PACK or MSC ADAMS, the greatest advantage of the system "SJKV" is a detailed knowledge
of algorithms on which the calculations are based. In case of the commercial MBS software,
"black-box-principle" from the user's point of view is often a source of problems, especially at
the verification of computational models.

In the program system "SJKV", the whole non-linear system vehicle–track is modelled as
a multi-body system; it means that all the bodies are considered rigid and coupled by means of
elastic and damping couplings. This dynamical model of the system vehicle–track is mathema-
tized on the base of *structural elements method*; therefore, one equation of motion belongs to

each degree of freedom of the multi-body system. So, the acceleration vector is generally given by equation:

$$\ddot{\mathbf{q}} = \mathbf{M}^{-1} \cdot \mathbf{L} \cdot \mathbf{F}, \tag{5}$$

where $\mathbf{M}$ is mass matrix, $\mathbf{L}$ is geometric matrix and $\mathbf{F}$ is vector of acting forces. These forces include forces of gravity, wheel/rail contact forces and coupling forces. In general case, characteristics of these couplings are non-linear. Components of the acceleration vector represent the acceleration of relevant bodies in directions of considered degrees of freedom (i.e. translation and rotation of these bodies).

Solving of mathematical model of the system is based on *finite differences method*. In this way, deflections and velocities in following time integration step can be calculated by means of deflections and accelerations in two previous steps; precondition for using of this method is a constant value of the acceleration during the time integration step Δt. So, the deflection vector and the velocity vector in the time step $i + 1$ are given as:

$$\mathbf{q}_{i+1} = 2 \cdot \mathbf{q}_i - \mathbf{q}_{i-1} + \ddot{\mathbf{q}}_i \cdot (\Delta t)^2, \tag{6}$$

$$\dot{\mathbf{q}}_{i+1} = \frac{1}{2 \cdot \Delta t} \cdot (3 \cdot \mathbf{q}_{i+1} - 4 \cdot \mathbf{q}_i + \mathbf{q}_{i-1}). \tag{7}$$

Special attention is paid to solving of wheel/rail contact in the program system "SJKV". Forces acting in the wheel/rail contact can be divided into 3 categories — wheel forces Q acting in vertical direction, guiding forces Y acting in lateral direction and longitudinal creep forces T_x. The lateral and longitudinal forces Y and T_x, which intermediate the adhesion joint between wheels and rails, are computed by means of algorithm proposed by Polách [4]. For purposes of using in the simulations, the wheel/rail contact is described by means of *characteristics of the wheel/rail contact geometry*. These characteristics are computed on the basis of wheel and rail profiles, track gauge, rail inclination and wheelset gauge by means of original program system "KONTAKT 5" and include:

- *position of the contact points on wheels and rails* in dependency on lateral movement of the wheelset y_w,

- *delta-r function* which gives actual roll radius difference of the left and right wheel of the wheelset in dependency on lateral wheelset movement y_w; it is given as:

$$\Delta r = r_L - r_R = f(y_w), \tag{8}$$

- *tangent-γ function* which is given by rake angles of contact planes on the left and right wheel in dependency on lateral wheelset movement y_w and which is an important parameter for the force interaction between the wheelset and the track; it is given as:

$$\tan \gamma = \tan \gamma_L - \tan \gamma_R = f(y_w), \tag{9}$$

- *equivalent conicity* which characterize intensity of wheelset centring effect. This quantity depends on amplitude of lateral wheelset motion y_0. Value of the equivalent conicity for the wheelset amplitude of $y_0 = 3$ mm is often used as comparative value of various wheelset/track contact pairs; this method of assessment of the wheel/rail contact geometry during the testing of running and guiding behaviour of railway vehicles is defined in standards [2] and [6], as well.

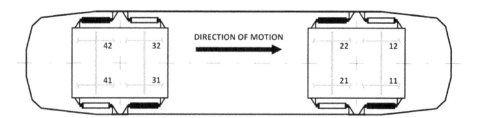

Fig. 4. Considered arrangement of classic (white) and active (black) yaw dampers on the locomotive

As it was said, mathematical description of the dynamical model is the basis of the MBS system "SJKV". Therefore, *input data* for simulations include above all dimensional and mass parameters of the model, characteristics of couplings and characteristics of contact geometry. Next important input data are given by alignment of the track, track irregularities and vehicle speed. More detailed description of the system "SJKV" is presented in [7].

3.2. Modification of the dynamical model of the locomotive

For purposes of assessment of active yaw dampers influence on guiding behaviour, dynamical model of the locomotive Škoda 109E had to be modified. Therefore, modelling of the system of active yaw dampers was the most significant modification in this case. New version of the program system named "SJKV-L3A" was created in this way.

Dynamical model of the vehicle consists of 7 rigid bodies, i.e. 4 wheelsets, 2 bogies and vehicle body. Smaller bodies (as for example springs, dampers or components of traction drive) are reduced to the considered bodies. This degree of simplification of physical reality seems to be fully sufficient because results of simulations performed with using of the system "SJKV" (see for example [9]) show relatively good agreement with results of measurements. Dynamical model of the track is created with reduced rail mass which belongs to each wheel (see [7]); it comprises 8 additional rigid bodies. In this case, the whole system vehicle–track has 50 degrees of freedom. Couplings between the bodies (i.e. springs, wheelset guiding, rotational stiffness bogie–vehicle body etc.) are considered elastic and damping; coupling forces are computed on the basis of deflections and velocities of the rigid bodies, i.e. deformations of the couplings and velocities of these deformations. Characteristics of considered couplings are usually given by producer of the vehicle or measured on test stand (see for example [10]).

Locomotive Škoda 109E is standardly equipped with 8 yaw dampers, i.e. 4 yaw dampers per bogie. In the framework of this work, 4 classic yaw dampers (i.e. 2 pieces per bogie) were replaced with active dampers in dynamical model. System of the active yaw dampers, which is technically based on the system ADD, was considered. It means that these dampers generate constant (pull or pressure) force, if the vehicle runs through small-radius curves, and help so with the radial steering of bogies. Small-radius curves are identified by means of angles between the bogies and the vehicle body, using the displacement sensors integrated into the active dampers. In Fig. 4 there is shown considered arrangement of classic and active dampers on the locomotive.

Arrangement of the active yaw dampers is designed in such a way that all these dampers generate either pull or pressure force at the same time. During the passing through curve, the active yaw dampers on the front bogie act in the opposite direction in comparison with moment against the bogie rotation; on the rear bogie, moment of the active dampers forces and the moment against the bogie rotation act in the same direction. So, total moment about vertical axis between the first/second bogie and the vehicle body is given as:

$$M_{I/II} = F_{act} \cdot 2W_{st} \mp \gamma \cdot \beta_{I/II}, \tag{10}$$

where F_{act} is force generated by the active yaw damper, $2W_{st}$ is lateral distance between yaw dampers, γ is rotational stiffness of secondary suspension and β is angle about vertical axis between the relevant bogie and the vehicle body during the passing through curve. In consequence of active elements actuation, additional moment about vertical axis acts on vehicle body, as well. Therefore, its magnitude must be compensated by means of additional lateral forces in secondary suspension — i.e. by deflection of flexi-coil springs, possibly by lateral forces acting on bogie bump stops.

3.3. Simulations of the locomotive equipped with the system of active yaw dampers

Simulations of guiding behaviour of the locomotive were performed for passing through small-radius curves (so-called "testing area No. 4" according to [2]) which is usually critical from the point of view of the limit value of quasistatic guiding force. Concretely, curves with radius of $R = 250$ m, 300 m and 350 m were investigated; superelevation in all the curves had value of $p = 150$ mm. For purposes of this work, only quasistatic calculations (i.e. simulations on ideal track without irregularities) were performed. However, this simplification should not have any significant influence on accuracy of results. As it was proved (see for example [8]), in case of assessment of the quasistatic guiding forces, results of the quasistatic calculations differ less than 1 % from mean values obtained by means of time-consuming statistic processing of results of dynamic calculations (i.e. simulations on real track with measured irregularities). The vehicle speed was chosen in such a way that it corresponded to unbalanced lateral acceleration of $a_q = 1.1$ m $\cdot$ s^{-2} — it means: $V = 81.7$ km/h in the $R250$-curve, $V = 89.5$ km/h in $R300$-curve and $V = 96.7$ km/h in $R350$-curve. This value of unbalanced lateral acceleration follows from requirements of the standards [2] and [6] on maximum testing vehicle speed in small-radius curves.

Conditions of wheel/rail contact geometry were given by measured (slightly worn) wheel profile S1002 and theoretical rail profile 60E1 with inclination 1 : 20; this contact pair is characterized with low value of equivalent conicity. Value of the friction coefficient in wheel/rail contact was changed in the range of 0.2 up to 0.5. For purposes of assessment of active yaw dampers influence on quasistatic guiding forces magnitudes, value of the force F_{act} generated by active elements was changed in the range of 5 up to 20 kN, as well.

4. Simulation results

As it was said in previous chapter, influence of the force generated by active yaw dampers as well as influences of the friction coefficient in wheel/rail contact and the curve radius on the guiding behaviour of the locomotive Škoda 109E were observed and subsequently analysed. Simulation results reached with locomotive equipped with the system of active yaw dampers were compared with quasistatic guiding forces of locomotive without them.

In bar chart in Fig. 5 there are shown magnitudes of the quasistatic guiding forces reached on outer wheels of the 1st and 3rd wheelset (i.e. guiding wheels in both bogies) in a curve with radius $R = 300$ m at the speed $V = 89.5$ km/h (i.e. $a_q = 1.1$ m $\cdot$ s^{-2}) for various values of friction coefficient in wheel/rail contact. It is evident that the system of active yaw dampers has a significantly positive influence on quasistatic guiding forces acting on guiding wheels. In comparison with the 1st wheelset, decreasing of the quasistatic guiding force magnitude on the

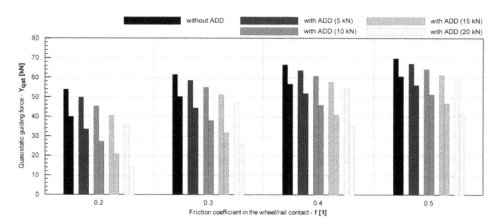

Fig. 5. Quasistatic guiding forces acting on outer wheels of the 1st and 3rd wheelset in curve with radius $R = 300$ m for various values of friction coefficient in wheel/rail contact and for various setting-up of maximum force generated by active yaw dampers

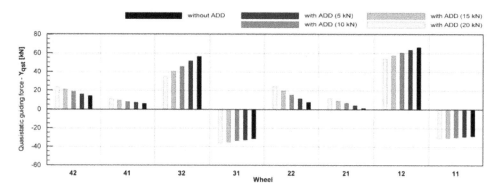

Fig. 6. Quasistatic guiding forces acting on constituent wheels in curve with radius $R = 300$ m for value of friction coefficient in wheel/rail contact $f = 0.40$ and for various setting-up of maximum force generated by active yaw dampers

3rd wheelset in dependency on increasing force generated by active elements is more significant. This fact can be explained with positive effect of the moment against bogie rotation in case of the rear (second) bogie — see equation (10). In case of normal friction conditions (it means that value of friction coefficient in wheel/rail contact belongs into the range of 0.3 up to 0.4), value of the force generated by active elements of $F_{act} = 10$ kN seems to be sufficient to meet the condition of the limit value of quasistatic guiding force $Y_{qst,lim} = 60$ kN which should not be exceeded in curves with mean value of radius $R_m = 300$ m ($+50$ m/-20 m) according to [2].

Influence of active yaw dampers on distribution of quasistatic guiding forces on constituent wheels of locomotive is shown for value of friction coefficient $f = 0.40$ and curve radius $R = 300$ m in bar chart in Fig. 6. It is evident that equation (1) must be always valid — in consequence of ADD actuation, magnitudes of guiding forces acting on wheels No. 12 and 32 (guiding wheels in both bogies) decrease; however, magnitudes of guiding forces acting on wheels of the 2nd and 4th wheelset increase. Numbering of wheels, which is used in bar charts in Fig. 6 up to 8, is explained in Fig. 4; the first number marks number of wheelset, the second number marks number of wheel of this wheelset. Considered motion of the locomotive has direction from the left to the right (see Fig. 4); so, the guiding wheels in the bogies in right hand curve are the wheels No. 12 and 32.

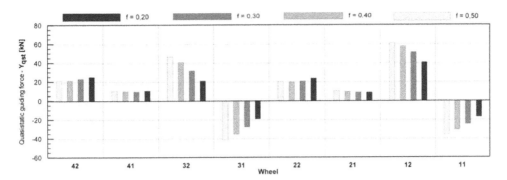

Fig. 7. Quasistatic guiding forces acting on constituent wheels in curve with radius $R = 300$ m for value of force generated by active yaw dampers $F_{act} = 15$ kN and for various values of friction coefficient in wheel/rail contact

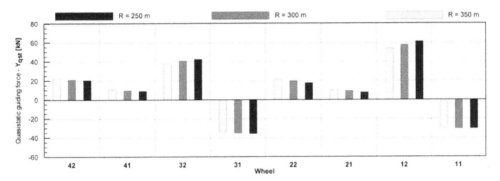

Fig. 8. Quasistatic guiding forces acting on constituent wheels in various curves for value of force generated by active yaw dampers $F_{act} = 15$ kN and for friction coefficient $f = 0.40$

Bar chart in Fig. 5 also confirms the fact that value of friction coefficient in wheel/rail contact has a very significant influence on magnitudes of quasistatic guiding forces; see for example [3], as well. Influence of friction coefficient on distribution of these forces on constituent wheels is demonstrated in Fig. 7. It is apparent that friction coefficient influences especially the forces acting on wheels of guiding wheelsets (i.e. the 1st and 3rd wheelset); equilibrium according to equation (1) must be retained again.

Influence of curve radius on quasistatic guiding forces was observed, as well. This influence is demonstrated for concrete conditions (given by values of friction coefficient and force generated by active elements) in bar chart in Fig. 8. In all cases the vehicle speed corresponded to unbalanced lateral acceleration $a_q = 1.1$ m$\cdot$s^{-2}. In comparison with influences of active yaw dampers and friction coefficient, this effect is relatively small.

5. Conclusion

This paper deals with a possibility of reduction of quasistatic guiding forces during the passing through curves by means of system of active yaw dampers. Except the analysis of the influence of these active elements, effects of friction coefficient in wheel/rail contact and curve radius were observed by means of computer simulations.

On the basis of reached results it is possible to say that the system of active yaw dampers seems to be an effective way how to reduce magnitudes of quasistatic guiding forces and consequently wear of rails in curves and wheels, as well. Lower degree of wear leads to lower operational costs, too; this effect, which is very important for carriers, is more significant if the fee for using of railway infrastructure depends on the effects of the vehicle on the track. In spite of this, using of active yaw dampers is not usual nowadays. Implementation of active elements into the running gear of rail vehicle puts very high requirements on safety; therefore, authorisation process of these systems is very complicated and expensive.

However, performed simulations show that implementation of the system of active yaw dampers (ADD) into secondary suspension of the locomotive Škoda 109E can cause more significant improvement of its guiding behaviour than the hydraulic bogie coupling which was tested on this vehicle recently (see [9]). For purposes of meeting the conditions of the European Standard EN 14363 [2] on the maximum value of quasistatic guiding force in the area of small-radius curves (so-called "testing area No. 4"), the system ADD, which is able to generate the maximum force $F_{act} = 10$ up to 15 kN, seems to be sufficient. It is important to say that many other conditions (for example the friction coefficient in wheel/rail contact or moment against the bogie rotation) have substantial influence on vehicle guiding behaviour, as well. However, application of active elements into the running gear of locomotives could represent the way how to design more effective railway vehicles in the future.

Acknowledgements

This work was supported by internal grant project of the University of Pardubice No. 51030/20/ SG510001 and by project of the Ministry of Education, Youth and Sports No. 1M0519 — Research Centre of Rail Vehicles.

References

[1] Breuer, W., Active torsional damper — an innovative damper concept in operational use, (in German), Eisenbahntechnische Rundschau 4 (2007), p. 186–189.

[2] ČSN EN 14363:2006. Railway applications — Testing for the acceptance of running characteristics of railway vehicles — Testing of running behaviour and stationary tests, (in Czech), Czech Institute for Normalisation, 2006.

[3] Michálek, T., Analysis of the influence of the friction coefficient on the guiding behaviour, Conference proceedings of the 13th conference Applied Mechanics (2011), p. 147–150.

[4] Polách, O., A Fast Wheel-Rail Forces Calculation Computer Code, Proceedings of the 16th IAVSD Symposium, Vehicle System Dynamics Supplement 33 (1999), p. 728–739.

[5] Schneider, R., Swaying-compensation and active wheelset steering "Flexx Tronic" from Bombardier, (in German), Eisenbahn-Revue International 4 (2010), p. 174–181.

[6] UIC Code 518:2009. Testing and approval of railway vehicles from the point of view of their dynamic behaviour — Safety — Track fatigue — Running behaviour, 4th edition, International Union of Railways (UIC), 2009.

[7] Zelenka, J., Running and guiding behaviour of two-axled diesel-electric locomotives CZ LOKO, (in Czech), New Railway Technique 6 (2009), p. 15–23.

[8] Zelenka, J., Michálek, T., A New Method of the Assessment of Rail Vehicles Guiding Behaviour in Small-radius Curves, International Journal of Applied Mechanics and Engineering 15 (2) (2010) 511–519.

[9] Zelenka, J., Špalek, P., Simulations of the guiding behaviour of the locomotive Class 380, assessment of the bogie coupling influence, (in Czech), Proceedings of the 19th conference with international participation Current Problems in Rail Vehicles (2009), p. 195–198.

[10] Zelenka, J., Vágner, J., Hába, A., Experimental verification of possibilities of determination of flexi-coil springs lateral stiffness, (in Czech), Vědeckotechnický sborník ČD 31 (2011).

[11] Liebherr — InnoTrans 2010, Berlin, promotional material.

9

Energy transformation and flow topology in an elbow draft tube

D. Štefan[a,*], P. Rudolf[a], A. Skoták[b], L. Motyčák[b]

[a] Faculty of Mechanical Engineering, Brno University of Technology, Technická 2896/2, 616 69 Brno, Czech Republic
[b] CKD Blansko Engineering, a.s., Čapkova 2357/5, 678 01 Blansko, Czech Republic

Abstract

Paper presents a computational study of energy transformation in two geometrical configurations of Kaplan turbine elbow draft tube. Pressure recovery, hydraulic efficiency and loss coefficient are evaluated for a series of flow rates and swirl numbers corresponding to operating regimes of the turbine. These integral characteristics are then correlated with local flow field properties identified by extraction of topological features. Main focus is to find the reasons for hydraulic efficiency drop of the elbow draft tube.

Keywords: draft tube, efficiency, pressure recovery, flow topology

1. Introduction

Draft tubes are diffusers placed at the outlet of the hydraulic turbine runner, see Fig. 1. Their main purpose is transformation of the residual kinetic energy into pressure energy with maximum efficiency. Although geometrically relatively simple the internal flow can be quite complex due to adverse pressure gradient, possible boundary layer separation, swirling and streamline curvature. The elbow draft tubes are usually used for a vertical arrangement of Kaplan turbines. The most problematic part of the draft tube with negative influence on flow properties is the elbow. Paper summarizes results of a computational study of energy transformation in two geometrically different configurations of Kaplan turbine elbow draft tube.

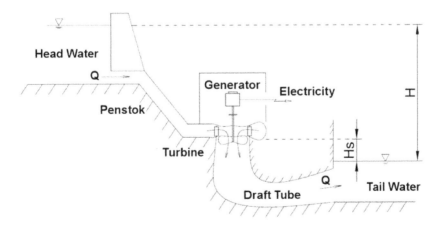

Fig. 1. Longitudinal cross-section view of hydraulic power plant

*Corresponding author. e-mail: y101274@stud.fme.vutbr.cz.

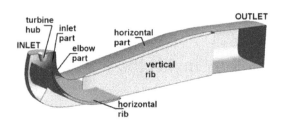

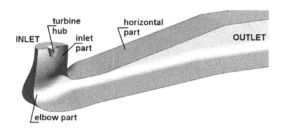

Fig. 2. Longitudinal cross-section view of DFT-A   Fig. 3. Longitudinal cross-section view of DFT-B

Draft tube geometrical configuration A (DFT-A) – The elbow draft tube for vertical Kaplan turbine includes one horizontal rib (in elbow part) and one vertical rib (in horizontal part), Fig. 2. This draft tube is installed in hydro-power plant Střekov, Czech Republic.

Draft tube geometrical configuration B (DFT-B) – The elbow draft tube for vertical Kaplan turbine without any ribs, Fig. 3. This draft tube is result of shape modification of DFT-A for one specific velocity profile (different than velocity profile used in this work) [12].

Part of turbine runner hub (at the inlet of the draft tube), see Fig. 2 and Fig. 3, has been included and modelled as a stationary wall for both configurations of the draft tube.

The numerical computation has been carried out in commercial CFD software ANSYS FLUENT r.12. employing Reynolds Averaged Navier-Stokes equations to solve computational domain by finite volume method. We used Realizable $k - \varepsilon$ model of turbulence (RKE) with non-equilibrium wall function to perform steady state solution. The RKE model of turbulence and steady state of computation are chosen because of suitable capturing of flow properties and computational demands for evaluation, where dynamics properties of flow are not extracted. Each draft tube has been computed for a series of flow rates corresponding to operating regimes of the turbine. The best efficiency point of the turbine (100 % Q_{BEP}) corresponds to the mass flow $Q_m = 440$ kg/s. Inlet velocity profiles in Fig. 5 are the result of numerical computation of the new design of Kaplan turbine runner for hydro-power plant Střekov and data were obtained from ČKD Blansko Engineering.

Investigation of the three-dimensional separation, carried out by software ANSYS CFD-POST, and its influence on energy transformation is discussed. Skin friction lines, surface streamlines and critical points are used to identify the topological features of the flow.

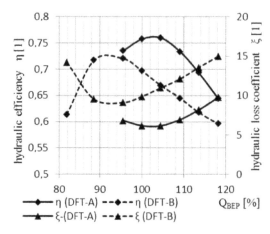

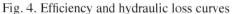

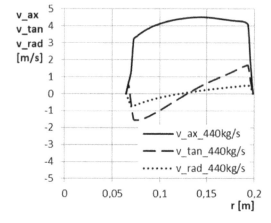

Fig. 4. Efficiency and hydraulic loss curves   Fig. 5. Inlet velocity profile for CFD computation, flow rate 100 % Q_{BEP}

2. Energy transformation

The hydraulic efficiency (1), pressure recovery factor (2) and hydraulic loss coefficient (3) have been evaluated in several cross-sections for series of flow rates and swirl numbers

$$\eta = \frac{p_{s(2)} - p_{s(1)}}{p_{d(1)} - p_{d(2)}}, \tag{1}$$

$$c_p = \frac{p_{s(2)} - p_{s(1)}}{\frac{1}{2}\rho\bar{v}_{(1)}^2}, \tag{2}$$

$$\xi = \frac{2}{\bar{v}_{(2)}^2}\left[\left(\frac{\alpha_{(1)}\bar{v}_{(1)}^2 - \alpha_{(2)}\bar{v}_{(2)}^2}{2}\right) + \frac{p_{s(1)} - p_{s(2)}}{\rho}\right], \tag{3}$$

where p_s is static pressure, p_d is dynamic pressure, $\bar{v}$ is mean velocity and α is Coriolis number. The number in parentheses represents $1 = $ inlet and $2 = $ outlet. The global development of these integral characteristics (1–3) is plotted in Fig. 6 for DFT-A and in Fig. 7 for DFT-B.

2.1. Energy transformation in DFT-A

Changes of DFT-A cross-sectional area are nearly linear from inlet to outlet of the draft tube. For flow rates 95.5 % Q_{BEP} and 100 % Q_{BEP} decrease of pressure recovery factor at the elbow part is observed and as will be shown in section 2.1, it is caused by flow separation under horizontal rib. For 109.1 % Q_{BEP}, the flow separation is significantly reduced, therefore, any considerable decrease of pressure recovery factor is not apparent.

The other decrease in pressure recovery factor is evident between cross-sections 9 and 11 for all three flow rates and is caused by the beginning of the vertical rib. The volume of vertical rib reduces cross-sectional area and thereby increases velocity and frictional losses. This effect is a consequence of the design so therefore it occurs for every flow rate. The smallest back-flow region at horizontal part of draft tube for whole range of flow rates is observed for 109.1 % Q_{BEP}. This is the reason for steeper slope of the pressure recovery curve than for other operating points. The highest value of DFT-A efficiency (1) is derived from efficiency curve plotted in Fig. 4 and corresponds to flow rate 103 % Q_{BEP}.

2.2. Energy transformation in DFT-B

The shape modification of DFT-B has had appreciable influence on the development of cross-sectional area at the elbow part of draft tube, see area between cross-section 3 and 6 in Fig. 7. The flow cross-section at the end of elbow part has been reduced because of separation risk on the inner curved wall. This modification was done for some past design of the draft tube with slightly different inlet velocity profile. It was anticipated that it might be beneficial also in combination with new runner design, which has to some degree altered outlet blade angles and hence also the outlet velocity components. Unfortunately, this assumption was not confirmed, flow inside the draft tube is rather sensitive to inlet boundary conditions and no positive impact for the DFT-B modification was observed.

DFT-B draft tube features rather steep increase of pressure recovery factor at the inlet part, which is connected with large wall divergence leading into reduction of velocity and abrupt transformation of kinetic energy into pressure energy. This is an undesirable effect, because too low kinetic energy of flow stream brings risk of flow separation at the downstream parts of the draft tube. In Fig. 13, it is shown that a large part of kinetic energy is dissipated just at the inlet part of draft tube and as will be shown in section 6.1, the flow separation occurs at the elbow part of the draft tube. The highest value of DFT-B efficiency (1) is derived from efficiency curve plotted in Fig. 4 and corresponds to flow rate 92 % Q_{BEP}.

3. Back-flow regions

The back-flow regions block flow area, increase velocity and cause higher hydraulic losses.

DFT-A: The first significant back-flow region is situated under the horizontal rib and occupies right (for flow rate lower than that of the best efficiency point of turbine, Fig. 8) or left (for flow rate higher than that of the best efficiency point of turbine, Fig. 9) side of the draft tube. The highest suppression of this region is reached for flow rate 109.1 % Q_{BEP}, Fig. 10. The second back-flow region is situated on the top of horizontal part of the draft tube and is the most suppressed between operating points 104.5 % Q_{BEP} and 109.1 % Q_{BEP}. The highest hydraulic efficiency of DFT-A is reached for 103 % Q_{BEP} when the back flow regions are not the smallest. It shows influence of higher hydraulic losses caused by surface friction on the ribs when the flow rate increases.

DFT-B: Two back-flow regions occur. The first one is situated at the end of elbow part and second one at the horizontal part of draft tube. For flow rate 104.5 % Q_{BEP}, they are the largest and connected together, as shown in Fig. 10. The largest suppression of these regions is observed close to flow rate 95.5 % Q_{BEP}, Fig. 11.

4. Contours of dissipation

The dissipation function D for computation of turbulent flow by RANS equation is defined as follows

$$D = 2 \iiint\limits_{V} \left\{ (\mu + \mu_t) \left[\left(\frac{\partial v_x}{\partial x}\right)^2 + \frac{1}{2}\left(\frac{\partial v_x}{\partial y} + \frac{\partial v_y}{\partial x}\right)^2 + \frac{1}{2}\left(\frac{\partial v_x}{\partial z} + \frac{\partial v_z}{\partial x}\right)^2 + \right. \right.$$
$$\left. \left. \left(\frac{\partial v_y}{\partial y}\right)^2 + \frac{1}{2}\left(\frac{\partial v_y}{\partial z} + \frac{\partial v_z}{\partial y}\right)^2 + \left(\frac{\partial v_z}{\partial z}\right)^2 \right] \right\} \, \mathrm{d}V. \tag{4}$$

For both draft tubes, the contours of dissipation (4) were computed at several cross-sections. Areas with the highest value of dissipation are coloured from black to white/vanish (over-range of colormap). The range of the colormap has been set and reduced so that only the dissipation inside of the volume is visible, because the highest value of dissipation occurs in boundary layer regions.

DFT-A: Several important areas with high value of dissipation are observed. The first one is at the inlet part of the draft tube. Considerable dissipation is caused by high velocity gradient of stream coming out from turbine runner. The second area is under the horizontal rib and corresponds with the back-flow region, compare Fig. 12 and Fig. 8. This result confirms that the back-flow regions are highly dissipative. The third area is located behind the trailing edge of the horizontal rib and is caused by sweeping of boundary layer into the main stream.

DFT-B: The main area of high dissipation is located close to the inner bend radius, Fig. 13. It is caused by very high velocity gradients induced by flow inside the elbow part. As in the case of DFT-A, significant dissipation caused by stream coming out from the turbine runner is observed at the inlet part of the draft tube.

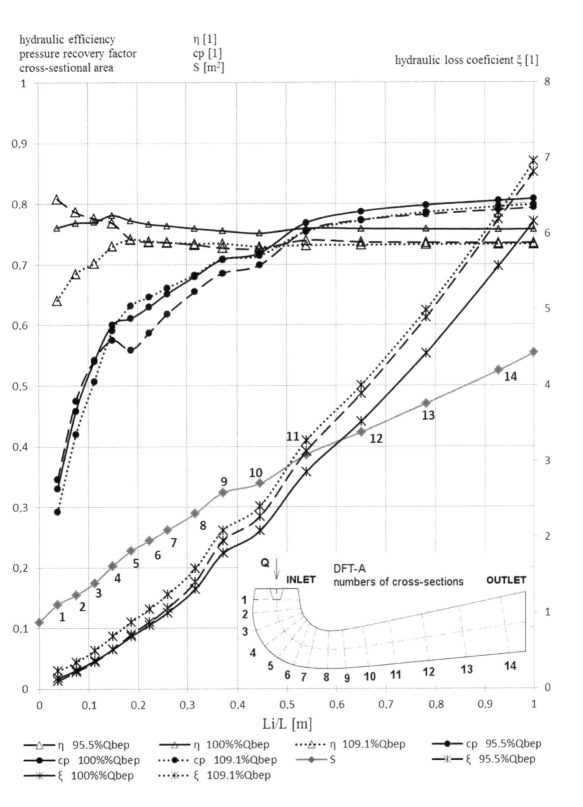

Fig. 6. Hydraulic efficiency, pressure recovery factor and loss coefficient evaluated in several cross-sections for three flow rates 95.5 % Q_{BEP}, 100 % Q_{BEP} and 109.1 % Q_{BEP} in case of DFT-A

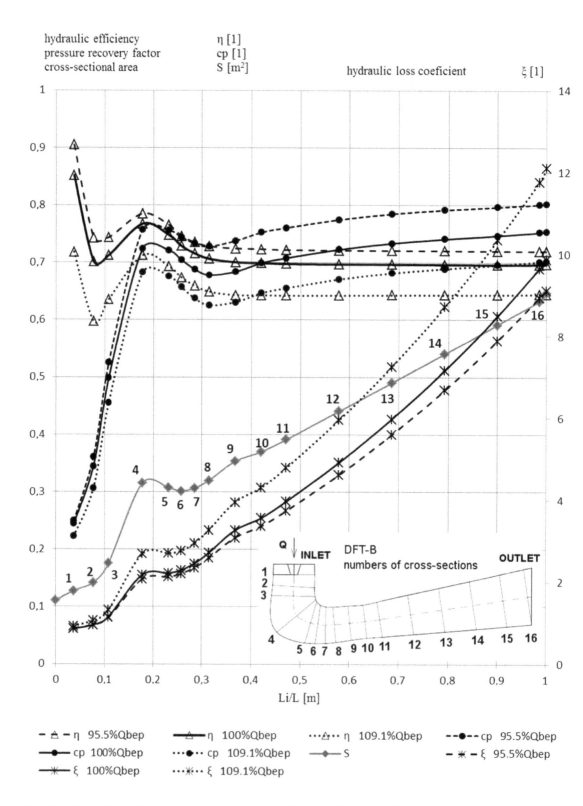

Fig. 7. Hydraulic efficiency, pressure recovery factor and loss coefficient evaluated in several cross-sections for three flow rates 95.5 % Q_{BEP}, 100 % Q_{BEP} and 109.1 % Q_{BEP} in case of DFT-B

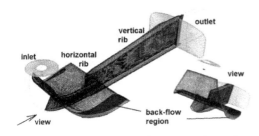

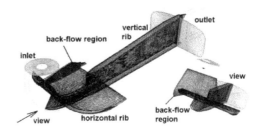

Fig. 8. Back-flow regions (in grey colour) in case of DFT-A, flow rate 104.5 % Q_{BEP}

Fig. 9. Back-flow regions (in grey colour) in case of DFT-A, flow rate 109.1 % Q_{BEP}

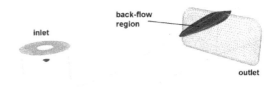

Fig. 10. Back-flow regions in case of DFT-B flow rate 104.5 % Q_{BEP}

Fig. 11. Back-flow regions in case of DFT-B flow rate 95.5 % Q_{BEP}

Fig. 12. Contours of dissipation in case of DFT-A

Fig. 13. Contours of dissipation in case of DFT-B

5. Topology of the three-dimensional separation

As mentioned in Tobak and Peak [13] and also in Depardon et al. [2], the hypothesis proposed by Legendre in 1956 brought a mathematical framework for description of the three-dimensional flow separation. The hypothesis is based on shear-stress patterns and critical points situated in the flow field and corresponding with the three-dimensional separated flow.

Fig. 14. Skin friction lines and surface streamlines in case of DFT-A

Fig. 15. Skin friction lines and surface streamlines in case of DFT-B

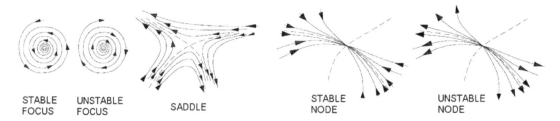

STABLE   UNSTABLE   STABLE   UNSTABLE
FOCUS   FOCUS   SADDLE   NODE   NODE

Fig. 16. Types of critical points

The three types of critical points in the flow field are identified, see Fig. 16. **Focus F** is a point where the vortex filament of the three-dimensional separation core is concentrated. Surface friction lines go into (stable) or out (unstable) of the centre of focus. **Saddle point S** is a place where surface friction lines converge from one side and diverge to the other side. **Node N**: Surface friction lines go into (stable) or out (unstable) of the centre of node. Node usually lies near or directly onto solid surfaces.

For the exact determination of critical points in particular flow cross-section computation of velocity gradient tensor eigenvalue (5) is used [3,4]

$$v_{ij}|_{x_{i,s}} = \begin{bmatrix} \frac{\partial v_x}{\partial x} & \frac{\partial v_x}{\partial y} \\ \frac{\partial v_y}{\partial x} & \frac{\partial v_y}{\partial y} \end{bmatrix}. \tag{5}$$

Each critical point (saddle, stable and unstable node, stable and unstable focus) is defined by sign of the eigenvalue of the tensor (5).

In case of elbow draft tube, several regions with risk of the three-dimensional separation occur. The beginning of the three-dimensional flow separation lies onto draft tube body surface where boundary layer separates and is carried out by main stream. The vortex region originates from this separation and causes blockage effect leading to the flow acceleration.

Exact visualization and evolution of separation core (so-called dividing surface) in flow field is rather difficult, especially when dealing with 3D data. One of the possibilities is to use the Sujudi-Haimes algorithm [9]. The computational algorithm looks for the points in the velocity field where a single real eigenvector exists and this is parallel to the velocity vector [5,6]. There are also other approaches to visualize separation surfaces emanating from critical points, see [7,10]. Some of them are specifically focused on swirling flows, see [1].

In this work, prediction of the core evolution is solved only by visualization method based on searching of foci in particular draft tube cross-sections (surface streamlines) and on the solid surfaces (skin friction lines), Fig. 14 and Fig. 15. This method is simplification of finding possible occurrence of core but not very suitable for tracking the spatial evolution of the core. Setting the cross-section orientation represents a very difficult task, because foci are properly visible only on cross-section which is almost exactly perpendicular to the separation core. It is also advised to observe back-flow regions, which are related to global flow separation and lead to the hydraulic efficiency drop.

5.1. Global flow separation

In case of elbow diffusers, the main part with risk of the three-dimensional separation is the bend. For rectangular curved diffuser, the authors of [8] investigated three kinds of global flow separations: *massive*, *typical* and *simple*. The typical idea of global flow separation: "The global flow separation begins, where the flow is separated from the wall and formed the back-

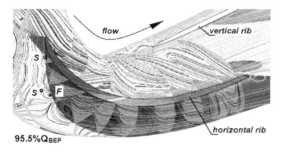

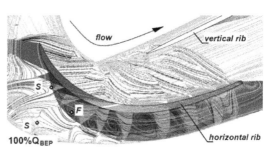

Fig. 17. Right view on elbow part of DFT-A for flow rate 95.5 % Q_{BEP}

Fig. 18. Right view on elbow part of DFT-A for flow rate 100 % Q_{BEP}

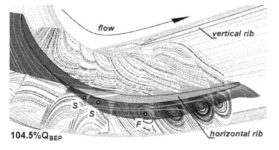

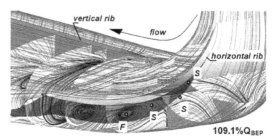

Fig. 19. Right view on elbow part of DFT-A for flow rate 104.5 % Q_{BEP}

Fig. 20. Left view on elbow part of DFT-A for flow rate 109.1 % Q_{BEP}

flow regions" is stated in [8]. The global flow separation is highly dissipative phenomenon, increases hydraulic losses and reduces hydraulic efficiency.

In case of turbulent flow in the draft tube (unsteady, three-dimensional with rotational character), the flow topology of the separation is very complex and directly corresponds with shape of the draft tube. This statement is especially characteristic of the draft tube containing ribs. Types of the three-dimensional separation corresponding to each of the investigated draft tubes will be shown in sections 5 and 6.

5.2. Local flow separation

In contrary to the case of global flow separation, the local flow separation is not related to back-flow regions. Hence no significant negative effect, as lower efficiency due to higher dissipation, has been documented. The example of local type of three-dimensional separation is observed in section 5.2 for the DFT-A draft tube.

6. Three-dimensional separation in DFT-A draft tube

6.1. Separation under horizontal rib

This separation is mainly caused by leading edge of horizontal rib which suppresses stream rotation at the inlet of the draft tube. This separation is in combination with back-flow region that means it is the global flow separation which decreases energy transformation and deteriorates efficiency.

Flow rate: (Figs. 17–20)

- **95.5 %** Q_{BEP}: The back-flow region is very large and starts from leading edge of the horizontal rib, see Fig. 17, and develops downstream to the draft tube. The global flow separation is represented by saddle point S in combination with focus F.

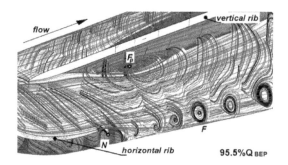

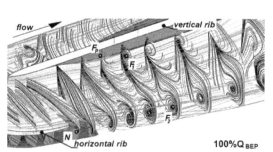

Fig. 21. Right view on horizontal part of DFT-A for flow rate 95.5 % Q_{BEP}

Fig. 22. Right view on horizontal part of DFT-A for flow rate 100 % Q_{BEP}

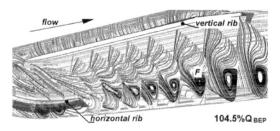

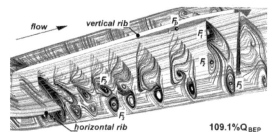

Fig. 23. Right view on horizontal part of DFT-A for flow rate 104.5 % Q_{BEP}

Fig. 24. Right view on horizontal part of DFT-A for flow rate 109.1 % Q_{BEP}

- **100 % Q_{BEP}:** The back-flow region and global flow separation are smaller by getting closer to the best efficiency point of the draft tube, see Fig. 18.
- **104.5 % Q_{BEP}:** This flow rate is close to the best efficiency point. The global flow separation is considerably reduced but still observed, see Fig. 19.

6.2. Separation at horizontal part of draft tube

The global and the local types of flow separation occur at the horizontal part of draft tube. The global flow separation with large back-flow region is observed for higher and lower flow rates out from the best efficiency point. The local flow separation is caused by boundary layers sweeping from trailing edge of horizontal rib. These boundary layers are then carried out by main stream and formed into separation core.

Flow rate: (Figs. 21–24)

- **95.5 % Q_{BEP}:** The large back-flow area surrounds the top of vertical rib. The global flow separation starts in F_p onto roof of draft tube body and winds into main stream. The local flow separation is represented by node N and focus F, Fig. 21.
- **100 % Q_{BEP}:** The separation with beginning in F_p is smaller than previous one but still considerable back-flow region occurs. Separation core in Fig. 22 is wound through the focus F_1. Node N and focus F_2 represent the local flow separation.
- **104.5 % Q_{BEP}:** This flow rate is close to the best efficiency point of the draft tube. The global flow separation in Fig. 23 is almost reduced. On the other hand, the local flow separation with core in focus F is getting stronger.
- **109.1 % Q_{BEP}:** High extensive local flow separation with two cores goes through the foci F_2 and F_3 as seen in Fig. 24. The small global flow separation with beginning in F_p and separation core going through F_1 is located near the top of vertical rib.

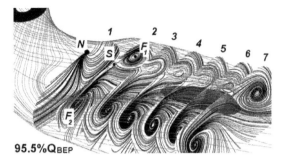

95.5%Q_BEP

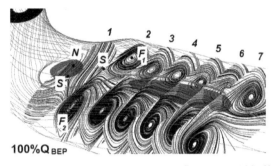

100%Q_BEP

Fig. 25. Top view on DFT-B, flow rate 95.5 % Q_{BEP}

Fig. 26. Top view on DFT-B, flow rate 100 % Q_{BEP}

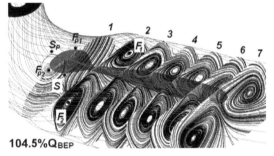

104.5%Q_BEP

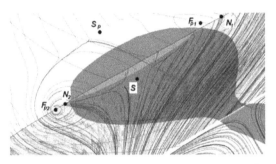

Fig. 27. Top view on DFT-B, flow rate 104.5 % Q_{BEP}

Fig. 28. Detailed view on global flow separation beginning, flow rate 104.5 % Q_{BEP}

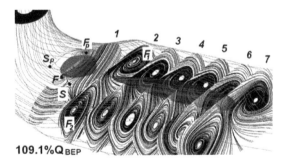

109.1%Q_BEP

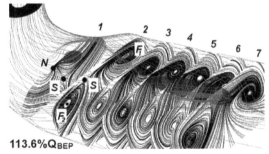

113.6%Q_BEP

Fig. 29. Top view on DFT-B, flow rate 109.1 % Q_{BEP}

Fig. 30. Top view on DFT-B, flow rate 113.6 % Q_{BEP}

Only the right channel of horizontal part is shown for demonstration. The same kinds of the three-dimensional separations occur also in the left channel.

7. Three-dimensional separation in DFT-B draft tube

The main region of elbow draft tube (without ribs) where could be formed the three-dimensional separated flow is at the inner radius of the bend.

7.1. Separation at the end of elbow part of draft tube

Flow rate: (Figs. 25–30)

- **95.5 % Q_{BEP}:** The unstable node N in combination with saddle S forms the beginning of separation. The back-flow region is very small and is situated far away from separation beginning. Hence this separation should be local type of separation, see Fig. 25.

- **100 %** Q_{BEP}: The unstable node N in combination with saddle S forms the beginning of global flow separation because the small back-flow region is formed at the end of elbow part, see Fig. 26.

- **104.5 %** Q_{BEP}: The beginning of large global flow separation is represented by combination of saddle S_p with foci F_{p1} and F_{p2} on solid surface. From these foci separation cores start, which go through foci F_1 and F_2 situated in flow cross sections 2–7, see Fig. 27 and Fig. 28.

- **109.1 %** Q_{BEP}: The global flow separation is still extensive, but diminishes and only one focus F_p on draft tube body occurs in Fig. 29.

- **113.6 %** Q_{BEP}: The flow topology in Fig. 30 is similar to the flow rate 100 % Q_{BEP} as shown in Fig. 26. At the end of elbow part is formed the small back-flow region which leads to global flow separation.

The detailed view on the beginning of global-flow separation for flow rate 104.5 % Q_{BEP} is shown in Fig. 28. The two separation cores start in foci F_{p1} and F_{p2}, which are in combination with saddle S_p. Nodes N_1 and N_2 and saddle S characterize the separation in flow cross-section.

8. Swirling flow

The swirling flow at the draft tube inlet is mainly caused by tangential velocity of stream exiting from the turbine runner. The swirl number S_n [11] is used for description of the swirling flow

$$S_n = \frac{\int v_{ax} v_{tan} r \, dS}{R \int v_{ax}^2 \, dS} \,, \tag{6}$$

where v_{ax} is axial velocity, v_{tan} is tangential velocity, r is radial coordinate and R is inlet radius of the draft tube. The swirl numbers computed by (6) in the inlet cross-section are plotted in Fig. 31. It can be seen that from flow rate 104.5 % Q_{BEP} to higher the flow rotates in opposite direction than turbine runner. Fig. 32 shows hydraulic efficiency curves for each draft tube against values of the swirl numbers. The best efficiency point for DFT-A is obtained for lower swirl number than in the case of DFT-B. This difference shows the significant influence of horizontal guide rib on flow in draft tube in relation to global flow separation caused by this rib. Result for DFT-A indicate that low swirling flow at the inlet of the draft tube is important

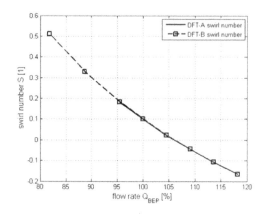

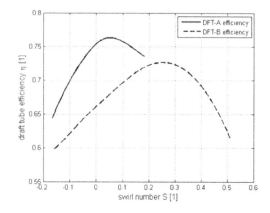

Fig. 31. Swirl number S_n versus flow rate Q_m Fig. 32. Draft tube efficiency η vs. swirl number S_n

for smoother splitting of water stream by leading edge of the horizontal rib. Guide rib acts like inhibition for swirling flow exiting from the runner. Thus, for the highest or lowest value of the swirl number, the back-flow region situated under horizontal rib and three-dimensional separation grows as well. On the other hand, the higher swirling flow with positive value of swirl number (the swirl rotates in same direction as turbine runner) is required for flow rate corresponding with the best efficiency point of DFT-B. Higher swirling flow pushes the boundary layer against the walls, which reduces the risk of separation mainly in the elbow part of draft tube.

9. Conclusion

The main result of this work is referring to the global flow separation influence on the draft tube efficiency. The global flow separation in combination with the back-flow regions is a highly dissipative phenomenon. When flow rate moves out of the best efficiency point of the draft tube, then extension of the global flow separation is observed for both draft tubes. Insignificant role on draft tube efficiency is also hydraulic losses caused by surface friction, mainly in case of DFT-A with two ribs.

The efficiency maximum of DFT-A is obtained for flow rate 103 % Q_{BEP} and swirl number $S_n = 0.05$. The global flow separation is the most considerably reduced between flow rates 104.5 % Q_{BEP} and 109.1 % Q_{BEP} and for swirl number close to $S_n = 0$. These results show compromise between frictional losses which grow with increasing flow rate, and non-swirling flow ensuring the lowest three-dimensional separation. But this relation is only valid for case of DFT-A or the similar elbow draft tubes with guide rib.

In case of DFT-B the swirl flow does not have undesirable effect, on the contrary the swirling flow with $S_n = 0.25$ in the best efficiency point of draft tube prevents risk of separation. It is seen that for flow rate 104.5 % Q_{BEP} and $S_n = 0$ the largest global flow separation is reached with important influence on the draft tube efficiency. This separation is in large amount caused by non-swirling flow.

It is difficult to describe exact character of the local type of flow separation in draft tubes because it does not correspond with back-flow regions. Thus, it becomes difficult to predict evolution of separation core. Present investigation indicates that local type of flow separation has no evident influence on the draft tube efficiency.

The results of the flow topology confirm that the patterns of skin friction lines onto solid surfaces are good indicators for quality of flow in particular region of the flow field. Identification of the skin friction patterns is rather sensitive tool to observe flow variations in the draft tube interior, if the inlet velocity profile is changed. In other words skin friction lines enable convenient observation how design modifications are reflected in flow topology. Combining the information about details of the flow field with integral parameters like hydraulic efficiency or loss coefficient, it offers full picture enabling optimization of the draft tube shape or the runner blade outlet angles.

Acknowledgements

Authors gratefully acknowledge support of the research by Czech Science Foundation under project No. 101/09/1715 "Cavitating vortical structures induced by rotation of liquid" and by Ministry of Trade and Industry under project No. 2A-1TP1/108 "Performance increase and operational range broadening for upgraded low head hydraulic power plants".

References

[1] Bisgaard, A. W., Structures and bifurcations in fluid flows with application to vortex breakdown and wakes, Ph.D. thesis, Technical University of Denmark, Kongens Lyngby, 2005.

[2] Depardon, S., Lasserre, J. J., Boueilh, J. C., Brizzi, L. E., Borée, J., Skin friction pattern analysis using near-wall PIV, Experiments in Fluids 39 (2005) 805–818.

[3] Helman, J., Hesselink, L., Representation and display of vector field topology in fluid flow data sets, Computer 22(8) (1989) 27–36.

[4] Kenwright, D. N., Henze, C., Levit, C., Feature extraction of separation and attachment lines, IEEE transactions on Visualization and Computer Graphics 5(2) (1999) 135–144.

[5] Mauri, S., Kueny, J. L., Avellan, F., Werlé-Legendre separation in a hydraulic machine draft tube, Journal of Fluids Engineering 126 (2004) 976–980.

[6] Mauri, S., Numerical simulation and flow analysis of an elbow diffuser, Ph.D. thesis, EPFL Lausane, 2002.

[7] Peikert, R., Sadlo, F., Topologically relevant stream surfaces for flow visualization, Proceedings of the Spring Conference on Computer Graphics, Budmerice, Slovakia, 2009, pp. 43–50.

[8] Sedlář, M., Příhoda, J., Investigation of flow phenomena in curved channels of rectangular cross-section, Engineering Mechanics 14 (2007) 387–397.

[9] Sujudi, D., Haimes, R., Identification of swirling flow in 3-D vector fields. Department of Aeronautics and Astronautics Massachusetts Institute of Technology Cambridge, MA 02139, pp. 1–8.

[10] Surana, A., Jacobs, G. B., Haller, G., Extraction of separation and attachment surfaces from three-dimensional steady shear flows, AIAA Journal 45(6) (2007) 1 290–1 302.

[11] Susan-Resiga, R., Ciocan, G. D., Anton, I., Avellan, F., Analysis of the swirling flow downstream a Francis turbine runner, Journal of Fluids Engineering 128 (2006) 177–189.

[12] Svozil, J., Rudolf, P., Kaplan turbine draft tube optimization, Research report VUT-EU13303-QR-01-10, Brno, 2010. (in Czech)

[13] Tobak, M., Peake, D. J., Topology of three-dimensional separated flows, Annual Review of Fluid Mechanics 14 (1982) 61–85.

10

Springback analysis of thermoplastic composite plates

Z. Padovec[a,*], M. Růžička[a], V. Stavrovský[a]

[a]*Department of Mechanics, Mechatronics and Biomechanics, CTU in Prague, Faculty of Mechanical Engineering, Technická 4, 166 07, Prague, Czech Republic*

Abstract

Residual stresses, which are set in the fiber reinforced composites during the laminate curing in a closed form, lead to dimensional changes of composites after extracting from the form and cooling. One of these dimensional changes is called "springback" of the angle sections. Other dimensional changes are warpage of flat sections of composite or displacement of single layers of composite for example. In our case four different lay-ups were analysed (three symmetrical and one unsymmetrical). An analytical model which covers temperature changes, chemical shrinkage during curing and moisture change was used. Also a FEM analysis was done for predicting the springback, and both calculations were compared with the measured data from manufacturer.

Keywords: springback, thermoplastic composite, analytical calculation, FEM

1. Introduction

Incorporating thermoplastic matrices (PEEK etc.) in carbon fibre-reinforced composites results considerably in higher toughness and impact resistance than have traditional thermoset based composites. In addition, the thermoplastic materials have significant advantages during fabrication and allow applying optimized metal working technology (stamping). However, the high temperature at which the thermoplastic composite must be processed does suggest an increased significance of thermally induced stresses and distortions in a product finished. Therefore it is desirable to be able to predict distortions accurately, reducing the trial and error time when producing a new component [5].

Change in composite dimensions is related to many parameters as: part angles, thicknesses, lay-ups, flange length, but also tool materials, tool surface or cure cycles [1]. When the composite L (or U) section is extracted from the form that was cooled to the room temperature, the change in the angle of the part (e.g. Fig.1) can be observed.

Tool angles have to be modified to affect this problem. The tool design is based on either the "rules of thumb", from the past experience, or on the trial-and-error. For the angular parts, the compensation is normally between 1 and $2,5°$. The most common problem found, using a standard factor, is that the springback may vary with the lay-up, material, cure temperature, etc. Therefore, what worked once does not necessarily work next time. The main cause of springback was the mismatch in thermal expansion along and across the fibers in a laminate. But there are also other causes which will be discussed in the next analysis [4].

---

*Corresponding author. e-mail: zdenek.padovec@fs.cvut.cz.

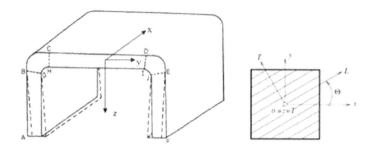

Fig. 1. Distortion of moulded U-section [3]

2. Analytical calculation

Analytical solution of the springback of the layered composite plate is based on the derivation of the springback for the unidirectional composite plate. The springback (SB) is defined as a difference between the final angle of the product and the tool angle divided by the tool angle

$$SB = \frac{\gamma' - \gamma}{\gamma} \tag{1}$$

when the angles can be computed from the lengths of arcs, thickness and their changes due the temperature, moisture and matrix shrinkage during cure (e.g. Fig. 2).

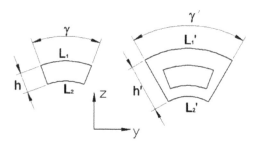

Fig. 2. To the derivation of springback [2]

Total change in the angle can be written as

$$\Delta\gamma = \Delta\gamma_t + \Delta\gamma_h + \Delta\gamma_c = \gamma\frac{\varepsilon_y^t - \varepsilon_z^t}{1 + \varepsilon_z^t} + \gamma\frac{\varepsilon_y^h - \varepsilon_z^h}{1 + \varepsilon_z^h} + \gamma\frac{\varepsilon_y^c - \varepsilon_z^c}{1 + \varepsilon_z^c}, \tag{2}$$

where $\Delta\gamma_t$ is temperature part of the angle change, $\Delta\gamma_h$ is change in the angle due to hygroscopic effect and $\Delta\gamma_c$ is change in the angle due to shrinkage effect during the cure cycle. Strains (from temperature, moisture absorption and shrinkage) marked with y index stand for longitudinal directions, coefficients marked with z stand for transversal (trough thickness) directions. Temperature and moisture absorption effects are notoriously known so in next paragraph will be described just the resin shrinkage effect during curing.

This structural effect of volumetric change of matrix due to the crystallization will be important in composites with thermoplastic matrices that change their phase from amorphous to crystalline one during heating and curing. During curing and crystallization, the semi-crystalline matrices shrink due to crowding of the mass (the crystals have higher density then the amorphous phase). The shrinkage of semi-crystalline matrix due to this effect can be significantly higher than that due to the temperature change. It is evident that amorphous polymers haven't got the effect of volumetric change due to the recrystallization, so the influence of temperature change is crucial [4].

Let's have a laminated composite plate with $i = 1, \ldots, N$ layers with thickness H ($H = \sum_i^N h_i$). Springback derivation for the layered composite plate stands on derivation of the thickness change due to temperature, moisture and shrinkage effect. Detailed derivation is presented in [2] or [4]. Final equations of the thickness change are

$$
\begin{aligned}
\Delta H^t = \sum_{k=1}^{N} & \left\{ [S_{13}\ S_{23}\ S_{36}]\,[T_q]_k \left((z_k - z_{k-1})\,[\bar{Q}]_k \left(\begin{bmatrix} \varepsilon_x^0 \\ \varepsilon_y^0 \\ \gamma_{xy}^0 \end{bmatrix} - \Delta T \begin{bmatrix} \alpha_x \\ \alpha_y \\ \alpha_{xy} \end{bmatrix}_k \right) + \right. \right. \\
& \left. \left. \frac{z_k^2 - z_{k-1}^2}{2}\,[\bar{Q}]_k \begin{bmatrix} \kappa_x \\ \kappa_y \\ \kappa_{xy} \end{bmatrix} \right) + (z_k - z_{k-1})\,(\Delta T (\alpha_3)_k) \right\},
\end{aligned} \tag{3}
$$

$$
\begin{aligned}
\Delta H^h = \sum_{k=1}^{N} & \left\{ [S_{13}\ S_{23}\ S_{36}]\,[T_q]_k \left((z_k - z_{k-1})\,[\bar{Q}]_k \left(\begin{bmatrix} \varepsilon_x^0 \\ \varepsilon_y^0 \\ \gamma_{xy}^0 \end{bmatrix} - \Delta c \begin{bmatrix} \beta_x \\ \beta_y \\ \beta_{xy} \end{bmatrix}_k \right) + \right. \right. \\
& \left. \left. \frac{z_k^2 - z_{k-1}^2}{2}\,[\bar{Q}]_k \begin{bmatrix} \kappa_x \\ \kappa_y \\ \kappa_{xy} \end{bmatrix} \right) + (z_k - z_{k-1})\,(\Delta c (\beta_3)_k) \right\},
\end{aligned} \tag{4}
$$

$$
\begin{aligned}
\Delta H^c = \sum_{k=1}^{N} & \left\{ [S_{13}\ S_{23}\ S_{36}]\,[T_q]_k \left((z_k - z_{k-1})\,[\bar{Q}]_k \left(\begin{bmatrix} \varepsilon_x^0 \\ \varepsilon_y^0 \\ \gamma_{xy}^0 \end{bmatrix} - \begin{bmatrix} \Phi_x \\ \Phi_y \\ \Phi_{xy} \end{bmatrix}_k \right) + \right. \right. \\
& \left. \left. \frac{z_k^2 - z_{k-1}^2}{2}\,[\bar{Q}]_k \begin{bmatrix} \kappa_x \\ \kappa_y \\ \kappa_{xy} \end{bmatrix} \right) + (z_k - z_{k-1})\,(\Phi_3)_k \right\},
\end{aligned} \tag{5}
$$

where N is a number of layers, S_{ij} are elements of the compliance matrix, T_q is a transformation matrix, z is a coordinate of the layer, $\bar{Q}$ is 2D plain stress matrix in x, y coordinate system (see Fig. 1) , ΔT is a change in temperature, Δc change in moisture, α_{ij} is a coefficient of thermal expansion, β_{ij} is a coefficient of moisture absorption and φ_{ij} is a coefficient of chemical shrinkage during recrystallization.

The total change in the laminate thickness can be computed as

$$
\Delta H^{thc} = \Delta H^t + \Delta H^h + \Delta H^c. \tag{6}
$$

Generalized strain in z direction can be computed from Eq. (6) as

$$
\varepsilon_z^{thc} = \frac{\Delta H^{thc}}{H}. \tag{7}
$$

According to the classical lamination theory (CLT, e.g. [5]), the strain and curvatures of the middle plane of laminated composite plate, loaded with unit forces and moments, can be obtain. Accordingly, it can be stated that strain and curvatures, arised from the temperature, moisture and resin shrinkage changes, will produce mechanical forces and moments. According to CLT

$$
\left\{ \begin{array}{c} N_x \\ N_y \\ N_{xy} \\ M_x \\ M_y \\ M_{xy} \end{array} \right\} = \left[\begin{array}{cccccc} A_{11} & A_{12} & A_{16} & B_{11} & B_{12} & B_{16} \\ A_{12} & A_{22} & A_{26} & B_{12} & B_{22} & B_{26} \\ A_{16} & A_{26} & A_{66} & B_{16} & B_{26} & B_{66} \\ B_{11} & B_{12} & B_{16} & D_{11} & D_{12} & D_{16} \\ B_{12} & B_{22} & B_{26} & D_{21} & D_{22} & D_{26} \\ B_{16} & B_{26} & B_{66} & & & \\ D_{16} & D_{26} & D_{66} & & & \end{array} \right] \left(\left\{ \begin{array}{c} \varepsilon_x^0 \\ \varepsilon_y^0 \\ \gamma_{xy}^0 \\ \kappa_x \\ \kappa_y \\ \kappa_{xy} \end{array} \right\} - \left\{ \begin{array}{c} N_x^{thc} \\ N_y^{thc} \\ N_{xy}^{thc} \\ M_x^{thc} \\ M_y^{thc} \\ M_{xy}^{thc} \end{array} \right\} \right), \qquad (8)
$$

where A_{ij}, B_{ij} and D_{ij} are the generally known elements of membrane stiffness, bending-extension coupling stiffness and bending stiffness matrices, and quantities N_i^{thc}, M_i^{thc} are defined by equations

$$
N_i^{thc} = \int Q_{ij}\varepsilon_j^{thc}\,dz, \qquad (9)
$$

$$
M_i^{thc} = \int Q_{ij}\varepsilon_j^{thc}z\,dz, \qquad (10)
$$

with the fact that integration boundaries are from $-H/2$ to $H/2$. N_i^{thc} and M_i^{thc} have the same dimension as N_i and M_i and they are called the resultants of the thermohygrocrystallic unit internal forces and moments. Q is 2D plain stress matrix in L, T coordinate system (see Fig. 1). The plane strain $\varepsilon_i^{0,thc}$ and relative change in curvature κ_i^{thc} will arise due to N_i^{thc} and M_i^{thc} at absence of N_i and M_i and they can be calculated by

$$
\left\{ \begin{array}{c} \varepsilon_x^{0,thc} \\ \varepsilon_y^{0,thc} \\ \gamma_{xy}^{0,thc} \\ \kappa_x^{thc} \\ \kappa_y^{thc} \\ \kappa_{xy}^{thc} \end{array} \right\} = \left[\begin{array}{cccccc} a_{11} & a_{12} & a_{16} & b_{11} & b_{12} & b_{16} \\ a_{12} & a_{22} & a_{26} & b_{12} & b_{22} & b_{26} \\ a_{16} & a_{26} & a_{66} & b_{16} & b_{26} & b_{66} \\ b_{11} & b_{12} & b_{16} & d_{11} & d_{12} & d_{16} \\ b_{12} & b_{22} & b_{26} & d_{21} & d_{22} & d_{26} \\ b_{16} & b_{26} & b_{66} & d_{16} & d_{26} & d_{66} \end{array} \right] \left\{ \begin{array}{c} N_x^{thc} \\ N_y^{thc} \\ N_{xy}^{thc} \\ M_x^{thc} \\ M_y^{thc} \\ M_{xy}^{thc} \end{array} \right\}, \qquad (11)
$$

where a_{ij}, b_{ij} and d_{ij} are elements of inverse matrix to the whole A, B, B, D matrix. If the lay-up of the composite is symmetric (according to CLT there is no connection between normal forces and moments or between axial strains and curvatures) then the springforward of the complete composite can be calculated by Eq. (2) which was derived for one layer.

For the unsymmetrical lay-up, an additional term must be included. Eq. (2) rewritten to the springback form will look [2]

$$
SB = \frac{\Delta\gamma}{\gamma^{thc}} = R_y\kappa_y^{thc} + \frac{\varepsilon_y^{thc} - \varepsilon_z^{thc}}{1 + \varepsilon_z^{thc}}, \qquad (12)
$$

where κ_y^{thc} is the change in curvature and R_y is the radius which affected the straight part of the plate.

When the cylindrical shell of the diameter $D = 2R_y$ is analyzed, this springback effect must be included in the thermal force-strain relationship of laminated plates. This is accomplished by making following replacement in Eq. (11) [2]

$$
\kappa_y^{thc} \Rightarrow \kappa_y^{thc} + \frac{1}{R_y}\left(\varepsilon_y^{thc} - \varepsilon_z^{thc}\right). \qquad (13)
$$

In the case of double curved shells (with two main curvatures radiuses R_x, R_y) we have to make also this replacement in Eq. (11) [2]

$$\kappa_x^{thc} \Rightarrow \kappa_x^{thc} + \frac{1}{R_x}\left(\varepsilon_x^{thc} - \varepsilon_z^{thc}\right). \tag{14}$$

3. Calculation for the given plates

The main goal of this work is to make a tool, to predict springback angle for given laminated plate with single or double curvature with defined number of layers made from defined combination of fibre and matrix. The springback angle is necessary to predict because of manufacturing more precious parts and also for good correction of the compression mould. The manufacturer of the composite plates (Letov Letecká Výroba, s. r. o.) has provided some measured data from the manufacturing process. The data was for the layered composite plates made from carbon fibre and PPS matrix with $V_f = 49\%$.

Lay up's and radiuses in y direction for the single curvature plates are:

- 30 layers $[(0,90)/(\pm45)]_7/(0,90)$ $R_y = 5\,\text{mm}$

- 32 layers $[[(0,90)/(\pm45)]_4]_s$ $R_y = 5\,\text{mm}$

- 36 layers $[[(0,90)/(\pm45)]_4/(0,90)]_s$ $R_y = 6\,\text{mm}$

- 40 layers $[(0,90)/[(0,90)/(\pm45)]_4/(0,90)]_s$ $R_y = 7\,\text{mm}$

Lay up's and radiuses in x and y directions for the double curvature plates are:

- 30 layers $[(0,90)/(\pm45)]_7/(0,90)$ $R_x = 2\,811{,}5\,\text{mm}$, $R_y = 5\,\text{mm}$

- 32 layers $[[(0,90)/(\pm45)]_4]_s$ $R_x = 2\,811\,\text{mm}$, $R_y = 5\,\text{mm}$

- 36 layers $[[(0,90)/(\pm45)]_4/(0,90)]_s$ $R_x = 2\,810{,}5\,\text{mm}$, $R_y = 6\,\text{mm}$

- 40 layers $[(0,90)/[(0,90)/(\pm45)]_4/(0,90)]_s$ $R_x = 2\,810\,\text{mm}$, $R_y = 7\,\text{mm}$

Single and double curvature plates can be seen in Fig. 3.

Fig. 3. Investigated plates with single and double curvature [4]

Eqs. (7) and (11) were used for computing coefficients of thermal expansion and recrystallization for the analyzed composite plates. Computed coefficients can be seen in Table 1. The hygroscopic effect is not considered ($\Delta c = 0$).

The comparison between the data calculated by the analytical model, the data from the manufacturer and the data from numerical simulation are shown in Figs. 6 and 7.

Table 1. Coefficients of thermal expansion and recrystallization for the analyzed lay-ups

Lay-up	α_x $[\mathrm{K}^{-1}]$	α_y $[\mathrm{K}^{-1}]$	α_z $[\mathrm{K}^{-1}]$	Φ_x $[-]$	Φ_y $[-]$	Φ_z $[-]$
$[(0,90)/(\pm45)]_7/(0,90)$	$4{,}82\mathrm{e}^{-6}$	$4{,}82\mathrm{e}^{-6}$	$3{,}91\mathrm{e}^{-5}$	$1{,}38\mathrm{e}^{-3}$	$1{,}38\mathrm{e}^{-3}$	$1{,}23\mathrm{e}^{-2}$
$[[(0,90)/(\pm45)]_4]_s$	$4{,}98\mathrm{e}^{-6}$	$4{,}98\mathrm{e}^{-6}$	$3{,}88\mathrm{e}^{-5}$	$1{,}43\mathrm{e}^{-3}$	$1{,}43\mathrm{e}^{-3}$	$1{,}22\mathrm{e}^{-2}$
$[[(0,90)/(\pm45)]_4/(0,90)]_s$	$4{,}70\mathrm{e}^{-6}$	$4{,}70\mathrm{e}^{-6}$	$3{,}93\mathrm{e}^{-5}$	$1{,}34\mathrm{e}^{-3}$	$1{,}34\mathrm{e}^{-3}$	$1{,}24\mathrm{e}^{-2}$
$[(0,90)/[(0,90)/(\pm45)]_4/(0,90)]_s$	$4{,}48\mathrm{e}^{-6}$	$4{,}48\mathrm{e}^{-6}$	$3{,}98\mathrm{e}^{-5}$	$1{,}27\mathrm{e}^{-3}$	$1{,}27\mathrm{e}^{-3}$	$1{,}25\mathrm{e}^{-2}$

4. Numerical calculation

A finite element model was created and analyzed for thermo-elastic deformations using ABA-QUS solver. For the calculation of plates with the single curvature was created model of the one quarter of the plate, for the plates with the double curvature was done model of one half. The model was solved with the hexahedron incompatible mode elements and two types of material properties. The first was defined by the orthotropic linear elastic properties and the thermal expansion material coefficients for the unidirectional lamina, the second one by the linear elastic material properties and the orthotropic thermal expansion for the whole composite. For the case of the first material model the local orientation of material axis in element has to be done with the method "OFFSET TO NODES" because of the change in material orientation due to changing radius. The theoretical material constants were computed by using CLT. The chemical shrinkage during the curing process was modeled by using "fake" coefficients of thermal expansion which represented the values of the chemical shrinkage. When the change in the temperature was $1°$, we obtained deformations according to the chemical shrinkage effect. The model of the single curvature plate can be seen in Fig. 4 and the double curvature plate can be seen in Fig. 5.

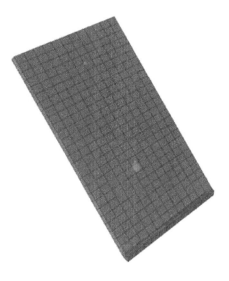

Fig. 4. FE model of single curvature plate

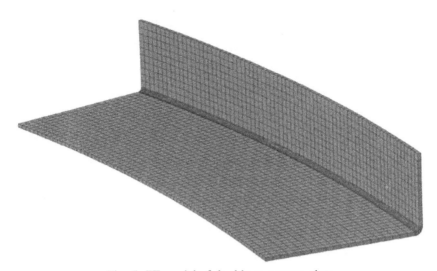

Fig. 5. FE model of double curvature plate

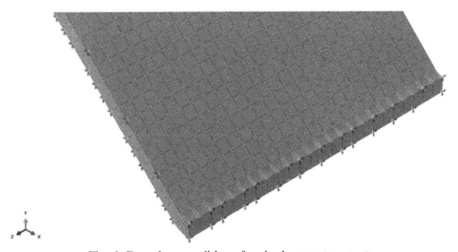

Fig. 6. Boundary conditions for single curvature part

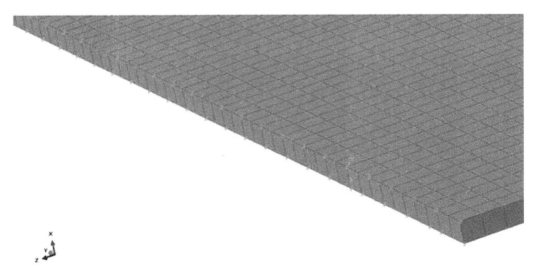

Fig. 7. Boundary conditions for double curvature part

Boundary conditions for the case of single curvature part were symmetrical and displacement in y direction was forbidden to highlighted node in Fig. 6. For the case of double curvature part, boundary conditions are shown in Fig. 7. Displacements in x and z are forbidden for marked area and also displacement in y direction was forbidden to highlighted node. Elements which were used in mesh of both parts were C3D8I, which are improved versions of C3D8 elements (there is no shear locking and volumetric locking is much reduced). These elements have an additional degree of freedom that enhances the ability to model a displacement gradient through the element (these elements act like quadratic elements but the computational cost is lower). It means that this element can be used in all instances, in which linear elements are subjected to bending (our case) and it allows the use of just one element in through thickness direction. A drawback to these elements are their sensitivity to element distortions which end up making the elements too stiff [6].

5. Results comparison, conclusions and future work

The comparison of the results computed by analytical and numerical solution with the data measured by manufacturer is shown in Figs. 8 and 9. Both figures show good agreement between measured data and springback angle predicted by analytical solution. Numerical model which used thermomechanical data computed by CLT shows the same values of springback angle for all types of the symmetrical lay-up. Numerical model which used thermomechanical data for the whole composite shows the same trend as the analytical results — springback angle for the symmetrical lay-up grows with the number of the layers (the third column in each data set — the growth is not very clear because of the chosen scale).

The analytical solution was put into the Matlab code. This program allows calculation of the springback angle for several combinations of fiber and matrix which can be chosen by user. User can also choose between the calculation of composite properties by using CLT or by input of properties directly. User also set up the lay-up of whole composite plate, radiuses of the

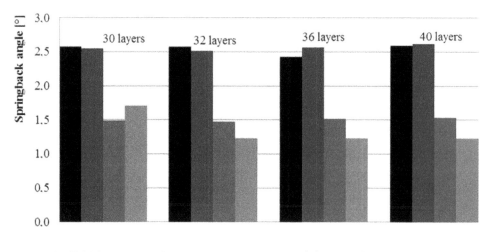

■MEASURED ■ANALYTICALY COMPUTED ■FEM-whole composite ■FEM-lamination theory

Fig. 8. Comparison of the analytical, numerical and experimental results for every type of investigated lay-up — plates with the single curvature

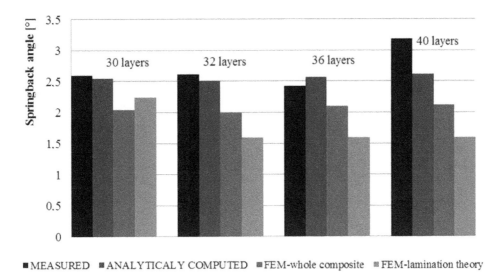

Fig. 9. Comparison of the analytical, numerical and experimental results for every type of investigated lay-up — plates with the double curvature

single and double curvature plates and the part angle of the plate. The analytical solution is faster than the numerical and the program doesn't need any special knowledge of software as FE solution does.

The future work in this area will focus on improving the FE results and their comparison with analytical and experimental (using other types of elements, mesh density, etc.). The material model which deals with the undulation of the fibers will be implemented into current Matlab code for comparison with the straight fibers model (because the reinforcement of the composite is 5H satin fabric). Also the influence of different types of the thermoplastic matrices on the springback angle will be focused in the future.

Acknowledgements

This work was supported by the Grant Agency of the Czech Republic number 101/08/H068 and project of Ministry of Industry and Trade of the Czech Republic number FR-TI1/463.

References

[1] Albert, C., Fernlund, G., Spring in and warpage of angled composite laminates, Composite Science and Technology 62 (2002) 1896–1912.

[2] Kollár, L. P., Springer, G. S., Mechanics of Composite Structures, Cambridge University Press, Campbridge, 2003.

[3] O'Neill, G. M., Rogers, T. G., Spencer, J. M., Thermally induced distortions in the moulding of laminated channel sections, Mathematical Engineering in Industry 2 (1988) 65–72.

[4] Padovec, Z., Růžička, M., Stavrovský, V., Chlup, H., Springforward phenomenon of angular sections of composite materials — Analytical, numerical and experimental approach, Proceedings of the 16th International Conference on Composite Structures, Porto, University of Porto, 2011, pp. 1–2.

[5] Stavrovský, V., Růžička, M., Padovec, Z., Chlup, H., Thermal and mechanical behaviour of the angle sections of composite plates with single and double curvature, Proceedings of the 18th International Conference on Composite Materials, Jeju Island, The Korean Society of Composite Materials, 2011, pp. 1–5.

[6] Brown, J., Characterization of MSC/NASTRAN & MSC/ABAQUS Elements for Turbine Engine Blade Frequency Analysis,
Available online on http://www.mscsoftware.com/support/library/conf/auc97/p02097.pdf.

11

Contact of screw surfaces with displaced axes

V. Machulda[a], J. Švígler[a],*

[a]Faculty of Applied Sciences, UWB in Pilsen, Univerzitní 22, 306 14 Plzeň, Czech Republic

Abstract

A couple of conjugate screw surfaces, which create a three-member mechanism with a higher kinematic pair, constitutes an important part of mechanical systems. One of these surfaces is created as an envelope of the second surface. In an ideal arrangement surface axes are parallel and the contact between both surfaces is curvilinear. After displacement of the surface axes, which is induced in technical equipments by force and heat loading, the original parallel arrangement of the surface axes is changed into an incorrect, spatial or alternatively parallel displaced, position. Consequently the original curvilinear contact changes into a contact at a point. This change is significant from the view of contact analysis. Determination of the contact point of conjugate surfaces, which are in the incorrect position, based on the kinematic way is presented in this contribution.

Keywords: conjugate screw surfaces, kinematic method, contact point, axes displacement

1. Introduction

Let us consider two conjugate surfaces. In theory the arrangement of surface axes is usually parallel and the contact between conjugate screw surfaces is curvilinear. However owing to external influences the original parallel position of surface axes acquires an incorrect configuration. As a consequence the original curvilinear contact is changed into a contact at a point. This change influences force loading between the surfaces and their relative motion. The original relative rolling motion changes into a spatial motion that can be described with kinematic screw, twist [4]. Ascertaining the contact between conjugate surfaces is described for example in [1,3]. In [1] the solution is based upon differential approach and in [3] a spatial problem is simplified. For using the solution from [3] it is necessary to consider a minor displacement of axes. In this contribution the kinematic method of determining a contact point of conjugate surfaces is presented. The kinematic method, which is based on Disteli principle [2], is highly illustrative and provides an effective solution of the problem. The kinematic way is very suitable for arbitrarily generated surfaces, thus not only for the conjugate screw surfaces, which are only a special case of general surfaces. The presented solution can be used for any displacement of axes. The solution is divided into two steps. In the first step a parallel displacement of surface axes is considered. In the second step general displacement is taken into consideration. In this paper the matrix notation of the kinematic analysis is made in homogeneous coordinates in this paper. Let be remarked an arrangement of axes of considered surfaces with curvilinear contact can be, in a theoretical case, arbitrary.

---

*Corresponding author. e-mail: svigler@kme.zcu.cz.

2. Creation of conjugate surfaces

2.1. Creating surface σ_2

Let us consider profile p_2 of creating surface σ_2 as an arc of circle, which is described in coordinate system $R_{2\gamma}$, Fig. 1, with equation

$$
_{R_{2\gamma}}\boldsymbol{r}_L^{\sigma_2}(\chi) = \begin{bmatrix} _{R_{2\gamma}}x_L^{\sigma_2} \\ _{R_{2\gamma}}y_L^{\sigma_2} \\ _{R_{2\gamma}}z_L^{\sigma_2} \\ 1 \end{bmatrix} = \begin{bmatrix} r_S \sin \Phi - r_K \sin \chi \\ -r_S \cos \Phi + r_K \cos \chi \\ 0 \\ 1 \end{bmatrix} , \tag{1}
$$

where $_{R_{2\gamma}}\boldsymbol{r}_L^{\sigma_2}$ is the position vector of point $L \in p_2$ and r_S, r_K, Φ, χ are geometric parameters. Surface $\sigma_2(\psi_2, \chi)$, which is two parametric manifold, is created by screw motion of profile p_2 along surface axis o_2. This screw motion, which is equivalent to transformation from coordinate system $R_{2\gamma}$ to R_2, is described with equation

$$
_{R_2}\boldsymbol{r}_L^{\sigma_2}(\psi_2, \chi) = \boldsymbol{T}_{R_{2\gamma},R_2} {}_{R_{2\gamma}}\boldsymbol{r}_L^{\sigma_2}(\chi) = \begin{bmatrix} \cos \psi_2 & -\sin \psi_2 & 0 & 0 \\ \sin \psi_2 & \cos \psi_2 & 0 & 0 \\ 0 & 0 & 1 & \delta_2 \\ 0 & 0 & 0 & 1 \end{bmatrix} {}_{R_{2\gamma}}\boldsymbol{r}_L^{\sigma_2}(\chi), \tag{2}
$$

where $_{R_{2\gamma}}\boldsymbol{r}_L^{\sigma_2}(\chi)$ is the position vector of profile point $L \in \sigma_2$, $\delta_2 = r_{W2}\psi_2 \tan \gamma$ is the displacement along the surface axis o_2, r_{W2} is the radius of rolling cylinder and γ is the helix angle on the rolling cylinder. With the use of (2) the creating surface σ_2 is, in coordinate system R_2, defined as

$$
_{R_2}\boldsymbol{r}_L^{\sigma_2}(\psi_2, \chi) = \begin{bmatrix} r_S \sin (\Phi + \psi_2) - r_K \sin (\chi + \psi_2) \\ -r_S \cos (\Phi + \psi_2) + r_K \cos (\chi + \psi_2) \\ \delta_2 \\ 1 \end{bmatrix} . \tag{3}
$$

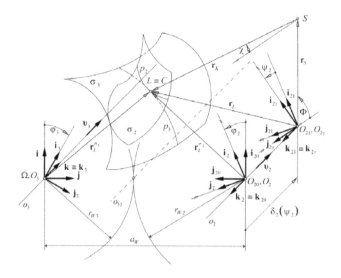

Fig. 1. Creation of conjugate surfaces σ_2 and σ_3

2.2. Conjugate surface σ_3

Creating conjugate surface σ_3 is based, as mentioned above, on Disteli principle. A solution presupposes the fact, that contact point $L \equiv L^{\sigma_2} \equiv L^{\sigma_3} \equiv C$ is a common point of both surfaces for an instantaneous position of angular displacement φ_2. In the basic coordinate system $R \equiv (i, j, k)$, see Fig. 1, the contact point has to comply with two conditions

$$_R r_L^{\sigma_3} = _R r_L^{\sigma_2}, \tag{4}$$

$$n_L \cdot v_{L32} = 0, \tag{5}$$

where $_R r_L^{\sigma_3}$ is the position vector of point $L \in \sigma_3$, $_R r_L^{\sigma_2}$ is the position vector of point $L \in \sigma_2$, n_L is the normal vector in point L and v_{L32} is the vector of the relative velocity in point L. The normal vector n_L can be defined as a cross product of the tangent vectors to the profile p_2 in point L and to the helix through point L. The vector of the relative velocity can be defined as difference between the vector of velocity of surface σ_3 and the vector of velocity of surface σ_2 in the same point L.

With the use of transformation matrices equation (4) can be written as

$$_{R_3} r_L^{\sigma_3} = T_{R,R_3} T_{R_{20},R} T_{R_2,R_{20}} \, _{R_2} r_L^{\sigma_2} \tag{6}$$

and substituting (3) into (6) we obtain

$$_{R_3} r_L^{\sigma_3}(\psi_2, \varphi_2, \chi) = \begin{bmatrix} r_S \sin(\Phi + \varphi_2 + \varphi_3 + \psi_2) - r_K \sin(\chi + \varphi_2 + \varphi_3 + \psi_2) + a_W \sin \varphi_3 \\ r_S \cos(\Phi + \varphi_2 + \varphi_3 + \psi_2) - r_K \cos(\chi + \varphi_2 + \varphi_3 + \psi_2) + a_W \cos \varphi_3 \\ \delta_2 \\ 1 \end{bmatrix}. \tag{7}$$

Equation (5) represents the condition of perpendicularity of normal vector n_L and vector of the relative velocity v_{L32} in contact point $L \equiv C$ of surfaces σ_2 and σ_3. Equation (5) can be rewritten, on condition that the profile p_2 is created by an arc of a circle, into form

$$(i_{32} + 1) \cdot r_K \sin(\chi - \Phi) + a_W \sin(\varphi_2 + \chi + \psi_{2L}) = 0, \tag{8}$$

where a_W is the distance between axes o_2, o_3 and $i_{32} = \frac{\omega_3}{\omega_2} = \frac{\varphi_3}{\varphi_2}$ is a gear ratio. Equation (8) represents functional relationship $f(\varphi_2, \chi) = 0$. Using equations (7) and (8) conjugate surface σ_3 is determined as two dimensional manifold $f(\psi_2, \chi) = 0$.

3. Contact of surfaces with displaced axes

The variation of the position of surface axes entails a displaced position of conjugate surfaces, which must be taken into consideration. As a result of this phenomenon an alternation of the surface contact takes place. Theoretical accurate contact, referred to as "correct contact", changes into incorrect contact. The incorrect contact is solved for two cases of arrangement of displaced axes. In the first case, a parallel displacement of axes is studied. In the second step, a general displacement of axes is taken into consideration.

3.1. Parallel displacement of axes

The incorrect contact of screw surfaces is caused by parallel displacement of the axes which were inserted, by fixed axis o_3, in the axis o_2. Subsequently axis o_2 is shifted in new, deformed, position o_2^Δ and surface σ_2 is displaced in position σ_2^Δ. Surfaces σ_2^Δ and σ_3 were divided into separated cross sections. The determination of the contact point of the surfaces is solved using cross sections with way which is indicated further at a given instantaneous time. Let us consider plane τ where the contact of profiles $p_3 \equiv \sigma_3 \cap \tau$ and $p_2^\Delta \equiv \sigma_2^\Delta \cap \tau$ is solved. The solution is based, Fig. 2, on the determination of the point of intersection of straight line m and profile p_3, $X \equiv p_3 \cap m$, where m is tangent to circle k_E which passes through chosen point $E \in p_2^\Delta$. Creating surface σ_2^Δ is gradually turned, angle φ_2^τ, around axis o_2^Δ until $E \equiv X$ i.e. $_R r_K \equiv {_R} r_X \equiv {_R} r_E \equiv {_R} r_L^{p_3}$ is achieved.

General point K of tangent m is determined with equation

$$r_K = r_E + p m, \tag{9}$$

where m is the unit vector of tangent m and p is a parameter. Numerical solution is made in space R.

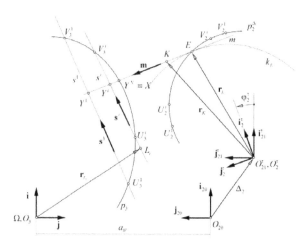

Fig. 2. Determination of contact point of profile p_3 and p_2^Δ in cross section τ

The difficulty of the solution consists in the fact that conjugate surface σ_3, which is created, as mentioned above, as two parametric manifold, is defined through the parametric equation of creating surface σ_2. It means that a general point of surface σ_3, $X \in \sigma_3 \wedge X \equiv m \cap p_3$, is not defined. As a result of the discrete determination of conjugate surface σ_3 the problem of intersection of profile p_3 and line m is solved using iterative method. In the iterative step intersection point $Y^i \equiv m \cap s^i$, $i = 1 \div n$, is determined. Line s^i is a secant of profile p_3 and it is defined by points U_3^i, V_3^i. With the use of equation (7) and its transformation into coordinate system R these points are defined as

$$_R r_{A_3^i} = \begin{bmatrix} r_S \sin\left(\Phi + \varphi_2^{A_2^i} + \varphi_3^{A_2^i} + \psi_\tau - \varphi_{3K0}\right) - \\ r_K \sin\left(\chi^{A_2^i} + \varphi_2^{A_2^i} + \varphi_3^{A_2^i} + \psi_\tau - \varphi_{3K0}\right) + a_W \sin\left(\varphi_3^{A_2^i} - \varphi_{3K0}\right) \\ r_S \cos\left(\Phi + \varphi_2^{A_2^i} + \varphi_3^{A_2^i} + \psi_\tau - \varphi_{3K0}\right) - \\ r_K \cos\left(\chi^{A_2^i} + \varphi_2^{A_2^i} + \varphi_3^{A_2^i} + \psi_\tau - \varphi_{3K0}\right) + a_W \cos\left(\varphi_3^{A_2^i} - \varphi_{3K0}\right) \\ \delta_\tau \\ 1 \end{bmatrix}, \tag{10}$$

where $A_2^i \equiv U_2^i, V_2^i$ are points on profile $p_2^\Delta \in \sigma_2^\Delta$, $A_3^i \equiv U_3^i, V_3^i$ are corresponding points on profile p_3 and i determines the number of an iterative step. Angles φ_j^A, $j = 2, 3$ are defined by

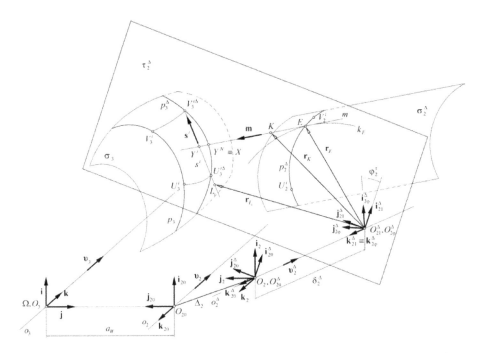

Fig. 3. Determination of contact point of profile p_3^Δ and p_2^Δ in cross section τ_2^Δ

equation (8) and angle φ_{3K0} determines the position of surfaces σ_3 and σ_2^Δ. With the use of (10) line s^i, Fig. 2, can be defined as

$$\boldsymbol{r}_{L_i} = \boldsymbol{r}_{U_3^i} + q(\boldsymbol{r}_{V_3^i} - \boldsymbol{r}_{U_3^i}) = \boldsymbol{r}_{U_3^i} + q\boldsymbol{s}^i, \tag{11}$$

where $L_i \in s^i$, $\boldsymbol{s}^i$ is the directional vector of a secant s^i and q is a parameter. Using equations (9) and (11) we obtain vector equation for determination of point of intersection $Y^i \in s^i$ with line m as

$$\boldsymbol{r}_E + p\boldsymbol{m} - \boldsymbol{r}_{U_3^i} - q\boldsymbol{s}^i = 0. \tag{12}$$

Interval bisection method is used for the determination of parameters χ^{A_i} of points U_3^i and V_3^i in the next iteration step. The iterative process continues until $U_3^N \equiv V_3^N \equiv X$. In the next step surface σ_2^Δ is gradually turned around axis o_2^Δ until $p = 0$, where p is a parameter determined by equation (12). Then $E \equiv X \wedge X \equiv m \cap p_3$. This algorithm is applied on all points of creating surface σ_2^Δ and the contact of surfaces σ_3 and σ_2^Δ is achieved under condition

$$\boldsymbol{r}_L^{\sigma_2^\Delta} \equiv \boldsymbol{r}_E \equiv \boldsymbol{r}_X \equiv \boldsymbol{r}_L^{\sigma_3} \wedge \boldsymbol{n}_L^{\sigma_3} \equiv \boldsymbol{n}_L^{\sigma_2^\Delta} \wedge \varphi_2^\tau = \min \varphi_2^\tau \tag{13}$$

and then $L \equiv C$.

3.2. General displacement of axes

In this case the arrangement of axes is generally skewed. Similarly, as in the case of parallel displacement of axes, axis o_3 is fixed and axis o_2 is shifted into new, deformed, position o_2^Δ and surface σ_2 is displaced in position σ_2^Δ, Fig. 3. The determination of the contact point C of surfaces σ_3 and σ_2^Δ is based on the solution for the parallel displaced axes. The main distinction consists in the fact that plane τ_2^Δ, where contact of profiles $p_3^\Delta \equiv \sigma_3 \cap \tau_2^\Delta$ and $p_2^\Delta \equiv \sigma_2^\Delta \cap \tau_2^\Delta$ is solved, is not perpendicular to axis o_3 and profile p_3^Δ differs from profile p_3 of surface σ_3

in the cross section which is perpendicular to axis o_3. Analogous to the case with parallel displaced axes the solution is divided into two steps. In the first step the determination of point $X \equiv m \cap p_3^{\tau_2^\Delta} \wedge X \equiv Y^N, Y^N \in \{Y^i\}$, where m is tangent to circle k_E which passes through chosen point $E \in p_2^\Delta$ and $Y^i \equiv m \cap s_i$, is achieved. In the second step creating surface σ_2^Δ is gradually turned around axis o_2^Δ until $E \equiv X \wedge X \equiv m \cap p_3^\Delta \equiv Y^N$. The solution is made in coordinate system $R_{20}^\Delta \equiv (i_{20}^\Delta, j_{20}^\Delta, k_{20}^\Delta)$.

Analogous to the previous case, an iterative method was used for the determination of point $X \equiv Y^N$. General point K of tangent m is determined by equation

$$r_K = r_E + pm, \tag{14}$$

where m is the unit vector of line m and p is a parameter. For determination of secant s_i of profile p_3^Δ points $U_3^{i\Delta}, V_3^{i\Delta} \in \sigma_3 \wedge U_3^{i\Delta}, V_3^{i\Delta} \in \tau_2^\Delta, i = 1 \div n$, are necessary to use. The position vectors of these points are determined by equation

$$R_{20}^\Delta r_{A_3^{i\Delta}} = \begin{bmatrix} R_{20}^\Delta x_{A_3^{i\Delta}} \\ R_{20}^\Delta y_{A_3^{i\Delta}} \\ R_{20}^\Delta z_{A_3^{i\Delta}} \\ 1 \end{bmatrix} = T_{R,R_{20}^\Delta} \begin{bmatrix} \cos \Delta\psi_3 & -\sin \Delta\psi_3 & 0 & 0 \\ \sin \Delta\psi_3 & \cos \Delta\psi_3 & 0 & 0 \\ 0 & 0 & 1 & \Delta\delta_3 \\ 0 & 0 & 0 & 1 \end{bmatrix} {}_R r_{A_3^i}, \tag{15}$$

where $A_3^{i\Delta} \equiv U_3^{i\Delta}, V_3^{i\Delta}, A_3^i \equiv U_3^i, V_3^i$ are points of profile p_3 which position vectors ${}_R r_{A_3^i}$ are determined by equation (10). $\Delta\delta_3 = r_{W3}\Delta\psi_3 \tan\gamma$ is the displacement along the surface axis o_3 and angle $\Delta\psi_3$ is determined by equation

$$R_{20}^\Delta z_{A_3^{i\Delta}} = \delta_2^\Delta. \tag{16}$$

Using points $U_3^{i\Delta}, V_3^{i\Delta}$ general point L_i of secant s_i is defined as

$$r_{L_i} = r_{U_3^{i\Delta}} + q(r_{V_3^{i\Delta}} - r_{U_3^{i\Delta}}) = r_{U_3^{i\Delta}} + qs^i, \tag{17}$$

where $L_i \in s^i$, s^i is the directional vector of secant s^i and q is a parameter.

Using equations (14) and (17) intersection $Y^i \in s^i$ with line m is determined as solution of equation

$$r_E + pm - r_{U_3^{i\Delta}} - qs^i = 0. \tag{18}$$

The rest of the solution to the iteration and process of gradual turning of surface σ_2^Δ are analogous as in the case of parallel displaced axes and contact of surfaces σ_3 and σ_2^Δ is obtained under condition

$$R_{20}^\Delta r_L^{\sigma_2^\Delta} \equiv R_{20}^\Delta r_E \equiv R_{20}^\Delta r_X \equiv R_{20}^\Delta r_L^{\sigma_3} \wedge n_L^{\sigma_3} \equiv n_L^{\sigma_2^\Delta} \wedge \varphi_2^\tau = \min \varphi_2^\tau \tag{19}$$

and then $L \equiv C$.

4. Application

4.1. Application to simple screw conjugate surfaces

For application, creating surface σ_2, Fig. 1, with geometric parameters $r_S = 45$ mm, $\Phi = \frac{2}{3}\pi$ rad, $r_K = 30$ mm, $\gamma = \frac{1}{4}\pi$ rad, $r_{W2} = a_W/(1 + 1/i_{32})$, where $a_W = 100$ mm and $i_{32} = 0.75$, was considered. Domains of the definition of surface parameters are $\chi \in \langle 0; \frac{\pi}{3} \rangle$ rad, $\psi_2 \in \langle 0; 1 \rangle$ rad.

Using equations (3), (7) and (8) creating surface σ_2 and conjugate surface σ_3 are determined. In the first case, the parallel displacement of axes was considered. Displacement of creating surface was stated $\Delta_2 = [1; 2; 0]^T$ mm. With the use of the algorithm mentioned in section 3.1, contact point of surfaces σ_2^Δ and σ_3 was determined for three positions of surfaces defined by angle $\varphi_{3K0} = -10°, 0°, 10°$. Dependence of min φ_2^τ, equation (19), on displacement δ_2 along axis o_2^Δ of surface σ_2^Δ is shown in Fig. 4. A variation of the position of contact point C of conjugate surfaces $\sigma_2^\Delta, \sigma_3$ depending on setting up conjugate surfaces is shown in Fig. 5. Fig. 6 presents a shifting of contact point C on surface σ_3, for surface arrangement mentioned above, which is situated in position $\varphi_{3K0} = 0°$.

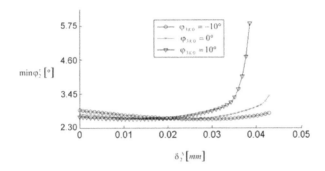

Fig. 4. Dependence of min φ_2^τ on δ_2^Δ

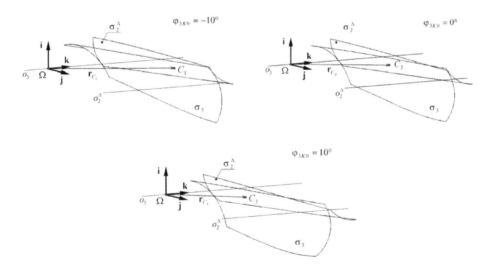

Fig. 5. Positions of contact point C of conjugate surfaces σ_2^Δ and σ_3

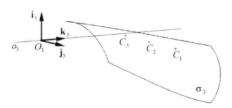

Fig. 6. Displacement of contact point C on surface σ_3

Let us consider the identical couple of conjugate surfaces as in the previous case but for a spatial arrangement. Skew arrangement of axes is determined by displacement of creating surface σ_2 in frontal plane $\Delta_2 = [1; 2; 0]^T$ mm of axis o_2 and the unit direction vector of axis $\boldsymbol{\nu}_2^\Delta = [0.0250; -0.0499; 0.9984]^T$. The algorithm described in section 3.2 was applied to considered case of conjugate surfaces and dependence of angle min φ_2^τ on the displacement along axis o_2^Δ of surface σ_2^Δ is shown in Fig. 7. A variation of contact point C depending on the rotary angle of conjugate surfaces is presented in Fig. 8 and shifting of the same contact point on surface σ_3 is shown in Fig. 9 where the position $C_1 \div C_3$ corresponds to the position of the same points in Fig. 8.

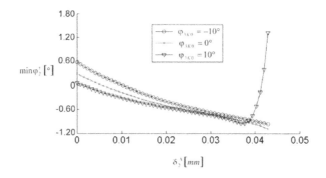

Fig. 7. Dependence of min φ_2^τ on δ_2^Δ

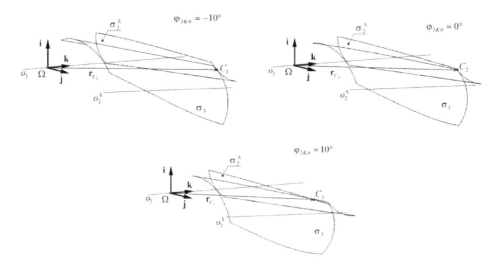

Fig. 8. Positions of contact point C of conjugate surfaces σ_2^Δ and σ_3

Fig. 9. Displacement of contact point C on surface σ_3

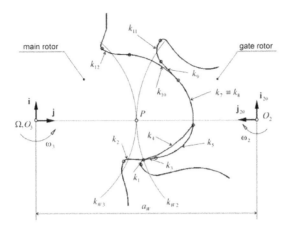

Fig. 10. Profiles of conjugate tooth surfaces

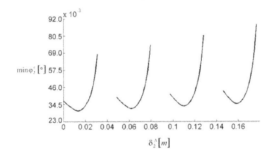

Fig. 11. Dependence of $\min \varphi_2^\tau$ on δ_2^Δ

4.2. Application to tooth surfaces of screw compressor

Tooth surfaces of the screw compressor with the profile SLF4 were considered in the case of application of the presented method to real technical equipment with conjugate surfaces. The main parameters of the compressor rotors are the following. The distance between axes $a_W = 85$ mm, gear ratio $i_{32} = 5/6$, helix angle on the rolling cylinders $\gamma = \pi/4$ rad and the length of toothed part of rotors $l = 193.8$ mm. Tooth surface σ_2 of the gate rotor, which is the creating surface, consists of several single surfaces that are interconnected continuously and smoothly. Profiles of the considered tooth surfaces in the frontal section are shown in Fig. 10. Creating curves of the profile of the gate rotor are arcs of circles k_1, k_3, k_7 and k_{11}, arc of trochoid k_5 and straight line k_9. A tooth profile of the main rotor consists of curves k_2, k_4, k_8, k_{10} and k_{12}, which are envelopes of mentioned creating curves of the gate rotor. For contact analysis the meshing parts of the profiles, which are defined by curves k_7, k_9 and k_{11} of the gate rotor and conjugate curves k_8, k_{10} and k_{12} of the main rotor, are considered. Skew arrangement of rotor axes was defined by displacement of the gate rotor in frontal plane $\Delta_2 = [1; 2; 0]^T$ mm and the unit direction vector of axis o_2^Δ of gate rotor $\boldsymbol{\nu}_2^\Delta = [0.010\,3; 0.005\,2; 0.999\,9]^T$. First of all, the contact of surfaces in a position defined by angle $\varphi_{3K0} = 0°$ was solved. Dependence of $\min \varphi_2^\tau$ on displacement δ_2^Δ along axis o_2^Δ is presented in Fig. 11 for this case. For analysis of the contact of tooth surfaces σ_3, σ_2^Δ during an operating cycle, which is given by a turning about one tooth pitch of the main rotor, the domain of definition of angular displacement of the main rotor is $\varphi_{3K0} \in \langle 0°, 72° \rangle$. Displacement of contact point C of the p-th pair of teeth on tooth

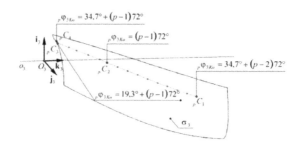

Fig. 12. Translation of contact point on surface σ_3 of the main rotor

surface σ_3 of the main rotor is shown in Fig. 12. Point $_pC_1$ is the point where the contact of the p-th pair of teeth starts. Point $_pC_2$ determines the position in which the contact point conforms to the starting position of both surfaces, Fig. 12, upon their creating. Point $_pC_3$ determines the first point in frontal cross-section, $_p\varphi_{3K0} = 19.3° + (p-1)72°$, where contact of tooth surfaces σ_3, σ_2^Δ of considered p-th pair is made and point $_pC_4$ is the last contact point of the p-th pair of teeth in frontal plane, $_p\varphi_{3K0} = 34.7° + (p-1)72°$.

5. Conclusion

As a result of external influences such as force, temperature and dynamic loading in mechanical systems the original arrangement of surfaces is shifted into a displaced position. This change causes a variation of the characteristics of the contact of surfaces and their relative motion. The originally relative rolling motion changes into the general spatial motion, which is given by relative twist, and the curve contact of surfaces changes into the contact at point. The presented method makes it possible to establish the instantaneous point of contact of conjugate surfaces which are in a deformed, incorrect position, in general. This circumstance is of extraordinarily important for contact analysis and dynamic analysis of mechanic systems or their parts in an operating state. For the afore mentioned reason the determination and specification of the incorrect contact of conjugate screw surfaces with displaced axes represents the first step for the following analysis. The presented method is suitable for arbitrary surfaces which are in contact.

Acknowledgements

The work has been supported by the research project MSM 4977751303.

References

[1] Bär, G., Explicit Calculation Methods for Conjugate Profiles, Journal for Geometry and Graphics 7(2) (2003) 201–210.
[2] Disteli, M., Über instantane Schraubengeschwindigkeiten und die Verzahnung der Hyperboloidräder, Zeitschrift für Mathematic und Physic Bd 53, 1904, p. 51–88. (in German)
[3] Kameya, H., Aoki, M., Nozawa, S., Three-dimensional curvature analysis on screw rotor and its applications, Proceedings of the International Conference on Compressors and their Systems 2005, London, John Willey & Sons, LTD, 2005, p. 467–474.
[4] Švejl, M., Applied kinematics spatial gear transmission, Doctoral thesis, Pilsen, 1967. (in Czech)

12

Car impact to pedestrian – fast 2D numerical analysis

L. Hynčík[a,*], H. Čechová[a]

[a]*New Technologies – Research Centre, University of West Bohemia, Univerzitní 8, 306 14 Plzeň, Czech Republic*

Abstract

The paper concerns a modelling approach for fast 2D car to pedestrian impact analysis. The pedestrian model is composed using the Lagrange equations with multipliers. The model consists of rigid bodies defining the major human body segments. The bodies are connected by rotational joints with non-linear response. The model is scalable based on the age and the gender. The car model is multi-segment composed as an open polygon. Between the pedestrian and the car, there are contacts defined and modelled explicitly by force-penetration dependence. For a given car profile design and a given human gender, age and percentile, the pedestrian impact consequences can be evaluated quickly by means of virtual numerical analysis.

Keywords: contact, pedestrian, impact, multi-body numerical simulation

1. Introduction

The traffic accidents are within a group of external consequences causing the third highest number of death in the Czech Republic, just after deaths caused by falls and suicides [15]. The numbers of deaths or fatally injured citizens prove that the traffic accidents and their consequences are still a serious problem to be solved. The statistics show that the number of accidents decreases slowly in the past years [14]. However, the decrease is necessary to be speed up regarding also the socioeconomic aspects of the problem [4].

2. Motivation

A lot of effort is devoted to both passive and active safety systems development where virtual human body models start to play significant role since they become powerful tool for supporting development of human-friendly and safe vehicles by the numerical way using computer simulation. Regarding that the correct biofidelic models are necessary to be developed and validated [7]. However such complete models usually spend a lot of computational time. Having a set of structural designs of a vehicle to be analysed, simple models are usually sufficient tool for the first approximation. Based on the standard injury criteria implemented in the simple models, they might predict global human behaviour and injuries in very short computational time. Hence, a lot of structural designs and impact scenarios might be analysed quickly in order to choose the best one for a deeper analysis.

3. Method

The paper shows the numerical modelling approach towards pedestrian impact analysis. Currently, two-dimensional (2D) modelling approach is presented.

---

*Corresponding author. e-mail: hyncik@ntc.zcu.cz.

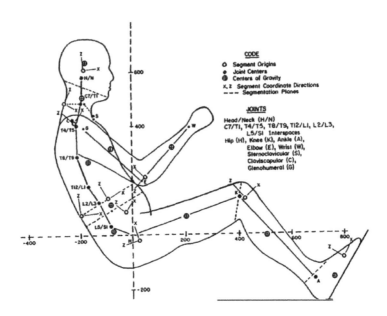

Fig. 1. Human body segmentation (taken from [12])

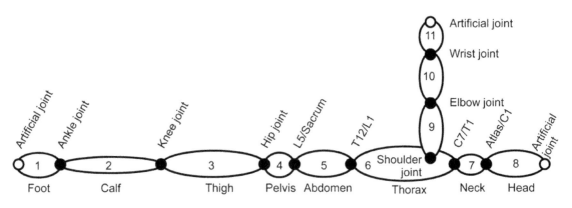

Fig. 2. Human body open-tree multi-body system in the horizontal position

3.1. Geometry and segmentation

The basic human body segmentation is taken from Robbins [12], see Fig. 1. The segmentation is as simple as possible to ensures the full articulability and correct kinematics of the human body under the loading. The anthropometry defined in Robbins [12] also presents the 50 %-tile male body including mass distribution, moments of inertia evaluation and joint stiffness definiton.

Based on the above-mentioned segmentation, the human body model is structured as a linked open-tree multi-body system (see Fig. 2) implemented using the Lagrange's equations with multipliers. The model is composed by 11 segments Ω_i, $i \in \{1, \ldots, 11\}$. The segments correspond to 11 rigid bodies, namely:

- Head
- Neck

- Thorax
- Abdomen
- Pelvis

- Left and right arm
- Left and right forearm
- Left and right hand

- Left and right thigh
- Left and right calf
- Left and right foot

Due to the planar (2D) approach, the couple segments (arm, forearm, hand, thigh, calf and foot) are considered as a one single segment.

3.2. Mass distribuion and joints

The position of each segment is geometrically defined by the centre $S_i, i \in \{1, \ldots, 11\}$ located between 2 joints $\boldsymbol{K}_{i,i-1}$ and $\boldsymbol{K}_{i,i+1}$ connecting the segment to the previous and the following segment in the tree system. The exception is for the head, the hand and the foot that are the outer segments. In this case, an artificial joint supplying the missing second joint is defined for the segment length measuring purposes. The segment centre

$$S_i = \frac{\boldsymbol{K}_{i,i-1} + \boldsymbol{K}_{i,i+1}}{2} \tag{1}$$

is an origin of the particular segment local coordinate system. The first local coordinate system axis is defined as

$$\bar{e}_1 = \frac{\boldsymbol{K}_{i,i+1} - \boldsymbol{S}_i}{\|\boldsymbol{K}_{i,i+1} - \boldsymbol{S}_i\|} \tag{2}$$

and the second one $\bar{e}_2$ is a unit vector perpendicular $\bar{e}_1$ constructing a clock-wise system. Each segment carries its mass m_i and moment of inertia $I_i, i \in \{1, \ldots, 11\}$ concentrated to its centre of gravity $T_i = \begin{bmatrix} {}^T\bar{x}_i, {}^T\bar{y}_i \end{bmatrix}, i \in \{1, \ldots, 11\}$ where the bar denotes coordinates in the segment local coordinate system $(\bar{e}_1, \bar{e}_2)$. The scheme is shown in Fig. 3.

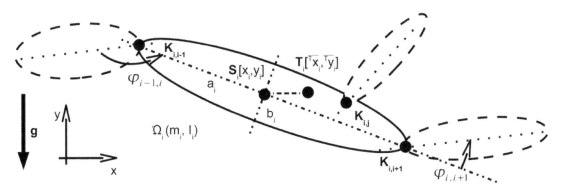

Fig. 3. Segment definition

Two neighbouring segments are connected by joints. The full model accommodates 10 joints, namely:

- Head/neck joint in the position between vertebrae Atlas bone and C1

- Neck/thorax in the position between vertebrae C7 and T1

- Thorax/abdomen in the position between vertebrae T12 and L1

- Abdomen/pelvis in the position between vertebrae L5 and Sacrum bone

- Left and right shoulder
- Left and right hip

- Left and right elbow
- Left and right knee

- Left and right wrist
- Left and right ankle

With respect to the Robbins segmentation [12], all joints in the current model are modelled as rotational ones accommodating non-linear rotational stiffness. Due to the planar approach, the couple joints (shoulder, elbow, wrist, hip knee and ankle) are concentrated into one single joint having double stiffness.

3.3. Anthropometry and stiffness scaling

The basic 50 %-tile human body anthropometry and mass distribution is based on [12] (for both male and female). The details of the scaling can be found in [8]. Due to the simple modelling approach and no detailed segment shape description, only the height based scaling is concerned. Bláha [2] defines the height from ground of all the above-mentioned joints in the human upright standing position, so the particular segment height can be obtained by substraction. The data measured by Bláha [2] depends on gender and age from 0 till 75 years.

Let us define a length vector $l_i = K_{i,i+1} - K_{i,i-1}$ for each segment Ω_i, $i \in \{1, \ldots, 11\}$. Concerning the size, each segment can be defined by its length $2a_i = \|l_i\|$ as the distance between the joints connecting the segment to the previous and the following segment in the tree system. Considering the upright standing position of the model, each segment height can be defined as $h_i = l_i e_2$ in the global coordinate system (e_1, e_2). The upright standing position of Robbins [12] is defined in [8].

Mathematically the anthropometry scaling means that the original size is multiplied by a scaling factor k_i defined as a ratio between the target height defined by [2] and the original height h_i of the segment in the upright standing positon. Concerning the head, the hand and the foot that are the outer segments the length is measured with help of the artificial joint defined above. In summary considering the uniform mass density distribution and the model upright standing position, the 3 basic points defining any segment (centre of gravity and both joints) can be scaled by multiplying by the scaling coeffitient k_i in the global coordinate system as $\tilde{x} = k_i x$. After the anthropometry scaling, all segments have to be reconnected in the joint positions. The mass and the moment of inertia of a particular segment change by changing its size. Considering the uniform mass density distribution, the mass and moment of inertia of each segment can be scaled as $\tilde{m} = k_i^3 m$ and $\tilde{I} = k_i^5 I$ [8], where $\tilde{m}$ and $\tilde{I}$ denotes the scaled values.

The segments are linked together using the rotational joints with non-linear stiffnesses defined in [12] for a 50 %-tile (average) subject. The joints flexibility also changes due to the age. The age dependent flexibility is also scaled based on so called flexindex [1]. Flexindex defines the flexibility of the whole human body by a single number (percentage to the average state) taking into account the flexibility of all major joints together [1]. The percentage is taken into account to multiply the all body joint ranges for the desired age.

3.4. Multi-body approach

Let us have the model structure described by rigid segments linked to an open-tree multi-body system as displayed in Fig. 3. The human model dynamical response due to the impact loading is investigated by the energy balance approach using the Lagrange's equations with multipliers. Let us have n segments, in our case $n = 11$, having the segment numbered by $i \in \{1, \ldots, 8\}$ from the foot through the tree system uptill the head and by $i \in \{6, 9, 10, 11\}$ from the thorax through the tree system uptill the hand. The motion of each segment Ω_i, $i \in \{1, \ldots, n\}$ is defined in plane by 3 independent coordinates where the first two defines the planar position of its centre of gravity and the third one the rotation around it. Let us define the vector

$$\boldsymbol{q} = [x_1, \dots, x_n, y_1, \dots, y_n, \varphi_1 \dots, \varphi_n]^T \tag{3}$$

including 33 generalized coordinates that are dependent by $m = 2(n-1) = 20$ constraints

$$\Phi_i = -x_i + x_{i+1} - (a_i - \bar{x}_i)\cos(\varphi_i) \tag{4}$$
$$- (a_{i+1} + \bar{x}_{i+1})\cos(\varphi_{i+1}) - \bar{y}_i \sin(\varphi_i) + \bar{y}_{i+1}\sin(\varphi_{i+1}), \tag{5}$$
$$\Phi_{n-1+i} = -y_i + y_{i+1} - (a_i - \bar{x}_i)\sin(\varphi_i) \tag{6}$$
$$- (a_{i+1} + \bar{x}_{i+1})\sin(\varphi_{i+1}) + \bar{y}_i \cos(\varphi_i) - \bar{y}_{i+1}\cos(\varphi_{i+1}), \tag{7}$$

where $\bar{x}_i$ and $\bar{y}_i$ are the coordinates of the segment centre of gravity in the local coordinate system (see Fig. 3) and $i \in \{1, \dots, 7\}$ from the ankle over the legs and spine to the head/neck joint,

$$\Phi_8 = = -x_6 + x_9 - (\bar{x}_{shoulder} - \bar{x}_6)\cos(\varphi_6) \tag{8}$$
$$- (a_9 + \bar{x}_9)\cos(\varphi_9) - \bar{y}_6 \sin(\varphi_6) + (\bar{y}_{shoulder} + \bar{y}_9)\sin(\varphi_9), \tag{9}$$
$$\Phi_{18} = = -x_6 + x_9 - (\bar{x}_{shoulder} - \bar{x}_6)\sin(\varphi_6) \tag{10}$$
$$- (a_9 + \bar{x}_9)\sin(\varphi_9) + (\bar{y}_6 - \bar{y}_{shoulder})\cos(\varphi_6) - \bar{y}_9\cos(\varphi_9), \tag{11}$$

for the shoulder joint where $[\bar{x}_{shoulder}, \bar{y}_{shoulder}]^T$ are the coordinates of the shoulder joint in the thorax local coordinate system and

$$\Phi_i = -x_i + x_{i+1} - (a_i - \bar{x}_i)\cos(\varphi_i) \tag{12}$$
$$- (a_{i+1} + \bar{x}_{i+1})\cos(\varphi_{i+1}) - \bar{y}_i \sin(\varphi_i) + \bar{y}_{i+1}\sin(\varphi_{i+1}), \tag{13}$$
$$\Phi_{n-1+i} = -y_i + y_{i+1} - (a_i - \bar{x}_i)\sin(\varphi_i) \tag{14}$$
$$- (a_{i+1} + \bar{x}_{i+1})\sin(\varphi_{i+1}) + \bar{y}_i \cos(\varphi_i) - \bar{y}_{i+1}\cos(\varphi_{i+1}), \tag{15}$$

where $i \in \{9, 10\}$ for the elbow and the wrist joints. The constraints link together the centres of gravity of all neighbouring couples of segments. So the whole system has $3n - m = 33 - 20 = 13$ degrees of freedom that are the initial position and rotation of the base body and 10 further relative rotations of the open-tree linked system of rigid bodies. The system of equations is conservative and the kinetic and the potential energies have the forms

$$E_k = \frac{1}{2}\sum_{i=1}^{n} m_i \left(\dot{x}_i^2 + \dot{y}_i^2\right) + \frac{1}{2}\sum_{i=1}^{n} I_i \dot{\varphi}_i^2, \tag{16}$$

$$E_p = g\sum_{i=1}^{n} m_i y_i + \frac{1}{2}\sum_{(i,j)} k_{ij}\left(\varphi_{ij}\right)\varphi_{ij}^2, \tag{17}$$

where the limits (i, j) in the second term of the potential energy mean all segment couples connected by joints, k_{ij} is the particular relative rotation dependent joint rotational stiffnes and $\varphi_{ij} = \varphi_j - \varphi_i$ is the relative rotation of the two neighbouring segments. Considering the Lagrange's equations with multipliers have the form

$$\frac{d}{dt}\left(\frac{\partial L}{\partial \dot{q}_i}\right) - \frac{\partial L}{\partial q_i} = \sum_{j=1}^{m} \lambda_j \frac{\partial f_j}{\partial q_i}, \quad i \in \{1, \dots, 3n\} \tag{18}$$

with the Lagrange's function $L = E_k - E_p$, we have the equations of motion in the form

$$\frac{\mathrm{d}}{\mathrm{d}t}\left(\frac{\partial E_k}{\partial \dot{q}_i}\right) + \frac{\partial E_p}{\partial q_i} = \sum_{j=1}^{m}\lambda_j\frac{\partial f_j}{\partial q_i} \quad i \in \{1,\ldots,3n\}. \tag{19}$$

Equations (18) can be rewritten using the matrix notation as

$$\begin{bmatrix} M & \Phi_q^T \\ \Phi_q & O \end{bmatrix}\begin{bmatrix} \ddot{q} \\ -\lambda \end{bmatrix} = \begin{bmatrix} r \\ \gamma \end{bmatrix}, \tag{20}$$

where M is the mass matrix, Φ_q is the matrix of constraint derivatives, O is the corresponding zero matrix, r is the external load matrix and γ is the matrix consisting of the rest after the constraint derivatives.

3.5. Contact definition

The model is defined by a linked open-tree multi-body system skeleton carrying the mass. For the purposes of pedestrian impact analysis, the contact forces play an important load. The contact forces express loading caused by the vehicle impact to the pedestrian. In order to model the contact well, each body of the multi-body system has to accommodate an external shape of the particular body segment. An ellipse attached to each segment is a good estimation for our purpose. Each segment ellipse is located by its centre to the segment centre (that is not necessarily its centre of gravity) and it is described by 2 semi-axes a_i and b_i as shown in Fig. 3. The scaling of the segment depth in the direction of the second semi-axis can be simply computed as $\bar{b}_i = k_i b_i$. The vehicle model is based on the approximation by rigid open polygon describing major vehicle frontal parts (bumper, bonnet and windshield) [10].

3.5.1. Collision detection

The shape of each segment $\Omega_i, i \in \{1,\ldots,n\}$ is geometrically defined by an ellipse. Considering homogeneous coordinates $X = [x, y, w]^T$, the ellipse can be written in the form

$$X^T A X = 0. \tag{21}$$

Any ellipse with the semi-axes a_i and b_i parallel to the global coordinate system axes has the matrix in the form

$$\bar{A}_i = \begin{bmatrix} \frac{1}{a_i^2} & 0 & 0 \\ 0 & \frac{1}{b_i^2} & 0 \\ 0 & 0 & -1 \end{bmatrix} \quad i \in \{1,\ldots,11\}. \tag{22}$$

The segments of the human body models are positioned and angled in the plane by their generalized coordinates $[x_i, y_i, \varphi_i]^T, i \in \{1,\ldots,n\}$ that can define transformation matrix R_i, $i \in \{1,\ldots,11\}$ that transforms (rotates and moves) the basic ellipse (22) to the particular position defined by $[x_i, y_i, \varphi_i]^T$ in the plane. Such positioned ellipse can be written in the form

$$A_i = R_i^T \bar{A}_i R_i, \quad i \in \{1,\ldots,n\}. \tag{23}$$

The vehicle model is composed of s straight lines. Any segment $\Omega_i, i \in \{1,\ldots,n\}$ defined geometrically by the particular ellipse defined by (23) can come into collision with one or more lines from the open polygon

$$X = P_l + (Q_l - P_l)t, \quad l \in \{1,\ldots,s\}, \quad t \in [0,1] \tag{24}$$

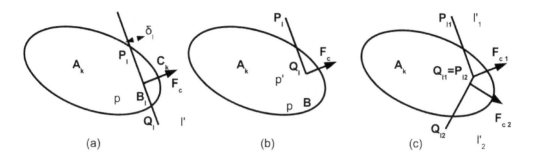

Fig. 4. Collision detection with contact force identification

that leads to a quadratic equation. Negative discriminant means no collision, zero or positive discriminant means contact or penetration.

Talking about contacts in multi-body dynamics, the energy absorption must be implemented in the internal contact variables since the rigidity of the contacting bodies. Furthermore, correct contact modelling in multi-body approach to biomechanics is crucial due to the high deformation of biological materials. Based on literature review, the continuous contact model, also referred to as compliant contact model, is chosen [5]. The basis of the continuous contact model is to measure the penetration of the contacting bodies, see Fig. 4a. This is done by defining the normal contact force F_n as an explicit function of local penetration δ and its rate as

$$F_n = F_n\left(\delta, \dot{\delta}\right). \tag{25}$$

For the total contact force, the simplified nonlinear spring-dashpot contact force model including tangential friction is used in the current paper as

$$\boldsymbol{F}_c = \left[F_n\left(\delta\right) - b\dot{\delta}\right]\left(\boldsymbol{n} - f\boldsymbol{t}\right), \tag{26}$$

where $\boldsymbol{n}$ and $\boldsymbol{t}$ are the penetration normal and tangent vectors respectively, f is the friction coefficient. The force on penetration dependance for various vehicle parts is take from [10]. The penetratration depth is checked couple by couple (segment k to open polygon line l, see Fig. 4c over the whole human body and over the whole open polygon and the particular penetration depth is computed as the distance between the vehicle structure open polygon line l and the parallel line tangent to the ellipse surface l'. Mathematically the line parallel to the open polygon line is constructed as the general line l' tangent to the ellipse with the tangent vector $|\boldsymbol{P}_l\boldsymbol{Q}_l|$. There are 2 solutions for such task, so it is necessary to choose the solution l' that is closer to the line $|\boldsymbol{P}_l\boldsymbol{Q}_l|$ otherwise the situation appeared on the ellipse opposite side. Introducing the line l', the contact point C_k is also detected. The contact force point of action (here line of action since all forces of the same value applied on a line to a rigid body have the same action) is the intersection of the vehicle structure open polygon line l and the perpendicular line going through the tangent point of contact p. Let us denote the intersection between lines p and l as the bottom point B_l. The force action line is constructed as the line p perpendicular to the line $|\boldsymbol{P}_l\boldsymbol{Q}_l|$ going through the contact point C_k.

In several cases, an edge contact might happen, see Fig. 4b. If there is the edge contact (see Fig. 4b and Fig. 4c) and the bottom point B_l is introduced outside the open polygon line (t fall outside the interval $[0, 1]$, see Fig. 4b), the bottom point comes to the edge point Q_l where the penetration is currently highest. The force action line becomes then p'.

If the segment k is in contact with more open polygon lines (e.g. l_1 and l_2, see Fig. 4c), all contacts are treated separately and separate contact forces $\boldsymbol{F}_{c1}$ and $\boldsymbol{F}_{c2}$ act on the particular segment. There is no reaction on the vehicle considering its higher mass. Concerning the particular segment k, the forces $\boldsymbol{F}_{ci}$, $i \in \{1, \ldots, s\}$ are translated to its centre of gravity including a moment caused by translation. The contact forces acting from the $i \in \{1, \ldots, s\}$ contact pairs on segment k is decomposed to the global coordinate system axis and togehter with the moment appears then in the right hand side vector r in (20) as

$$
\boldsymbol{r} = \begin{bmatrix} \vdots \\ \sum\limits_{i=1}^{l} \boldsymbol{F}_{c_{kl}} \boldsymbol{e}_1 \\ \vdots \\ m_k g + \sum\limits_{i=1}^{l} \boldsymbol{F}_{c_{kl}} \boldsymbol{e}_2 \\ \vdots \\ \sum\limits_{i=1}^{l} \mathbf{M}_{kl} \\ \vdots \end{bmatrix} . \tag{27}
$$

4. Results

Numerically, the complete simulation model is implemented and solved in the form

$$
\begin{bmatrix} \boldsymbol{M} & \boldsymbol{\Phi}_q^T \\ \boldsymbol{\Phi}_q & \boldsymbol{O} \end{bmatrix} \begin{bmatrix} \ddot{\boldsymbol{q}} \\ -\boldsymbol{\lambda} \end{bmatrix} = \begin{bmatrix} \boldsymbol{r} \\ \boldsymbol{\gamma} \end{bmatrix} \tag{28}
$$

that leads to the system of 33 second order non-linear ordinary differential equations

$$
\ddot{\boldsymbol{q}} = \boldsymbol{M}^{-1} \left[\boldsymbol{r} + \boldsymbol{\Phi}_q^T \left(\boldsymbol{\Phi}_q \boldsymbol{M}^{-1} \boldsymbol{\Phi}_q^T \right)^{-1} \left(\boldsymbol{\gamma} - \boldsymbol{\Phi}_q \boldsymbol{M}^{-1} \boldsymbol{r} \right) \right], \tag{29}
$$

where r is the right hand side including also the first order derivative expressions. The Baumgart's stabilization [6] for better convergence is used to stabilize the system. It leads to the final form of the solution as

$$
\ddot{\boldsymbol{q}} = \boldsymbol{M}^{-1} \left[\boldsymbol{r} + \boldsymbol{\Phi}_q^T \left(\boldsymbol{\Phi}_q \boldsymbol{M}^{-1} \boldsymbol{\Phi}_q^T \right)^{-1} \left(\boldsymbol{\gamma} - \alpha \dot{\Phi} - \beta^2 \Phi - \boldsymbol{\Phi}_q \boldsymbol{M}^{-1} \boldsymbol{r} \right) \right], \tag{30}
$$

where α and β are the Baumgart's stabilization constants. The system is numerically solved in MATLAB R2010a [11] computational environment using especially the *ode45* function for numerical integration and the *fzero* function for the contact point detection. Other mathematical operations concern standard matrix manipulation.

Fig. 5 shows a typical pedestrian kinematic state of the impact situation at time 75 ms after the impact in the developed software [9]. The impact takes into account a 5 %-tile female pedestrian impacted by the standard passengers' car defined by Kerrigan [10] with car initial velocity 30 km/h. Fig. 6 shows the head output, namely relative head to car position in plane, head velocity and head acceleration. The same output is stated in Fig. 7 for thorax. The accelerations are filtered by the Channel Frequency Class (CFC) [3]. In accordance with SAE J211,

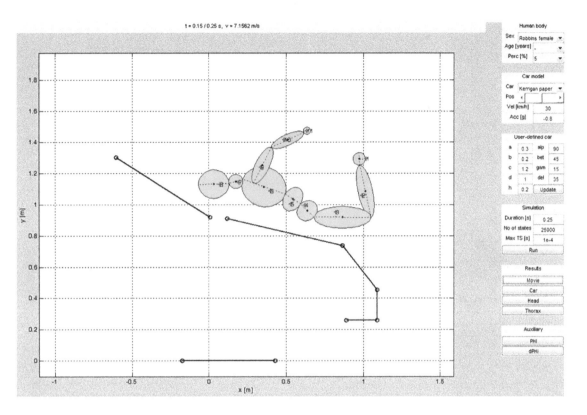

Fig. 5. Kinematic state

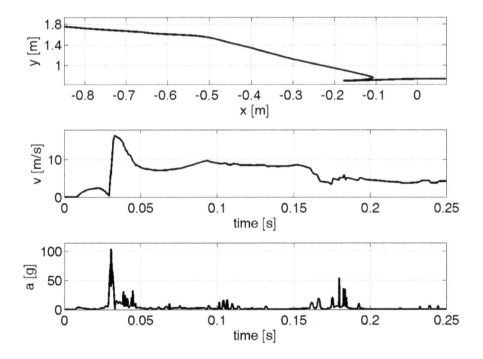

Fig. 6. Head output

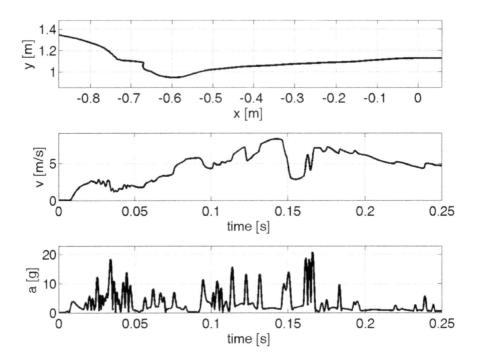

Fig. 7. Thorax output

a 4-channel Butterworth low pass filter with linear phase and special starting condition is used as a digital filter as

$$Y_i = a_0 X_i + a_1 X_{i-1} + a_2 X_{i-2} + b_1 Y_{i-1} + b_2 Y_{i-1}, \quad i \in 1, \ldots, N, \tag{31}$$

where X is the input signal, N is the number of samples and Y is the output filtered signal. The filter constants are calculated as [3]:

$$
\begin{aligned}
\omega_d &= 2\pi \cdot CFC \cdot 2.0775, & a_0 &= \frac{\omega_a^2}{1+\sqrt{2}\omega_a+\omega_a^2}, & b_1 &= \frac{-2(\omega_a^2-1)}{1+\sqrt{2}\omega_a+\omega_a^2}, \\
\omega_a &= \frac{\sin\frac{T\omega_d}{2}}{\cos\frac{T\omega_d}{2}}, & a_1 &= 2a_0, & b_2 &= \frac{-1+\sqrt{2}\omega_a-\omega_a^2}{1+\sqrt{2}\omega_a+\omega_a^2}, \\
& & a_2 &= a_0,
\end{aligned}
\tag{32}
$$

where CFC is the filter parameter (180 for the thorax according to ISO 6487 and 1000 for the head according to SAE J211 [3]) and T is the sampling rate, here it is the time interval between 2 followed saved time history states. The difference equation describes a 2-channel filter. To realize a 4-channel filter, the data of the 2-channel filter has to run twice; once forwards and once backwards, to prevent phase displacements. The filter constants ω_d are calculated in the ISO 6487 sample code differently to SAE J211 as

$$\omega_d = 2\pi CFC \cdot 1.25 \cdot \frac{5}{3} = 2\pi \cdot CFC \cdot 2.08\bar{3}. \tag{33}$$

The current simulation results in $HIC = 25846$ that is really high considering fatal head injuries but typical for such simulation and in $3ms = 56$ that is under the injury limit [13].

4.1. Injury description

Injury criteria meassures related to the acceleration of the particular segment are implemented in order to analyze the particular injury of a human body [13]. Currently it concerns the head injury criterion (*HIC*) and thethorax injury criterion (*3ms*). Whilst the $3ms$ criterion is the highest thorax acceleration lasting at least 3 miliseconds, the head injury criterion is stated as the integral

$$HIC = \max_{0 \leq t_1 \leq t_2 \leq T} \left(\frac{1}{t_2 - t_1} \int_{t_1}^{t_2} a_{head}(t)dt \right)^{2.5} (t_2 - t_1), \tag{34}$$

where $a_{head}(t)$ is the head centre of gravity acceleration during the loading, t_2 and t_1 are two arbitrary times during the acceleration pulse. Time values are to be provided in seconds. For example, the victim of an accident sustains a head injury if $HIC_{36} < 1\,000$, where the time window is $t_2 - t_1 \leq 36$ ms.

5. Discussion

The above-mentioned study employs the standard multi-body system modelling approach for the used in biomechanics, especially the impact biomechanics. The main advantage of the multi-body system approach is the computation speed. The pedestrian impact scenario takes usually considerable time comparing to the standards crash simulation since the global kinematics of the pedestrian due to the free space after the impact. Hence the fast computation of the global kinematics including basic injury description is profitable. Such simulation can be followed by a deeper injury analysis e.g. by detailed finite element models whose have much higher requirements on computational time. Moreover, merging the multi-body system approach with some deformable elements like flexible joints, continuous contact models or even finite element parts ensure even wider exploitation of the model in the field of safety system development. The current study shows the planar approach that can be simply extended to the full 3D spatial modelling. The multi-body skeleton of the model can also carry realistic human shapes modelled by finite elements surfaces.

6. Conclusion

The paper shows a multi-body approach towards modelling of the human body response caused by the external impact loading. The human body is segmented into 11 major parts that compose together an open-tree linked multi-body structure. Particular neighbouring segments are connected by rotatinal joints with non-linear stiffness. Each segment carries a mass and inertial properties and its shape is simplified by an ellipse in order to simulate a contact surface. Continuous contact model based on penetration dependent contact force for each penetrating segment couples is implemented in order to asure correct impact loading due to the external contact. Age scaling concerning anthropometry, mass distribution and joint flexibility is derived. Injury criteria meassures are implemented in order to analyze the particular injury of a human body. The model serves for fast analysis of a frontal pedestrian impact that might go before a detailed analysis using more complex virtual human body models or physical tests with dummies. The 2D approach seems to be a promising fast tool for simple analysis even including multiple body contact.

Acknowledgements

The work is supported by the project CG911-044-150 of the Czech Ministry of Transport.

References

[1] Araújo, C. G., Flexibility assessment: normative values for flexitest from 5 to 91 years of age. Arq. Bras. Cardiol., 90 (2008) 257–263.

[2] Bláha, P., et al., Anthropometry of the Czechoslovak population from 6 till 55 years, Czechoslovak spartakiade. 1985 (in Czech).

[3] Cichos, D., de Vogel, D., Otto, M., Schaar, O., Zölsch, S., Crash Analysis Criteria Description. 2006.

[4] Daňková, A., Economical aspect of traffic accidents. Traffic engineering, 2007 (in Czech).

[5] Gilardi, G., Sharf, I., Literature survey of contact dynamics modelling. Mechanism and Machine Theory, 37 (2002) 1 213–1 239.

[6] Hajžman, M., Polach, P., Application of stabilization techniques in the dynamic analysis of multibody systems. Applied and Computational Mechanics, 1 (2007) 479–488.

[7] Haug, E., Choi, H. Y., Robin, S., Beaugonin, M., Human models for Crash and impact simulation. Elsevier, North Holland, 2004.

[8] Hynčík, L., Nováček, V., Bláha, P., Chvojka, O., Krejčí, P., On scaling of human body models. Applied and Computational Mechanics, 1 (2007) 63–76.

[9] Hynčík, L., Čechová, H., Antroped2D: Software for 2d car to pedestrian frontal impact analysis.

[10] Kerrigan, J. R., Murphy, D. B., Drinkwater, D. C., Kam, C. Y., Bose, D., Crandall, J. R., Kinematic corridors for pmhs tested in full-scale pedestrian impact. University of Virginia Center for Applied Boimechanics, 2005.

[11] MathWorks, MATLAB R2010a.

[12] Robbins, D. H., Anthropometry of motor vehicle occupants. Technical report, 1983.

[13] Schmitt, K. U., Niederer, P., Walz, F., Trauma biomechanics, Introduction to accidental injury. Springer-Verlag, 2004.

[14] Skácal, L., Deep analysis of international comparison of traffic accidents in the Czech Republic. Transport Research Centre Brno, 2007 (in Czech).

[15] Czech Statistical Office, External death causes in the Czech Republic in years from 1994 till 2006 (in Czech).

13

Dynamic behavior of Jeffcott rotors with an arbitrary slant crack orientation on the shaft

R. Ramezanpour[a], M. Ghayour[a,*], S. Ziaei-Rad[a]

[a]*Department of Mechanical Engineering, Isfahan University of Technology 84156-83111 Isfahan, Iran*

Abstract

Dynamic behaviour of a Jeffcott rotor system with a slant crack under arbitrary crack orientations is investigated. Using concepts of fracture mechanics, flexibility matrix and stiffness matrix of the system are calculated. The system equations motion is obtained in four directions, two transversal, one torsional and one longitudinal, and then solved using numerical method. In this paper a symmetric relation for global stiffness matrix is presented and proved; whereas there are some literatures that reported nonsymmetrical form for this matrix. The influence of crack orientations on the flexibility coefficients and the steady-state response of the system are also investigated. The results indicate that some of the flexibility coefficients are greatly varied by increasing the crack angle from 30° to 90° (transverse crack). It is also shown that some of the flexibility coefficients take their maximum values at (approximately) 60° crack orientation.

Keywords: dynamic, rotor system, slant crack, compliance matrix, response spectrum

1. Introduction

Modern day rotors are designed for achieving higher revolutionary speed. On the other hand such systems have noticeable mass and thus considerable energy. It is obvious that any phenomenon that causes sudden release of this energy may lead to a catastrophic failure in such systems. Since 1980s, numerous researchers have studied the response of rotating systems with crack. Recently [3] investigated a simple Jeffcott rotor with two transverse surface cracks. It is observed significant changes in the dynamic response of the rotor when the angular orientation of one crack relative to the other is varied. A response-dependent nonlinear breathing crack model has been proposed in [2]. Using this model, they studied coupling between longitudinal, lateral and torsional vibrations. They observed that motion coupling together with rotational effect of rotor and nonlinearities due to their presented breathing model introduces sum and difference frequency in the response of the cracked rotor. Transient response of a cracked Jeffcott rotor through passing its critical speed and subharmonic resonance has been analysed by [4]. The peak response variations as well as orbit orientation changes have been also studied experimentally. In comparison to transverse crack, there are a few investigations on slant cracks. A qualitative analysis of a transverse vibration of a rotor system with a crack at an angle of 45 degrees toward the axis of the shaft has been presented in [5]. It has been concluded that the steady-state transversal response of the rotor system contain peaks at the operating speed, twice of the operating speed and their subharmonic frequencies. The transverse vibration of a rotor

---
*Corresponding author. e-mail: ghayour@cc.iut.ac.ir.

Nomenclature

$A, \bar{A}$	cross sectional area of crack and shaft (m^2)	K_I^i, K_{III}^i	opening and tearing mode of crack due to internal load "i" (N/m$^{3/2}$)
A'	cross sectional area of open part of crack surface (m^2)	$[K]_l$	local stiffness matrix
$[c]_l$	local flexibility matrix of cracked shaft	q_1	longitudinal force (internal reaction) (N)
$[C_s]$	total flexibility matrix of uncracked shaft	q_4, q_5	bending moments (internal reactions) (N · m)
E_{total}	total strain energy of a cracked shaft (N · m)	R_M	radius of the Mohr circle (m)
F_1, F_2, F_{III}	influential functions	T	torsional moment (N · m)
F_x, F_y	transversal forces (external loads) (N)	u	longitudinal displacement of center of disk (m)
F_z	longitudinal force (external load) (N)	U	strain energy of uncracked shaft (N · m)
G	modulus of rigidity (N/m^2)		
h	height of element strip (m)	W	strain energy due to crack (N · m)
$[H]$	transformation matrix	x, y	transversal displacements of center of disk (m)
I	moment of inertia for cross section (m^4)	α	rotor center displacement in rotational direction (torsional displacement of center of disk) (rad)
J	polar moment of inertia of disk (m^2)		
J_p	polar moment of inertia for cross section (kg · m^2)	β	rotation angle of element E2 (Mohr circle) (rad)
k_{ij}	cross-coupled stiffness (N/m, N/rad)	γ	crack depth (m)
k_x, k_y	stiffness in x and y direction (N/m)	η_0	location of elemental strip along η' direction (m)
k_u	stiffness in longitudinal direction (N/m)	θ	crack orientation angle (rad)
k_T	stiffness in torsional direction (N/rad)	σ_M	center of Mohr circle (N/m^2)
$[k]_g$	global stiffness matrix	σ_1', σ_2'	axial stress due on element E2 (after rotation) (N/m^2)
K_I, K_{III}	total opening and tearing mode of crack (N/m$^{3/2}$)		
		τ_1	shear stress on element E1 (before rotation) (N/m^2)
		τ_1'	shear stress on element E2 (after rotation) (N/m^2)

system with a slant crack under torsional vibration has been investigated in [6]. It has been considered that the transverse vibration of the rotor is to be closely related to the torsional vibration. A comparison between the response of transverse and slant cracks has been presented in [10]. They proposed use of mechanical impedance for crack detection. It is concluded that vibration behavior of a rotor with a slant crack is less sensitive to mechanical impedance. A simple Jeffcott rotor model of a rotor with a slant crack has been considered by [1]. It is observed a rotor with a slant crack is stiffer in lateral and longitudinal directions, but more flexible in torsion, compared to a rotor with a transverse crack. Recently [8] in his good review paper explained many crack models such as open crack model, switching crack model, second moment inertia model, breathing models and harmonic model approaches. The dynamics behaviour of a slant (45° crack angle) cracked rotor has been studied by [7]. Using Jeffcott rotor model, the equation of the motion extracted in four directions. Global stiffness of the system obtained from concepts of fracture mechanics and strain energy release rate. It is included that existence of the frequency of torsional excitation in longitudinal response and combined frequencies of the rotating frequency and frequency of torsional excitation in transversal response are good signs for slant crack detection.

In this paper, the dynamic behavior of a cracked Jeffcott rotor with a slant crack on the shaft is considered. Motion equations of the system that are obtained in four directions, two transversal, one torsional and one longitudinal, are solved by Runge-Kutta method. Using concepts of fracture mechanics, flexibility matrix and thus stiffness matrix of the system are calculated. Also the influence of crack orientations on the flexibility coefficients and subsequently on the amplitude of the frequency responses in several prominent frequencies is investigated. These results depict a better understanding for the dynamic behavior of slant cracked shaft under various crack orientations.

2. Equations of motion

Consider a Jeffcott rotor rotating at speed Ω (Fig. 1). The shaft is assumed to be massless and a disk of mass m is placed in the middle of the shaft. A view of cross section of the disk is shown in Fig. 2. In this figure XOY is the fixed coordinate, $\xi o\eta$ is the rotational coordinate with center o and $\xi'o'\eta'$ is rotational coordinate that is located at the center of the disk and attached to it. Point o' is the center of the disk, c is the disk center of mass, α is the angle represents the torsional vibration of the system and φ is the angle between center of mass and rotational coordinate.

In the following equations indices u and T denote the coefficient for torsional and longitudinal directions respectively. Using d'Alambert principle (Fig. 3), equation of the motion in four directions (two transversal, one torsional and one longitudinal) can be established as

$$m\ddot{x} + c\dot{x} + k_x x + k_{xy} y + k_{xT}\alpha + k_{xu}u = \tag{1}$$
$$-mg + me(\Omega + \dot\alpha)^2 \cos(\Omega t + \alpha + \varphi) + me\ddot\alpha \sin(\Omega t + \alpha + \varphi),$$
$$m\ddot{y} + c\dot{y} + k_{xy} x + k_y y + k_{yT}\alpha + k_{yu}u = \tag{2}$$
$$me(\Omega + \dot\alpha)^2 \sin(\Omega t + \alpha + \varphi) - me\ddot\alpha \cos(\Omega t + \alpha + \varphi),$$
$$J\ddot\alpha + c_T(\Omega + \dot\alpha) + k_{xT}x + k_{yT}y + k_T\alpha + k_{Tu}u = \tag{3}$$
$$M(t) + mge\sin(\Omega t + \alpha + \varphi) + me\ddot{x}\sin(\Omega t + \alpha + \varphi) - m\ddot{y}e\cos(\Omega t + \alpha + \varphi),$$
$$m\ddot{u} + c_u\dot{u} + k_{xu}x + k_{yu}y + k_{Tu}\alpha + k_u u = 0, \tag{4}$$

where J is the mass moment of inertia of the disk about o', c, c_T, and c_u are the damping coefficients in transversal, torsional and longitudinal directions. It should be mentioned that these equations are the same with those reported in [7]. Also, $M(t)$ is the torsional excitation and e is the eccentricity of the disk. According to (1)–(4) the stiffness matrix of the system can be extracted as:

$$F = \begin{bmatrix} k_x & k_{xy} & k_{xT} & k_{xu} \\ k_{xy} & k_y & k_{yT} & k_{yu} \\ k_{xT} & k_{yT} & k_T & k_{Tu} \\ k_{xu} & k_{yu} & k_{Tu} & k_u \end{bmatrix} \begin{pmatrix} x \\ y \\ \alpha \\ u \end{pmatrix} \Rightarrow [k]_g = \begin{bmatrix} k_x & k_{xy} & k_{xT} & k_{xu} \\ k_{xy} & k_y & k_{yT} & k_{yu} \\ k_{xT} & k_{yT} & k_T & k_{Tu} \\ k_{xu} & k_{yu} & k_{Tu} & k_u \end{bmatrix}. \tag{5}$$

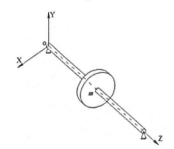

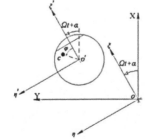

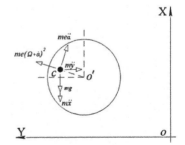

Fig. 1. A schematic of Jeffcott rotor

Fig. 2. Cross sectional view of crack at middle point of the shaft

Fig. 3. Forces exerted on the mass center of the disk

Existence of crack can affect the elements of this matrix. This will be shown in the following sections.

3. Flexibility of a rotor with slant crack

In this section using strain energy release rate method and Castigliano's theorem, the crack compliance matrix is calculated. It is known that the total strain energy of a cracked shaft is the sum of strain energy of uncracked shaft and strain energy caused by crack.

$$E_{\text{total}} = U_{\text{uncracked shaft}} + W_{\text{cracked shaft}}. \tag{6}$$

Consider a cracked shaft (Fig. 4) under four external loads, three forces exerted in principle directions and one torsional moment in Z directions. Thus, the strain energy of an uncracked shaft can be expressed as

$$U = \frac{F_x^2 l^3}{96EI} + \frac{F_y^2 l^3}{96EI} + \frac{T^2 l}{4GJ_p} + \frac{F_z^2 l}{4\bar{A}E}, \tag{7}$$

where G is the shear modulus and $\bar{A}$ is the cross sectional area of the shaft.

Suppose that internal reactions on an element of shaft containing crack, are two bending moments q_4 and q_5, one torsional moment T and one longitudinal force q_1 (Fig. 5). Thus the additional strain energy due to crack is a function of q_5, q_4, T and q_1.

According to (6) and also using Castigliano's theorem, the local flexibility of cracked shaft will be determine using following relation

$$\frac{\partial^2 E}{\partial F_i \partial F_j} = \frac{\partial^2 U}{\partial F_i \partial F_j} + \frac{\partial^2 W}{\partial F_i \partial F_j}. \tag{8}$$

If there is not exist a crack on the shaft, the flexibility due to crack is zero. Therefore, flexibility of the system will be equal to the flexibility of an uncracked shaft. Using (7), the first term in the right hand side of (8) can be determined.

The next step is to find relations between F_i and q_i. Using Figs. 6a and 6b, the following relations can be obtained

$$q_4 = \frac{F_y}{2}\left(\frac{l}{2}\right) = \frac{F_y l}{4}, \qquad q_5 = \frac{F_x}{2}\left(\frac{l}{2}\right) = \frac{F_x l}{4}. \tag{9}$$

Considering (9) and using the chain rule, (8) leads to flexibility matrix of cracked shaft, $[c]_l$ (see

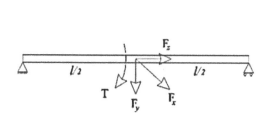

Fig. 4. A cracked shaft under external loads

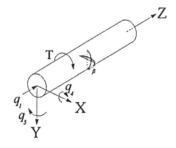

Fig. 5. Internal reactions on the crack

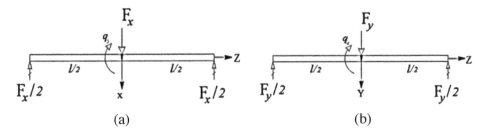

Fig. 6. Relation between external and internal loads: (a) between F_x and q_5, (b) between F_y and q_4

Appendix I)

$$[c]_l = \begin{bmatrix} \dfrac{l^2}{16}\left(\dfrac{\partial^2 W}{\partial q_5^2}\right) & \dfrac{l^2}{16}\left(\dfrac{\partial^2 W}{\partial q_5 \partial q_4}\right) & \dfrac{l}{4}\left(\dfrac{\partial^2 W}{\partial q_5 \partial T}\right) & \dfrac{l}{4}\left(\dfrac{\partial^2 W}{\partial q_5 \partial q_1}\right) \\ & \dfrac{l^2}{16}\left(\dfrac{\partial^2 W}{\partial q_4^2}\right) & \dfrac{l}{4}\left(\dfrac{\partial^2 W}{\partial q_4 \partial T}\right) & \dfrac{l}{4}\left(\dfrac{\partial^2 W}{\partial q_4 \partial q_1}\right) \\ & & \dfrac{\partial^2 W}{\partial T^2} & \dfrac{\partial^2 W}{\partial q_1 \partial T} \\ & \text{sym.} & & \dfrac{\partial^2 W}{\partial q_1^2} \end{bmatrix} + \qquad (10)$$

$$\text{diag}\left(\dfrac{l^3}{48EI}, \dfrac{l^3}{48EI}, \dfrac{l}{2GJ_p}, \dfrac{l}{2AE}\right).$$

Eq. (10) can be written in compact form as

$$[c]_l = [G_1][\Delta c_{ij}][G_2] + [C_s], \qquad (11)$$

where

$$\Delta c_{ij} = \dfrac{\partial^2 W}{\partial q_i \partial q_j}, \quad [G_1] = \begin{bmatrix} \dfrac{l}{4}, \dfrac{l}{4}, 1, 1 \end{bmatrix}, \quad [G_2] = \begin{bmatrix} \dfrac{l}{4}, \dfrac{l}{4}, 1, 1 \end{bmatrix}, \qquad (12)$$

$$[C_s] = \text{diag}\left(\dfrac{l^3}{48EI}, \dfrac{l^3}{48EI}, \dfrac{l}{2GJ_p}, \dfrac{l}{2AE}\right).$$

It is obvious that (11) leads to a symmetric matrix whereas some researchers reported non symmetric form for it [8,9]. Using to (11), one is able to determine local flexibility of a cracked shaft if additional strain energy due to crack can be determined. This is feasible using concepts of fracture mechanics. According to [7], strain energy due to crack is given by

$$W = \int_{A'} \dfrac{1}{E'} K_I^2 \, dA' + \int_A \dfrac{1}{E'}(1+\nu)K_{III}^2 \, dA, \qquad (13)$$

where for tearing mode (K_{III}) the total surface of crack, i.e. A, is used for integration while for the opening mode (K_I) part of the crack surface which remains open during the rotation, A', should be taking into account [7].

Crack surface at θ orientation is shown in Fig. 7. Using this figure (Fig. 7) the stress intensity factors for a slant crack at θ angle are given by
For q_1,

$$K_I^1 = \dfrac{q_1}{\pi R^2} \sin^2(\theta)\sqrt{\pi\gamma}F_1, \quad K_{III}^1 = \dfrac{q_1}{\pi R^2} \sin(\theta \cos(\theta)\sqrt{\pi\gamma}F_{III}. \qquad (14)$$

For q_4,

$$K_I^4 = \dfrac{4q_4 x_0}{\pi R^4} \sin^2(\theta)\sqrt{\pi\gamma}F_1, \quad K_{III}^4 = \dfrac{2q_4 x_0}{\pi R^4} \sin(\theta)\cos(\theta)\sqrt{\pi\gamma}F_{III}. \qquad (15)$$

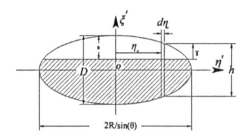

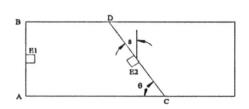

Fig. 7. The crack surface at orientation angle θ Fig. 8. The location of elements E1 and E2 used in Mohr circle

For q_5,

$$K_I^5 = \frac{4q_5\sqrt{R^2 - x_0^2}}{\pi R^4}\sin^2(\theta)\sqrt{\pi\gamma}F_2, \quad K_{III}^5 = \frac{2q_5}{\pi R^4}\sin(2\theta)\sqrt{\pi\gamma}F_{III}. \tag{16}$$

And finally for T,

$$K_I^T = \frac{2T\sqrt{R^2 - x_0^2}}{\pi R^4}\sin(2\theta)\sqrt{\pi\gamma}F_2, \tag{17}$$

$$K_{III}^T = \frac{-2T\sqrt{R^2 - x_0^2}}{\pi R^4}\cos(2\theta)\sqrt{\pi\gamma}F_{III}, \tag{18}$$

where

$$x_0 = \eta_0\sin(\theta). \tag{19}$$

According to [11]

$$F_1 = \sqrt{\frac{\tan(\lambda)}{\lambda}}\left[0.752 + 1.01\frac{\gamma}{\sqrt{R^2 - x_o^2}} + 0.37(1 - \sin(\lambda))^3\right]\frac{1}{\cos(\lambda)}, \tag{20a}$$

$$F_2 = \sqrt{\frac{\tan(\lambda)}{\lambda}}\left[0.923 + 0.199(1 - \sin(\lambda))^4\right]\frac{1}{\cos(\lambda)}, \tag{20b}$$

$$F_{III} = \sqrt{\frac{\tan(\lambda)}{\lambda}}, \quad \lambda = \frac{\pi\gamma}{4\sqrt{R^2 - x_o^2}}. \tag{20c}$$

Therefore, the total strain density functions are

$$K_I = \left(\frac{q_1}{\pi R^2}\sin^2(\theta)F_1 + \frac{4q_4x_0}{\pi R^4}\sin^2(\theta)F_1 + \frac{4q_5\sqrt{R^2 - x_0^2}}{\pi R^4}\sin^2(\theta)F_2 + \right.$$

$$\left. \frac{2T\sqrt{R^2 - x_0^2}}{\pi R^4}\sin(2\theta)F_2\right)\sqrt{\pi\gamma}, \tag{21}$$

$$K_{III} = \left(\frac{q_1}{\pi R^2}\sin(\theta)\cos(\theta) + \frac{2q_5}{\pi R^4}\sin(2\theta) - \frac{2T\sqrt{R^2 - x_0^2}}{\pi R^4}\cos(2\theta)\right)\sqrt{\pi\gamma}F_{III}. \tag{22}$$

Eq. (18) has appeared in different forms in literature [1,2]. Here, we have used Mohr circle to extract its correct form. To show this, let consider two elements E1 and E2 as depicted in Fig. 8. Element E1 coincides with element E2 after rotating it β degree counterclockwise where

$$\beta = \frac{\pi}{2} - \theta. \tag{23}$$

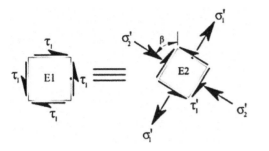

Fig. 9. The position of elements E1 and E2 after rotation β (CCW)

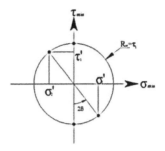

Fig. 10. The Mohr circle (center σ_M and radius R_M)

It is obvious that in relation (18), the stress intensity factor is due to the torsional moment T. Torsional moment T creates shear stress τ_1 equal to

$$\tau_1 = \frac{T\sqrt{R^2 - x_0^2}}{(\pi R^4/2)}. \tag{24}$$

Shear stress τ_1 on element E1 and stresses on element E2 are shown in Fig. 9.

From equalizing these to configurations, the center σ_M and radius R_M of the Mohr circle are zero and τ_1 respectively so according to Fig. 10, one can obtain

$$\sigma_1' = \tau_1 \sin(2\beta) = \tau_1 \sin(\pi - 2\theta) = \tau_1 \sin(2\theta), \tag{25}$$
$$\sigma_2' = -\tau_1 \sin(2\theta), \tag{26}$$
$$\tau_1' = \tau_1 \cos(2\beta) = -\tau_1 \cos(2\theta). \tag{27}$$

Thus, the correct form of the stress intensity factor for the third mode caused by T is given by

$$K_{III}^T = \frac{2T\sqrt{R^2 - x_0^2}}{\pi R^4}\cos(2\beta)\sqrt{\pi\gamma}F_{III} = \frac{-2T\sqrt{R^2 - x_0^2}}{\pi R^4}\cos(2\theta)\sqrt{\pi\gamma}F_{III}. \tag{28}$$

After calculating the local flexibility of a cracked rotor, local stiffness of the system can be calculated

$$[K]_l = [c]_l^{-1}. \tag{29}$$

The global stiffness matrix in the inertia coordinate system is

$$[K]_g = [H]^{-1}[K]_l[H], \tag{30}$$

where

$$[H] = \begin{bmatrix} \cos(\Phi) & \sin(\Phi) & 0 & 0 \\ -\sin(\Phi) & \cos(\Phi) & 0 & 0 \\ 0 & 0 & 1 & 0 \\ 0 & 0 & 0 & 1 \end{bmatrix}, \quad \Phi = \Omega t + \alpha. \tag{31}$$

For a 9.5 mm diameter shaft with a crack depth equal to its radius, the elements of the local flexibility matrix are evaluated for different crack orientations from $30°$ to $90°$ (transverse crack). In Fig. 11 the variations of these flexibilities versus CCLP[1] [2] and crack orientations ($30°$, $45°$, $60°$, $70°$, $80°$ and $90°$) are shown. It should be mentioned that the crack tip is divided into

---

[1]Crack closure line position

360 point for using CCLP method. It means that for CCLP = 180 the crack is fully open and CCLP = 0 or 360 exhibits a fully closed crack.

According to Fig. 11, increasing the value of crack angle increases the maximum value of $c(1,1)$ and $c(2,2)$. In fact the maximum value of these coefficients occurs when the crack is fully open (i.e. CCLP = 180). When the crack is fully open, in bending, the flexibility of the transverse crack is more than that of the slant crack and flexibility is a monotonic function of the crack angle. For fully open crack $c(3,3)$ for slant crack is higher than the transverse one. When the crack is fully closed, the value of $c(1,1)$ for slant crack is higher than the transverse one. However, there is no difference between the values of $c(2,2)$ for slant crack and transverse crack (for fully closed crack).

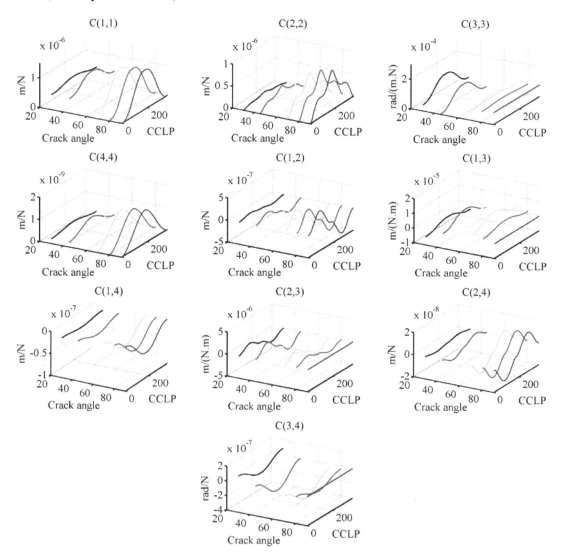

Fig. 11. Variation of the elements of local flexibility matrix versus CCLP and crack orientation from 30° to 90°

Fig. 11 shows that generally in torsion, slant crack is more critical than the transverse crack. Similar conclusion is presented in [1]. Whereas for closed crack, transverse one is flexible than slant one. It should be mentioned that for a fully closed crack, the slant crack with 45° orientation angle has no corresponding flexibility coefficient in torsion. This is due to the fact that according to relation (22), when θ is $\pi/2$, stress intensity factor in 3rd mode that caused by T will be zero. Therefore, among slant cracks with different orientations, $45°$ slant crack has the minimum value in torsion.

It should be noticed that $c(3,3)$ coefficient for a transverse crack is not sensitive to the amount of the open part of crack. In other words the value of $c(3,3)$ does not depend on the value of CCLP. $c(1,2)$ and $c(2,4)$ in CCLP $=0$, CCLP $=180$ and CCLP $=360$ are zero, but in other CCLPs are not zero.

The elements of $c(2,4)$ and $c(1,2)$ for open and closed cracks are not depended on the crack angle. In other words, $c(2,4)$ and $c(1,2)$ do not have any rule in coupling between the bending in different directions. If one considers the breathiong crack, the elements obtain their maximum at CCLP $=90$ and 270. Elements $c(4,4)$ and $c(1,4)$ have a trend similar to $c(1,1)$. For fully closed crack the value of $c(4,4)$ and $c(1,4)$ for slant crack are higher than the transverse one.

From Fig. 11 it can be seen that the coefficients $c(1,3)$ and $c(2,3)$ are zero for transverse crack. These elemnts cause coupling between torsional and transversal directions. This means that the effects of coupling for slant crack is more than the transverse one. Therefore, in general it is resonable that one expects there exist more frequencies in the spectrums of responses for the slant crack in comparison to those relate to the transverse one.

It is worth mentioning that from Fig. 11, one can observe that the maximum value of elements $c(1,3)$, $c(2,3)$ and $c(3,4)$ versus crack angle occurs at 60 degrees for open crack. However, for open crack $c(2,3)$ does not have any rule in coupling between torsional and bending vibration.

4. Vibration response of rotor system with slant crack

The parameters that are needed for solving (1)–(4) are tabulated here (Table 1).

Table 1. Solution parameters

Revolutionary speed	$\Omega = 500$ rpm	Disk mass	$m = 0.595$ kg
Torsional excitation freq.	$\omega_T = 0.6\Omega = 300$ rpm	Shaft length	$l = 0.26$ m
External torsional excitation	$M(t) = \sin(\omega_T t)$	Shaft diameter	$D = 9.5$ mm
Transversal damp coefficient	$c = 41.65$ kg/s	Disk diameter	$dp = 76$ mm
Torsional damp coefficient	$c_T = 0.009\,1$ kg $\cdot$ m^2/s	Initial phase angle	$\varphi = \pi/6$ rad
Longitudinal damp coefficient	$c_u = 146.203\,4$ kg/s	Poisson ratio	$\nu = 0.3$
Modulus of elasticity	$E = 210$ GPa	Eccentricity	$e = 0.164\,3$ mm

Solution of motion equations considering breathing model for the crack is very time consuming in comparison to open crack model. On the other hand, there are the same prominent characteristic frequencies for these two models [7]. Therefore, all calculations in this paper are about open crack model and its effects on the response of the system. Runge-Kutta method is used for solving the equations of the motion. Using this method, the response of the Jeffcott rotor with a slant crack under different crack angles is evaluated. Figs. 12–15 show the system responses for crack orientations $30°$, $45°$, $60°$ and $90°$ respectively. Theses responses are related to two transversal, one torsional and one longitudinal direction. It should be mentioned that response for other angles such as $60°$, $70°$ and $80°$ are obtained but are not presented here.

According to Fig. 12, for $30°$ slant crack, the spectrum of transversal (vertical and horizontal) responses contain Ω and 2Ω frequencies and their side bands ($\Omega \pm \omega_T$ and $2\Omega \pm \omega_T$). Fig. 13 show that there are Ω, 2Ω, $\Omega \pm \omega_T$, $2\Omega \pm \omega_T$ and 3Ω frequencies in the spectrum of

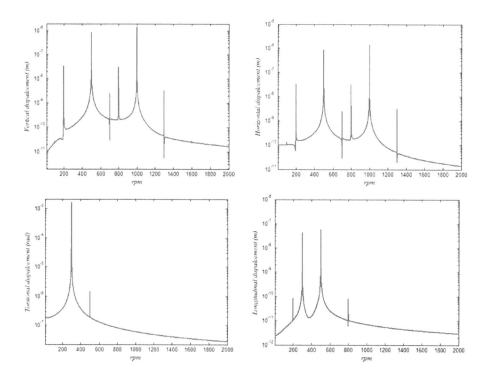

Fig. 12. Spectrum of the rotor response for 30° slant crack

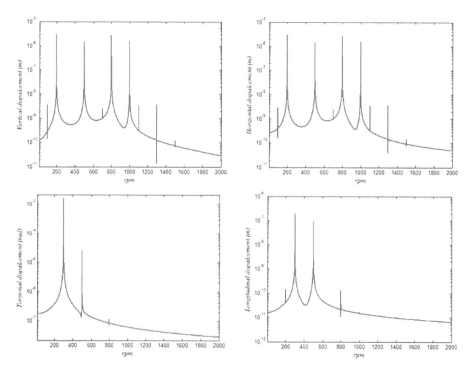

Fig. 13. Spectrum of the rotor response for 45° slant crack

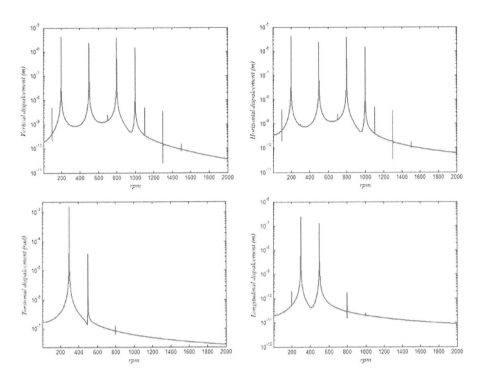

Fig. 14. Spectrum of the response for 60° slant crack

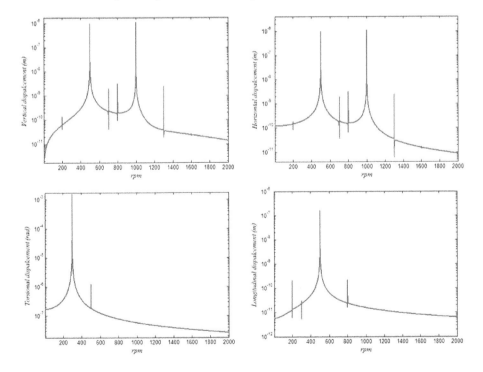

Fig. 15. Spectrum of the response for 90° slant crack

transversal responses for $45°$ slant cracks. Whereas there is only $\Omega, 2\Omega, \Omega \pm \omega_T$ and $2\Omega \pm \omega_T$ frequencies in the spectrum of transversal response for transverse crack. It is considerable that general schemas of spectrums of $30°$ and $90°$ slant cracks are almost the same and have sensible difference in compare to other spectrums. It can be explained that there are three coefficients ($c(1,3)$, $c(2,3)$ and $c(3,4)$) in the flexibility matrix that can cause coupling between torsional and other directions. Among these coefficients, there are two coefficients that can cause coupling between transversal and torsional response and they are $c(1,3)$ and $c(3,4)$. On the other hand for fully open crack $c(2,3)$ is zero.

According to Fig. 11, it can be seen that $c(1,3)$, for $30°$ slant crack and $90°$ slant crack are equal to each other and both of them are zero; therefor spectrums of transversal responses for both of them ($30°$ and $90°$ slant crack) have the same schema. It should be noticed that existence of combined frequencies such as $\Omega \pm \omega_T$ and $2\Omega \pm \omega_T$ in the spectrum of transversal response (for $30°$ and $90°$) are due to coupling phenomena that caused by eccentricity (that can be seen in the equations of the motion).

The spectrums of torsional responses (Figs. 12–15) for all crack angles contain Ω and ω_T frequencies. Also all the mentioned spectrums have $\Omega + \omega_T$ frequency except in spectrums with $30°$ and $90°$ slant cracks. Existence of Ω, ω_T and $\Omega \pm \omega_T$ frequencies in the spectrums of longitudinal responses (Figs. 12–15) is obvious. In all of these spectrums, the 2Ω frequency can be detected. However, as the peaks are very small in $30°$ and $90°$ slant crack, they are not easily detectable. It is clear that the amplitude of frequency response functions for different crack angles are not equal.

Fig. 16 compares these amplitudes at the prominent frequencies ($\Omega, 2\Omega, \Omega\pm\omega_T$ and $2\Omega\pm\omega_T$ for transversal, Ω and ω_T for torsional and longitudinal spectrums [7]).

According to Fig. 16, in Ω frequency, when the crack angle increases from $30°$ to $60°$, the amplitude of transversal responses increase to maximum, then increasing in the crack angle from $80°$ to $90°$ increases the amplitude. In 2Ω frequency, increasing in the crack angle from $30°$ to $45°$, increases the value of amplitude of the transversal responses and then from $45°$ to $90°$ the mentioned amplitude decreases. The $\Omega \pm \omega_T$ frequencies in the transversal responses have the same variations versus crack angles. In these frequencies, any increase in the crack orientations from $30°$ to $60°$, increases the amplitude of transversal responses and any increase in the crack angle from $60°$ to $90°$ decreases the amplitude of them. In Ω frequency the amplitude of torsional responses increases when the crack angle increases from $30°$ to $60°$. Whereas from $60°$ to $90°$, any increase in the crack angle, decreases the amplitude. In these spectrums and for ω_T frequency, any increase from $30°$ to $90°$ increases amplitude. In Ω frequency, any increase in the crack angle from $30°$ to $90°$, increases the amplitude of longitudinal responses. However in these spectrums and for ω_T frequency, any increase in crack angle from $30°$ to $60°$, increases the amplitude of these responses and for crack angles between $60°$ to $90°$ decreases them.

5. Conclusions

In this paper the dynamic behavior of a Jeffcott rotor system with a slant crack under arbitrary crack orientations is investigated. Using concepts of fracture mechanics, flexibility matrix and subsequently stiffness matrix of the system are evaluated and the influence of crack orientations on the flexibility coefficients is investigated. In this paper a symmetric relation for global stiffness matrix is presented and proved; whereas there are some literatures that reported non-symmetrical form for this matrix. It is shown that for fully open crack $c(1,1)$, $c(2,2)$, $c(4,4)$, and $c(1,4)$ coefficients are more for transverse crack rather than slant crack. For slant crack,

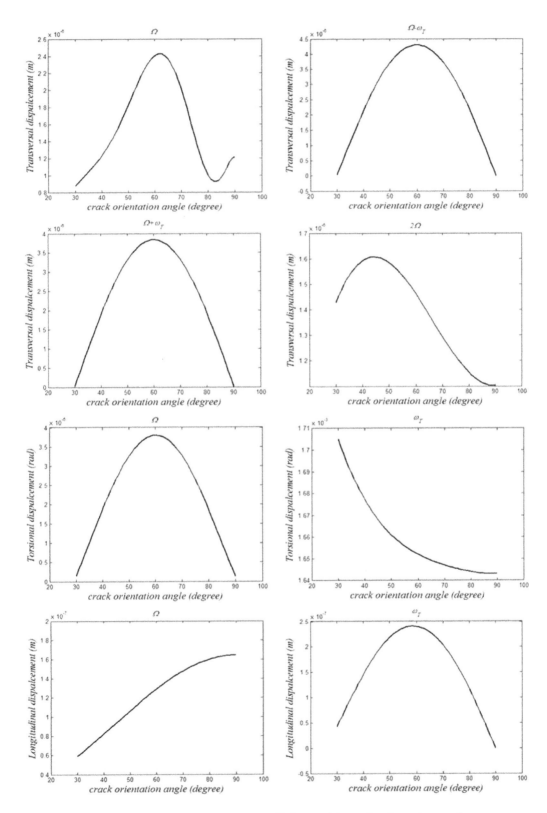

Fig. 16. Variation of flexibility coefficients with crack orientation angle

the more crack angle is, the more $c(3,3)$ coefficient will be. However for $60°$ slant crack $c(3,4)$ and $c(1,3)$ will be max and it shows that for $60°$ slant crack, the stiffness coupling between torsional direction and other directions increases. Therefor in the transversal response and in $\Omega \pm \omega_T$ frequencies, there is a maximum in the amplitude of the spectrum for $60°$ slant crack. In similar, there is a maximum in the amplitude of longitudinal response at $60°$ crack angle, because there is a maximum for $c(3,4)$ coefficient in this angle.

Also It is shown that the amplitude of transversal response in the Ω, $\Omega \pm \omega_T$, and 2Ω frequencies will be maximum at $60°$, $60°$ and $45°$ crack angles respectively. For $60°$ slant crack, the amplitude of torsional response in Ω frequency and the amplitude of longitudinal response in ω_T frequency are maximum.

Appendix A

Considering (9) and using chain rule we have

$$
\frac{\partial^2 W}{\partial F_x^2} = \frac{\partial}{\partial F_x}\left(\frac{\partial W}{\partial q_5} \cdot \frac{\partial q_5}{\partial F_x}\right) = \frac{\partial^2 W}{\partial q_5^2}\cdot\left(\frac{\partial q_5}{\partial F_x}\right)^2 + \frac{\partial W}{\partial q_5}\cdot\frac{\partial^2 q_5}{\partial F_x^2} = \tag{A-1}
$$
$$
\frac{\partial^2 W}{\partial q_5^2}\cdot\left(\frac{l}{4}\right)^2 = \frac{l^2}{16}\left(\frac{\partial^2 W}{\partial q_5^2}\right),
$$
$$
\frac{\partial^2 W}{\partial F_y^2} = \frac{\partial}{\partial F_y}\left(\frac{\partial W}{\partial q_4} \cdot \frac{\partial q_4}{\partial F_y}\right) = \frac{\partial^2 W}{\partial q_4^2}\cdot\left(\frac{\partial q_4}{\partial F_y}\right)^2 + \frac{\partial W}{\partial q_4}\cdot\frac{\partial^2 q_4}{\partial F_y^2} = \frac{l^2}{16}\left(\frac{\partial^2 W}{\partial q_4^2}\right),
$$
$$
\frac{\partial^2 W}{\partial F_z^2} = \frac{\partial}{\partial F_z}\left(\frac{\partial W}{\partial q_1} \cdot \frac{\partial q_1}{\partial F_z}\right) = \frac{\partial^2 W}{\partial q_1^2}\cdot\left(\frac{\partial q_1}{\partial F_z}\right)^2 + \frac{\partial W}{\partial q_1}\cdot\frac{\partial^2 q_1}{\partial F_z^2} = \frac{\partial^2 W}{\partial q_1^2},
$$
$$
\frac{\partial^2 W}{\partial T^2} = \frac{\partial^2 W}{\partial T^2}
$$

and

$$
\frac{\partial^2 W}{\partial F_x \partial F_y} = \frac{\partial}{\partial F_x}\left(\frac{\partial W}{\partial q_4} \cdot \frac{\partial q_4}{\partial F_y}\right) = \frac{\partial^2 W}{\partial q_5 \partial q_4}\cdot\frac{\partial q_5}{\partial F_x}\cdot\frac{\partial q_4}{\partial F_y} + \frac{\partial W}{\partial q_4}\cdot\frac{\partial^2 q_4}{\partial F_x \partial F_y} = \tag{A-2}
$$
$$
\left(\frac{l}{4}\right)\left(\frac{l}{4}\right)\left(\frac{\partial^2 W}{\partial q_5 \partial q_4}\right) = \frac{l^2}{16}\left(\frac{\partial^2 W}{\partial q_5 \partial q_4}\right),
$$
$$
\frac{\partial^2 W}{\partial F_x \partial F_z} = \frac{\partial}{\partial F_x}\left(\frac{\partial W}{\partial q_1} \cdot \frac{\partial q_1}{\partial F_z}\right) = \frac{\partial^2 W}{\partial q_5 \partial q_1}\cdot\frac{\partial q_5}{\partial F_x}\cdot\frac{\partial q_1}{\partial F_z} + \frac{\partial W}{\partial q_1}\cdot\frac{\partial^2 q_1}{\partial F_x \partial F_z} = \frac{l}{4}\left(\frac{\partial^2 W}{\partial q_5 \partial q_1}\right),
$$
$$
\frac{\partial^2 W}{\partial F_x \partial T} = \frac{\partial}{\partial F_x}\left(\frac{\partial W}{\partial T}\right) = \frac{\partial^2 W}{\partial q_5 \partial T}\cdot\frac{\partial q_5}{\partial F_x} = \frac{l}{4}\left(\frac{\partial^2 W}{\partial q_5 \partial T}\right),
$$
$$
\frac{\partial^2 W}{\partial F_y \partial F_z} = \frac{\partial}{\partial F_y}\left(\frac{\partial W}{\partial q_1} \cdot \frac{\partial q_1}{\partial F_z}\right) = \frac{\partial^2 W}{\partial q_4 \partial q_1}\cdot\frac{\partial q_4}{\partial F_y}\cdot\frac{\partial q_1}{\partial F_z} + \frac{\partial W}{\partial q_1}\cdot\frac{\partial^2 q_1}{\partial F_y \partial F_z} = \frac{l}{4}\left(\frac{\partial^2 W}{\partial q_4 \partial q_1}\right),
$$
$$
\frac{\partial^2 W}{\partial F_y \partial T} = \frac{\partial}{\partial F_y}\left(\frac{\partial W}{\partial T}\right) = \frac{\partial^2 W}{\partial q_4 \partial T}\cdot\frac{\partial q_4}{\partial F_y} = \frac{l}{4}\left(\frac{\partial^2 W}{\partial q_4 \partial T}\right),
$$
$$
\frac{\partial^2 W}{\partial F_z \partial T} = \frac{\partial}{\partial F_z}\left(\frac{\partial W}{\partial T}\right) = \frac{\partial^2 W}{\partial q_1 \partial T}\cdot\frac{\partial q_1}{\partial F_z} = \frac{\partial^2 W}{\partial q_1 \partial T}.
$$

Therefore using (A–1), (A–2) and (8), (14) is obtained.

Appendix B: Largest influence sixty degrees crack orientation

It is noticeable that the variation in the system response is due to the elements of the flexibility matrix. Therefore, any change in the system response is directly related to the flexibility matrix elements. In the following, we will show that, for instance, the maximum value of $c(3,4)$ will happen in an angle of approximately 60 degrees.

Let us consider the following figure:

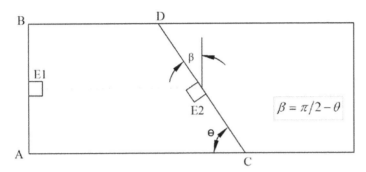

Fig. A–1. E1 and E2 elements for using in Mohr circle

According to this figure, it is evident that if element E1 is just under an axial load, the element E2 with $\beta = \theta = \pi/4$ will experience the maximum shear stress. Also, if the element E1 is under pure shear, the element E2 with $\beta = \theta = \pi/4$ is under axial stress only. However, for cases in which the element is under mixed loads, the maximum shear stress will not happen at an angle of 45 degrees.

Assume that element E1 is under pure shear stress. According to Fig. A–2, the tension and shear stresses for an element after rotation of β in CCW direction, is expressed as:

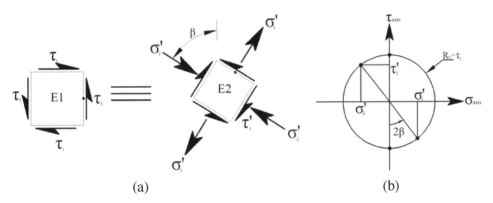

(a)   (b)

Fig. A–2. a) E1 element and E2 element after β rotation (CCW), b) Mohr circle with center σ_M and radius R_M

$$\sigma'_1 = \tau_1 \sin(2\beta) = \tau_1 \sin(\pi - 2\theta) = \tau_1 \sin(2\theta), \qquad \text{(A–3)}$$
$$\tau'_1 = \tau_1 \cos(2\beta) = -\tau_1 \cos(2\theta). \qquad \text{(A–4)}$$

In the above equations,

$$\tau_1 = \frac{TR}{(\pi R^4/2)} = \frac{2T}{\pi R^3}. \qquad \text{(A–5)}$$

In a similar way, consider the element E1 which is under uniaxial tension only.

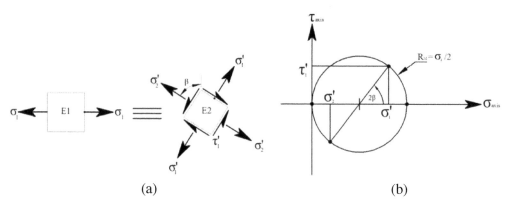

Fig. A–3. a) E1 element and E2 element after β rotation (CCW), b) Mohr circle with center σ_M and radius R_M

The shear and axial stresses for a rotated element with angle β in CCW direction can be expressed as:

$$\sigma'_1 = \frac{\sigma_1}{2} + \frac{\sigma_1}{2}\cos(2\beta) = \frac{\sigma_1}{2}(1 + \cos(\pi - 2\theta)) = \sigma_1 \sin^2(\theta), \tag{A–6}$$

$$\tau'_1 = \frac{\sigma_1}{2}\cos(2\beta) = \frac{\sigma_1}{2}\sin(2\theta) = \sigma_1 \sin(\theta)\cos(\theta). \tag{A–7}$$

In the above equations σ_1 is:

$$\sigma_1 = \frac{q_1}{\pi R^2}. \tag{A–8}$$

Therefore, the maximum tension and shear stresses for an element under combined loading is:

$$\sigma_{\max} = \frac{2T}{\pi R^3}\sin(2\theta) + \frac{q_1}{\pi R^2}\sin^2(\theta), \tag{A–9}$$

$$\tau_{\max} = -\frac{2T}{\pi R^3}\cos(2\theta) + \frac{q_1}{\pi R^2}\sin(\theta)\cos(\theta). \tag{A–10}$$

According to Eqs. (21) and (22) in calculating the elements of flexibility matrix, the squared of $\sigma_{\max}$ and $\tau_{\max}$ need to be calculated:

$$\sigma^2_{\max} = \left(\frac{q_1}{\pi R^2}\sin^2(\theta) + \frac{2T}{\pi R^3}\sin(2\theta) + \ldots\right)^2, \tag{A–11}$$

$$\tau^2_{\max} = \left(-\frac{2T}{\pi R^3}\cos(2\theta) + \frac{q_1}{\pi R^2}\sin(\theta)\cos(\theta) + \ldots\right)^2. \tag{A–12}$$

To determine the element $c(3,4)$, one should compute the second derivatives of of $\sigma^2_{\max}$ and $\tau^2_{\max}$ with respect to q_1 and T

$$\frac{\partial^2 \sigma^2_{\max}}{\partial q_1 \partial T} = \frac{\partial^2}{\partial q_1 \partial T}\left(\frac{q_1}{\pi R^2}\sin^2(\theta) + \frac{2T}{\pi R^3}\sin(2\theta) + \ldots\right)^2 = \tag{A–13}$$

$$2\left(\frac{2}{\pi R^3}\right)\left(\frac{1}{\pi R^2}\right)\sin^2(\theta)\sin(2\theta),$$

$$\frac{\partial^2 \tau^2_{\max}}{\partial q_1 \partial T} = \frac{\partial^2}{\partial q_1 \partial T}\left(-\frac{2T}{\pi R^3}\cos(2\theta) + \frac{q_1}{\pi R^2}\sin(\theta)\cos(\theta) + \ldots\right)^2 = \tag{A–14}$$

$$-2\left(\frac{2}{\pi R^3}\right)\left(\frac{1}{\pi R^2}\right)\sin(\theta)\cos(\theta)\cos(2\theta).$$

Therefore, the element $c(3, 4)$ of the flexibility matrix is proportional to function H in the following equation:

$$H(\theta) = \frac{\partial^2 \sigma_{\max}^2}{\partial q_1 \partial T} + \frac{\partial^2 \tau_{\max}^2}{\partial q_1 \partial T} = 2 \left(\frac{2}{\pi R^3}\right) \left(\frac{1}{\pi R^2}\right) F(\theta), \tag{A--15}$$

$$F(\theta) = \sin^2(\theta) \sin(2\theta) - \sin(\theta) \cos(\theta) \cos(2\theta). \tag{A--16}$$

The variation of $F(\theta)$ as a function of θ is shown in the following figure. The plot shows that the maximum occurs in an angle of 60 approximately.

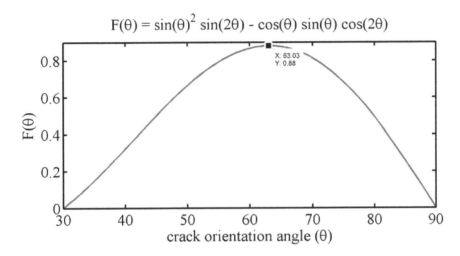

Fig. A--4. The variation of $F(\theta)$ versus of θ

References

[1] Darpe, A. K., Dynamics of a Jeffcott rotor with slant crack, Journal of Sound and Vibration 303 (2007) 1–28.

[2] Darpe, A. K., Gupta, K., Chawla, A., Coupled bending, longitudinal and torsional vibrations of a cracked rotor, Journal of Sound and Vibration 269 (2004) 33–60.

[3] Darpe, A. K., Gupta, K., Chawla, A., Dynamics of a two cracked rotor, Journal of Sound and Vibration 259 (3) (2003) 649–675.

[4] Darpe, A. K., Gupta, K., Chawla, A., Transient response and breathing behaviour of a cracked Jeffcott rotor, Journal of Sound and Vibration 272 (2004) 207–243.

[5] Ichimonji, M., Watanabe, S., The dynamics of a rotor system with a shaft having a slant crack, JSME, International Journal, Series III 31 (4) (1988) 712–718.

[6] Ichimonji, M., Kazao, S., Watanabe, S., Nonaka, S., The dynamics of a rotor system with a slant crack under torsional vibration, Nonlinear and Stochastic Dynamics 78 (1994) 81–89.

[7] Lin, Y., Chu, F., The dynamic behaviour of a rotor system with a slant crack on the shaft, Mechanical System and Signal Processing 24 (2010) 522–545.

[8] Papadopoulos, C. A., The strain energy release approach for modeling cracks in rotors: a state of the art review, Mechanical Systems and Signal Processing 22 (2008) 763–789.

[9] Papadopoulos, C. A., Dimarogonas, A. D., Stability of cracked rotors in the coupled vibration mode, Journal of Vibration Acoustics Stress and Reliability in Design-Transactions of the ASME 110 (3) (1988) 356–359.

[10] Sekhar, A. S., Mohanty, A. R., Prabhakar, S., Vibrations of cracked rotor system: transverse crack versus slant crack, Journal of Sound and Vibration 279 (2005) 1 203–1 217.

[11] Tada, H., Paris, P. C., Irwin, G. R., The stress analysis of cracks, Handbook, third edition, Professional Engineering Publishing, 2000.

14

Design of a large deformable obstacle for railway crash simulations according to the applicable standard

S. Špirk[a,*], V. Kemka[a], M. Kepka[a], Z. Malkovský[b]

[a]*UWB in Pilsen, Faculty of Mechanical Engineering, Research Centre of Rail Vehicles, Univerzitní 22, 306 14 Plzeň, Czech Republic*
[b]*VÚKV, a.s., Bucharova 1314/8, 158 00 Praha, Czech Republic*

Abstract

This article discusses the design of a deformable obstacle to be used in simulated rail and road vehicle collisions as prescribed by scenario 3 specified by standard ČSN EN 15227. To approve a vehicle in accordance with this standard, it is necessary to carry out numeric simulations of its collision with a large obstacle, following the standard's specification. A simulated impact of a rigid ball into the obstacle is used to calibrate the obstacle's properties, and the standard specifies limit deformation characteristics for that collision. The closer are the deformation characteristics observed in the test to the limit characteristics prescribed by the standard, the more favorable results can be expected when using the obstacle in actual numeric simulations of frontal impacts of rail vehicles. There are multiple ways to design a FEM model of the obstacle; this article discusses one of those. It shows that given a suitable definition of material properties, this particular approach yields quite favorable deformation characteristics.

Keywords: rail vehicles, railway, crash, crashworthiness, collision, FEM, simulation

1. Introduction

Since the introduction of standard ČSN EN 15227 [1], rail vehicle design has also been increasingly addressing passive safety. The main goal of this work consists in designing a deformation model of a tanker to be used in numeric simulations of frontal impacts of rail vehicles into large road vehicles on level crossings. The article discusses the topic of passive safety in rail vehicles, and focuses on the third collision scenario designing a FEM model of the obstacle. A novel approach to obstacle design consists in using a single type of isotropic material, which yields more favorable results than sheet metal shell obstacles used in simulating the 109E locomotive. The 109E is marking for the new Multi-system Universal locomotive developed by ŠKODA TRANSPORTATION [7]. It is reasonable to expect more favorable results when running numeric simulations according the third scenario. These simulations have been discussed for instance in *"Development of the crashworthy locomotive platform TRAXX"* [5] or *"Structural Crashworthiness standards Comparison: Grade-Crossing Collision Scenarios"* [6]. Fine details of obstacle calibration constitute the know-how for most designers, which is why that topic is not discussed in detail.

---

*Corresponding author. e-mail: spirks@fst.zcu.cz.

2. Passive safety in rail vehicles

Essential passive safety requirements applicable to rail vehicles are specified by standard ČSN EN 15227 (Railway applications – Crashworthiness requirements for railway vehicle bodies) [1]. "Measures outlined in this standard constitute the final stage of protection that does not take effect before all chances of accident prevention have failed." [2] It specifies risks and essential requirements that must be addressed in design so that the rail vehicle in question can withstand a crash. These conditions reflect typical accident scenarios. Four main pillars of passive safety are defined to meet appropriate requirements. Rail vehicles are classified into four categories based on their construction. Simply put, the first category includes train units and carriages, the second one contains subway vehicles, and the final two categories are for trams. Four crash scenarios are identified in parallel with these categories:

1. a front end impact between two identical train units,

2. a front end impact with a different type of railway vehicle,

3. train unit front end impact with a large road vehicle on a level crossing,

4. train unit impact into low obstacle (e.g. car on a level crossing, animal, rubbish).

By combining these categories and scenarios one receives a set of scenarios for the given vehicle category. In general, we can name five essential guidelines that must be followed when designing a rail vehicle:

1. reduce the risk of overriding,

2. absorb collision energy in a controlled manner,

3. maintain survival space and structural integrity of the occupied areas,

4. limit the deceleration,

5. reduce the risk of derailment and limit the consequences of hitting a track obstruction.

Passive safety requirements apply to a train as a whole, but the complex behavior of a train is virtually impossible to test; the actual train performance with respect to safety requirements is therefore verified by simulating reference collision scenarios. Numerical simulation can only be used to make reasonably precise predictions for structures subject to minor (limited) deformation. Structures undergoing major deformation must be tested full-scale to validate the output of the numerical simulation. For new designs at least one full-scale test of a part of the vehicle's structure is mandatory. The Approval Program employs a combination of testing (experiment) and numerical simulation (combination). The Validation Program is structured to verify that test results match simulation outcome. Validation also involves a Technical Report addressing all stages of the approval process, from material properties to component testing and overall simulation of the vehicle as a whole. In the end, behavior of the structure is assessed per individual criteria, referring to technical diagrams and reports. The ability to withstand impact is evaluated in three stages. At first, energy absorption properties of components and deformation zones are evaluated, then the numerical model is calibrated, and finally, simulations are performed for individual applicable collision scenarios. Specialized software designed to simulate the dynamics of high-speed collisions is used for modeling. A methodology should be employed for gradual verification of numerical models calibrated for various sub-sets of the whole train. It is necessary to pay attention when creating the model mesh and applying various finite element characteristics. Acceptance criteria are clearly specified by the standard. Numerical simulations of individual collision scenarios must use a numerical model corresponding precisely with the actual geometry of the structure, so that its compliance with overall requirements detailed in technical documentation can be verified.

3. Task details

3.1. Task analysis

The main goal consists in designing a deformable model of a tanker that meets the requirements of standard ČSN EN 15227. The resulting model is intended for use in simulating a collision as per scenario No. 3 — a train crashing into a large road vehicle on a level crossing. Being intended solely for numerical simulation, the model is not required to use realistic materials. Deformation characteristics of the obstacle are defined by the standard, addressing the essential properties influencing the simulated impact. This means that there is a minimum limit curve for the deformation characteristics. Obviously, the softer the characteristics, the more favorable are the impacts on the structure of the rail vehicle. Thus the key problem when designing a deformable obstacle model lies with specifying material properties.

3.2. Requirements for the deformable obstacle

Mandatory requirements for the numerical model of the large deformable obstacle are explicitly specified by the standard. The dimensions of the obstacle are shown in the diagram below.

The prescribed weight of the obstacle is 15 000 kg. Parts A and B may differ in weight, allowing for adjustments of the center of gravity, which must be positioned 1 750 mm above

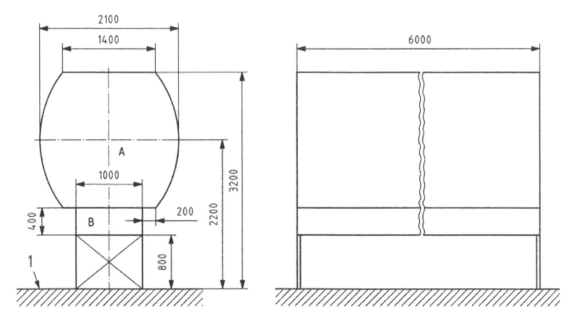

Fig. 1. Dimensions of the deformable obstacle [1]

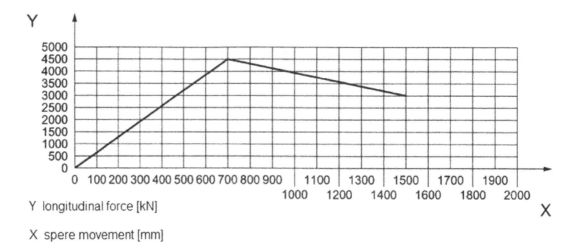

Y longitudinal force [kN]

X spere movement [mm]

Fig. 2. Stiffness of the Deformable Obstacle [1]

the top surface of the rail. Another important requirement calls for an even distribution of mass and stiffness along the length of the obstacle. The material of the obstacle must be chosen to achieve the required stiffness [4]. The dependence of force on displacement, determined by using a calibration model, must lie above the limit curve — see Fig. 2. For calibration, the obstacle must be impacted — in simulation — by a solid homogenous ball with a diameter of 3 m, weight of 50 000 kg and an initial speed of 30 m/s.

4. Numerical model

A simple numerical model is used to simulate a rigid ball impacting a deformable obstacle. The MSC Dytran software suite was used to run the simulation and MD Patran R2.1 was used for pre-processing and post-processing. The shape of the obstacle was designed by relying on the advantages offered by Unigraphics NX 6.0. The whole solution uses SI base units (m, kg, s, etc.). For a simple preview of the model please see below.

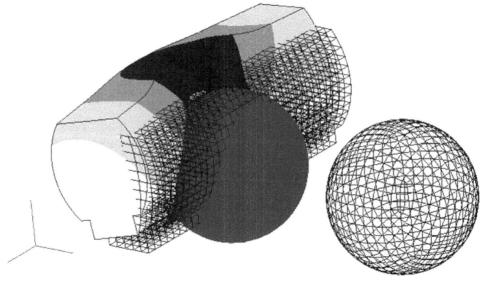

Fig. 3. Task preview

The mesh representing a ball consists of Quad4 finite elements with four vertices each. The mesh has been generated by the IsoMesh function with certain degeneration of the finite elements allowed in its properties. The mesh representing an obstacle has been generated by stretching out a 2D mesh representing the prescribed cross-section of the obstacle. A thickness of 0.001 m and an absolutely stiff material with an overall weight of 50 000 kg were assigned to finite elements forming the ball. The initial speed of the ball is specified by a vector. Since the ČSN EN 15227 standard calls explicitly for the ball's ability for displacement only in the lengthwise direction, suitable boundary values need to be applied to prevent the ball from moving in other directions. The deformation of the obstacle is not expected to be severe enough to actually result in self-contact, which is why only master-slave contact is being considered. No friction is considered as it is not required by the standard. The ball is the "master object". The number of time steps for the simulation has been determined by estimate initially and later modified to 8 000. Within those 8 000 steps the obstacle moves by approx. two meters, which is sufficient for the scenario specified by the standard. The critically important length of step 1 has been easily derived from the size of the smallest element, the Young's modulus and the density. The actual length of that step was based on the computation. The smallest step was 10 % shorter than the initial step.

$$\Delta t_{\mathrm{krit}} = \frac{L}{\sqrt{\frac{E}{\rho}}} = \frac{0.2}{\sqrt{\frac{4.8\mathrm{e}12}{416}}} = 1.86\mathrm{e} - 6, \tag{1}$$

Δt_{krit} critical time step [s],
L characteristic length [m],
E Young's modulus [Pa],
ρ mass density [kg/m^3].

5. Material

5.1. Material definition

Early simulations have shown that the required deformation characteristics can be achieved by using plastic material of type "piecewise-linear" (linear part-by-part). That was why we have chosen a constitutive model "ElasPlas (DYMAT24)" [3], which represents a material suitable for Lagrangian volume finite elements. "True Stress vs. Strain" was selected as the "Yield Model," which means — in principle — that the material is defined by the relation between true stress and true strain with respect to the initial cross-section. It is also worth remembering that although many materials may show a decreasing trend in the force/displacement chart (such as the area of significant displacement prior to structural failure in steel), the actual stress always increases. The selected material uses a set of 41 properties to specify this behavior, making sure that there are enough elements to yield a sufficiently smooth curve while keeping the demand for computing power at a reasonable minimum. Specific values are given in subsection 6.4. Besides that, the material is characterized by three values: density, Poisson number ($\mu = 0.3$) and the module of elasticity. A FEM-based tensile strength test simulation has shown that the solver in fact always derives the Young's modulus from the first position within the array, and uses the generated value in case of a discrepancy between the generated and the pre-set one. Thus, the Young's modulus value entered as a material property does not influence the results or the course of the simulation in any way.

5.2. Array definition method

The MD Patran application requires a separate array to be defined for any variable quantity. As explained in the previous paragraph, a dedicated array must be specified to describe the behavior of a plastic material. That array is generated by the Tabular Input function. "mat1" array has been defined using 41 separate stress-strain values . The application uses linear interpolation to fill in intermediate values. The CSV import feature was used to avoid the necessity of entering all 41 values manually. The simple structure of the CSV file format makes it possible to create a macro that converts CVS files used by Patran and MS Excel. Several dozen or even a few hundred simple FEM calculations must be performed to make the obstacle match the required deformation characteristics, which is why a macro was written to enable direct editing of the Patran database, avoiding the need to bring up the preprocessor prior to each run and make many unnecessary operations. Material properties are translated from Excel into Dytran input format and the computation may start immediately.

5.3. Defining material density

ČSN EN 15 227 standard specifies the required weight of the obstacle and the height of its center of gravity. To achieve both, the obstacle must consist of two materials demonstrating different densities. The first material is used at the bottom part of the obstacle, while the other type of material forms the top part. There are two formulae needed to satisfy the requirement of maintaining the total weight of 15000 kg and the position of the center of gravity at 1.75 m above the top of the rail. Both formulae already take into account the required shape of the obstacle, yielding the required density values upon adequate modification.

$$\rho_2 \cdot v_2 = m_c - \rho_1 \cdot v_1, \qquad \rho_1 \cdot y_1 \cdot v_1 + \rho_2 \cdot y_2 \cdot v_2 = T \cdot m_c, \tag{2}$$

$$\rho_2 = \frac{m_c - \rho_1 \cdot v_1}{v_2}, \qquad \rho_1 = \frac{m_c(T - y_2)}{y_1 \cdot v_1 - y_2 \cdot v_2}, \tag{3}$$

ρ_1 density of the bottom part of the obstacle [kg/m^3],
ρ_2 density of the top part of the obstacle [kg/m^3],
ν_1 volume of the bottom part of the obstacle [m^3],
ν_2 volume of the top part of the obstacle [m^3],
m_c weight of the whole obstacle [kg],
y_1 height of the bottom part's center of gravity above the top of the rail [m],
y_2 height of the top part's center of gravity above the top of the rail [m],
T height of the obstacle's center of gravity above the top of the rail [m].

5.4. Stress-strain characteristic of the material

After running many simulations of a rigid ball impacting a deformable obstacle, it became clear that a systematic approach must be taken to identifying the most suitable material. With material properties specified randomly, the deformation characteristics have never, even remotely, resembled the required curve. On top of that, even minor changes in stress-strain characteristic often resulted in order-of-magnitude changes in deformation characteristics, while other properties had no influence on the final result whatsoever. Based on experience gathered in early simulation runs, a curve was defined to depend on merely two parameters (Y, e), whose modification had an observed impact on deformation characteristics. The proposed shape of the curve builds on the assumption that there should be three distinguished stages in the evolution of stress-strain — a gradual ramp-up until the maximum level is reached, then a slow decrease and finally a stage with a very slow change. A geometric progression proved to be a very good generator of values for this purpose. The following is used as the actual stress-strain function:

$$n \in N, \quad (\sigma_n)_{n=0}^4 \in R, \quad (\varepsilon_n)_{n=0}^4 \in R, \quad e, Y \in R, \tag{4}$$

$$\varepsilon_n = n \cdot 5 \cdot 10^{-8}, \tag{5}$$

$$\sigma_0 = 0, \quad \sigma_1 = 1, \tag{6}$$

$$\forall \sigma_n \text{ where } n \in \langle 2; 12 \rangle \quad \sigma_{n+1} = [\sigma_n \cdot e] \cdot Y, \tag{7}$$

$$\forall \sigma_n \text{ where } n \in \langle 13; 23 \rangle \quad \sigma_{n+1} = [\sigma_n \cdot (1 + e^{25-n})] \cdot Y, \tag{8}$$

$$\forall \sigma_n \text{ where } n \in \langle 24; 40 \rangle \quad \sigma_{n+1} = [\sigma_n + 0.1] \cdot Y. \tag{9}$$

The fact that the actual stress-strain curve is now governed by no more than two parameters makes it possible to identify a type of material whose deformation characteristics match requirements. A careful evaluation of the stress-strain curve (see Fig. 5) reveals that the Young's modulus — and with it the stiffness of the material — increases until it reaches the maximum value, and then slowly deteriorates until reaching the minimum value, which it maintains. This gradual development ensures similarly gradual deformation characteristics. The "e" parameter determines the slope of the curve, and thus the maximum Young's modulus value. The "Y" parameter determines the limit value of stress wherein the curve stops dropping and the material stops offering much resistance to the penetrating body. No sophisticated algorithm was used to find suboptimal values Y and e. They were identified through a simple process explained in Fig. 4. At first, Y and e pairs were chosen, spread evenly across the whole range of reasonable

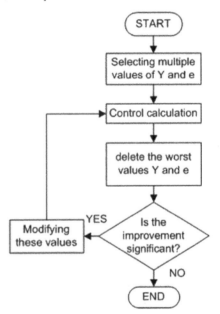

Fig. 4. Flow Chart detailing the identification of Y and e values

results expected. A test run was performed for each pair (see Fig. 4). Values that failed to produce expected results were deleted and new values were generated by modifying the remaining data points to maintain the full size of the data set. The new e and Y values where generated by enlargement and by reduction of suitable old e and Y values. Another test run was performed and the whole process repeated until there was no significant progression (approximately under 5 %) towards individual characteristic anymore. In the end, all results were evaluated.

There were three materials matching the required characteristics (Fig. 2) rather closely. Average force was used for their exact evaluation. All values of force acting between the two bodies were summed up across all steps and then divided by the number of steps. Material with the lowest average value was selected as the most suitable one. Although the ramp-up of force acting between the two bodies is rather steep, the maximum lies rather close to the required peak, and the remaining section of the curve matches the prescribed characteristics quite well. These characteristics were achieved with $e = 2.5$ and $Y = 13.3$.

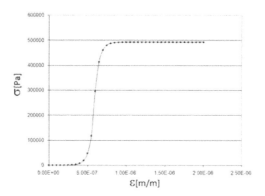

Fig. 5. The stress-strain curve

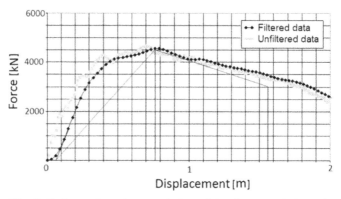

Fig. 6. Deformation characteristics of the fine-tuned obstacle

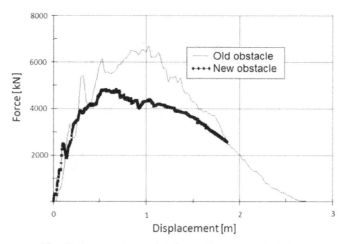

Fig. 7. Comparison of deformation characteristics

6. Evaluation of results

There are two ways of reading the development of force acting on the ball. It is possible either to analyze forces acting on the contact surface, or to derive the force from the deceleration of the impacting body. Both forces should according to Newton's third law be equal, which also allows us to provide partial verification of the result. Since the deceleration force is easier to express in the Dytran application in dependence on the object's trajectory, it will be the one used for our purpose. As the standard permits filtering of the resulting values, the development of force and trajectory in time was filtered through a 60 Hz low-pass filter applied by a Patran postprocessor. Filtering made the resulting characteristic smoother, allowing for even better convergence to characteristics prescribed by the standard. Filtered results were then exported into Excel, which was used to express the dependence of force on displacement as shown in the figure below. The light gray curve shows unfiltered values, while the dark curve represents the result of filtration. Two straight lines indicate limits prescribed by the standard. A phenomenon worth noting is the initial gradual ramp-up caused by software-based filtration. That is why the intersection of the horizontal axis with the linear trend of the ramp-up curve is considered the actual point of origin. The graph also shows quite clearly that the unfiltered curve breaches the mandatory limit at a certain point. That must be considered an advantage since the stiffness of the obstacle will be actually lower while the standard will still be adhered to.

As a secondary goal, the project was also aiming at designing a deformable obstacle with softer characteristics. That is why a comparison between the old and the new obstacle is in order (see Fig. 7). The light curve shows the characteristics for the original (old) obstacle — one with a sheet metal shell (see subsection 7.3) that was used in modeling collisions of the 109E locomotive [7]. A lighter curve closely copying the previous one represents results achieved by employing a similar obstacle provided by the Dyna application. Deformation characteristics of the new obstacle are shown in light gray and the respective curve is the lowest one in the chart. All curves show a distinguished ramp-up of force at first. In the case of the newly designed obstacle, though, the intensity of growth gradually diminishes and an overall view reveals a significant reduction of stiffness achieved by introducing the new obstacle, whose deformation characteristics are closest to those prescribed by the standard.

7. Others deformable obstacle design alternatives

7.1. Multiple isotropic materials

Different materials can be used for different finite elements and the required deformation characteristics can be achieved by applying materials with different stiffness. On one hand, employing a sufficient number of finite elements makes it possible to match characteristics required by the standard. On the other hand, the greater number of finite elements results in greater consumption of computing power. Compared to the previous option, the complexity of this one increases proportionally to the number of materials used in the model. This makes such approach suitable for automated optimization rather than safety testing. On top of that, it may sometimes be difficult to satisfy the standard's explicit requirement for constant stiffness in certain directions.

7.2. Honeycomb

Once again the model involves just one body consisting of a single type of material — a honeycomb. Honeycombs are anisotropic materials often used in vehicles to absorb kinetic energy by deformation. Simplicity is again the main advantage of the model — most software tools make it simple to specify a body made of a honeycomb-type material. Another obvious advantage consists in deformation characteristics of honeycombs. On the other hand, material properties must be defined by setting quite a large number of parameters to achieve deformation characteristics expected by the standard.

7.3. Isotropic material with a sheet metal shell

In this case the model consists of 3D bodies covered by a 2D shell. With different materials used for the inside and for the shell the model acts as a body with a metal sheet surface. This model makes it possible to simulate a rigid object with a soft core. An obvious disadvantage is the need to set and balance the properties of two different materials.

8. Conclusion

The article gives a brief overview of provisions prescribed by ČSN EN 15227 standard, followed by a more detailed description of requirements concerning deformable obstacles. Several alternative solutions are proposed and evaluated, and the most suitable approach is identified. The article details the development of a deformable obstacle to be used in simulating the front of a rail vehicle colliding with a large road vehicle on a level crossing. The obstacle must follow requirements specified by the standard while remaining as soft as possible, since the stiffness of the obstacle can be reasonably expected to significantly influence the collision. FEM-based simulations used in designing the model are explained in detail, starting with geometry designs and ending with the processing of results. The final section deals with identifying a suitable type of material that offers acceptable deformation characteristics.

The newly proposed type of material suitable for relatively soft obstacles is a major result of our work. The peak force recorded in deformation characteristics is approximately 30 % lower compared to the obstacle used in simulating the 109E locomotive. The material was found by exploiting an obvious empiric relationship between stress and strain. This is an approach that could be further improved in the future, but it is yet unclear whether that could lead to another improvement. It is certainly advisable to try and improve the deformation characteristics in the initial ramp-up phase, which is most important for the collision itself. It might be possible to automate the optimizing process and use multiple materials.

Another tool worth mentioning is the new algorithm used to interconnect Excel with the MD Patran preprocessor and the MSC Dytran solver. These scripts make it possible to modify material properties in the solver's database through Excel and run the computation immediately.

The properties of the newly designed obstacle were tested by simulation according to Scenario 3 and the tests have shown clearly that deformation characteristics of the new obstacle are better in comparison to the sheet metal shell type. This could allowed us, for instance, to reduce the weight of a train cockpit.

References

[1] Malkovský, Z., Nevěčný, K., ČSN EN 15227 Railway applications — Crashworthiness requirements for railway vehicle bodies. 1(2008) 23–27 (in Czech).

[2] Dostál, J., Heller, P., Railway vehicles II. 1 (2009) 187–195 (in Czech).

[3] MSC.software Corporation: Introduction to MSC Dytran, 1 (2005) 86–102.

[4] Kemka, V., Analysis of the large deformable obstacle, CADAM 2009 7 (2009) 29–30.

[5] Carl, F. B., Schneider, S., Wolter, W., Development of the crashworthy locomotive platform TRAXX, Berlin 5 (2005) 42–62.

[6] Llana, P., Structural Crashworthiness Standards Comparison: Grade-Crossing Collision Scenarios, Texas, USA, 3 (2009) 119–128.

[7] Jankovec, J., Numerical simulation of high-speed locomotive collision scenarios, SKODA RESEARCH Innovation, 1 (2007) 243–250 (in Czech).

15

Investigation of the effect of controllable dampers on limit states of rotor systems

J. Zapoměl[a,*], P. Ferfecki[b], J. Liberdová[a]

[a]Department of Mechanics, VŠB-Technical University of Ostrava, 17. listopadu 15, 708 33 Ostrava-Poruba, Czech Republic

[b]Centre of Excellence IT4Innovations, VŠB-Technical University of Ostrava, 17. listopadu 15, 708 33, Ostrava-Poruba, Czech Republic

Abstract

The unbalance and time varying loading are the principal sources of lateral vibrations of rotors and of increase of forces transmitted through the coupling elements into the stationary part. These oscillations and force effects can be considerably reduced if damping devices are added to the coupling elements placed between the rotor and its casing. The theoretical studies and practical experience show that to achieve their optimum performance their damping effect must be controllable. This article focuses on investigation of influence of controlled damping in the rotor supports on its limit state of deformation, fatigue failure and on magnitude of the forces transmitted into the stationary part. The analysed system is a flexible rotor with one disc driven by an electric DC motor and loaded by the disc unbalance and by technological forces depending on the rotor angular position. In the computational model the system vibration is governed by a set of nonlinear differential equations of the first and second orders. To evaluate the fatigue failure both the flexural and torsional oscillations are taken into account. The analysis is aimed at searching for the intervals of angular speeds, at which the rotor can be operated without exceeding the limit states.

Keywords: rotors, controllable damping, limit states, fatigue failure

1. Introduction

The time varying technological forces and unbalance loading are the main sources of lateral oscillations of rotors working in industrial devices. The excessive vibration reduces their service life and increases the noise and forces transmitted from the rotor through the coupling elements into the stationary part. In addition, large amplitude of the oscillations can lead to impacts between the rotor and its casing.

The damping devices added to the coupling elements can considerably reduce the rotor lateral vibration and magnitude of the transmitted forces. The theoretical analysis carried out in this article proves that to achieve their efficient work the damping force must be controllable to be possible to adapt their performance to the current operating conditions. This can be achieved by application of classical squeeze film dampers with a conical gap and adjustable position of their inner or outer rings [9], magnetorheological squeeze film dampers [1–4], electromagnetic damping devices [14, 15] and by further means. Efficiency of the dampers is evaluated here from the view point of the limit state of deformation and fatigue failure.

In this article the basic properties of the controllable damping elements are studied by means of attenuation of lateral vibration of a flexibly supported Jeffcott rotor. Results of the analysis

---

*Corresponding author. e-mail: jaroslav.zapomel@vsb.cz.

show that adaptation of the damping effect in the coupling elements to the current running speed arrives at significant extension of the velocity intervals, in which the rotor can be operated without exceeding the limit states. Consequently, the obtained experience is utilized for analysis of a rotor driven by an electric DC motor. The rotor is flexible and is subjected to a combined bending and torque loading. The computational simulations show that control of damping in the coupling elements reduces extent of the speed interval, in which amplitude of the rotor vibration exceeds the allowed value. The fatigue failure is evaluated by means of a safety factor determined from the point of view of unlimited rotor service life for all loading regimes.

2. Study of the effect of controlled damping on lateral vibration of a flexibly supported Jeffcott rotor

The analysed system is a Jeffcott rotor [6,7] mounted in flexible supports whose damping can be controlled. The rotor is assumed to be axisymmetric, external damping of the disc is isotropic and material damping of the shaft is neglected. Behaviour of the whole system is considered to be linear. The rotor turns at constant angular speed, is loaded by its weight and is excited by the disc unbalance. A simplified scheme of the studied Jeffcott rotor is drawn in Fig. 1.

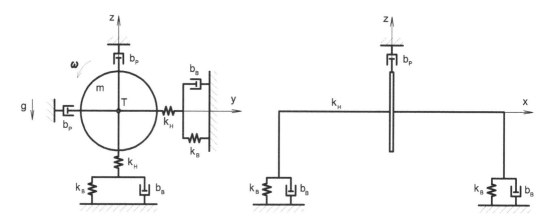

Fig. 1. Simplified scheme of the Jeffcott rotor flexibly supported

Its lateral vibration is governed by a set of four differential equations of the first and second orders

$$m\ddot{y} = -b_P\dot{y} - k_H(y - y_B) + m e_T \omega^2 \cos(\omega t + \psi_o), \tag{1}$$

$$m\ddot{z} = -b_P\dot{z} - k_H(z - z_B) + m e_T \omega^2 \sin(\omega t + \psi_o) - mg, \tag{2}$$

$$0 = -2b_B\dot{y}_B - 2k_B y_B - k_H(y_B - y), \tag{3}$$

$$0 = -2b_B\dot{z}_B - 2k_B z_B - k_H(z_B - z). \tag{4}$$

m is the disc mass, k_H, k_B are stiffnesses of the shaft and of each rotor support, b_P is the coefficient of the disc external damping, b_B is the controllable damping coefficient referred to each rotor support, e_T is eccentricity of the disc centre of mass, ω is the angular speed of the rotor rotation, g is the gravity acceleration, y, z, y_B, z_B are displacements of the disc and rotor journal centres in y and z directions respectively, t is the time, ψ_o is the phase shift of the disc centrifugal force and (˙), (¨) denote the first and second derivative with respect to time.

To solve two governing equations the y and z displacements are expressed from (3) and (4)

$$y = w_1 \dot{y}_B + w_0 y_B, \tag{5}$$

$$z = w_1 \dot{z}_B + w_0 z_B. \tag{6}$$

Coefficients w_0 and w_1 are defined by the relations

$$w_0 = 1 + \frac{2k_B}{k_H}, \tag{7}$$

$$w_1 = \frac{2b_B}{k_H}. \tag{8}$$

Consequently substitution of (5) and (6) and their first and second derivatives with respect to time into (1) and (2) arrives at a set of two linear differential equations of the third order

$$a_3 \dddot{y}_B + a_2 \ddot{y}_B + a_1 \dot{y}_B + a_0 y_B = N_C \omega^2 \cos \omega t - N_S \omega^2 \sin \omega t, \tag{9}$$

$$a_3 \dddot{z}_B + a_2 \ddot{z}_B + a_1 \dot{z}_B + a_0 z_B = N_S \omega^2 \cos \omega t + N_C \omega^2 \sin \omega t - mg. \tag{10}$$

The individual coefficients in Eqs. (9) and (10) are given by the following relationships

$$a_0 = 2k_B, \tag{11}$$

$$a_1 = b_P \left(1 + \frac{2k_B}{k_H} \right) + 2b_B, \tag{12}$$

$$a_2 = m \left(1 + \frac{2k_B}{k_H} \right) + \frac{2b_P b_B}{k_H}, \tag{13}$$

$$a_3 = m \frac{2b_B}{k_H}, \tag{14}$$

$$N_C = m e_T \cos \psi_o, \tag{15}$$

$$N_S = m e_T \sin \psi_o \tag{16}$$

and ($\dddot{}$) denotes the third derivative with respect to time.

The right hand sides of differential Eqs. (9) and (10) make it possible to assume their steady state solutions in the form of two harmonic functions of time

$$y_B = r_{BC} \cos \omega t - r_{BS} \sin \omega t, \tag{17}$$

$$z_B = r_{BS} \cos \omega t + r_{BC} \sin \omega t + r_{BG}, \tag{18}$$

where r_{BC}, r_{BS} and r_{BG} are the coefficients.

Substitution of (17) and (18) into (5) and (6) yields relations for the steady state time histories of the y and z displacements of the disc centre

$$y = (w_0 r_{BC} - \omega w_1 r_{BS}) \cos \omega t - (w_0 r_{BS} + \omega w_1 r_{BC}) \sin \omega t, \tag{19}$$

$$z = (w_0 r_{BS} + \omega w_1 r_{BC}) \cos \omega t + (w_0 r_{BC} - \omega w_1 r_{BS}) \sin \omega t + w_0 r_{BG}. \tag{20}$$

Eqs. (17)–(20) show that the steady state trajectories of the rotor journals and disc centres are circular shifted in the direction of the gravity acceleration. It holds for their radii

$$r_B = \sqrt{r_{BC}^2 + r_{BS}^2}, \tag{21}$$

$$r = \sqrt{(w_0 r_{BC} - \omega w_1 r_{BS})^2 + (w_0 r_{BS} + \omega w_1 r_{BC})^2}. \tag{22}$$

r_B, r are radii of the orbits of the rotor journal and disc centres respectively.

Displacements y_B and z_B, given by relations (17) and (18), and their first, second and third derivatives with respect to time are substituted into Eq. (10). Consequently the corresponding absolute term and the coefficients of proportionality of the cosine and sine ones on the left and right hand sides of the obtained equation are compared. This manipulation arrives at a set of three linear algebraic equations whose solving gives

$$r_{BG} = -\frac{mg}{a_0}, \tag{23}$$

$$r_{BC} = \frac{-a_3 N_S \omega^5 - a_2 N_C \omega^4 + a_1 N_S \omega^3 + a_0 N_C \omega^2}{a_3^2 \omega^6 + (a_2^2 - 2a_1 a_3)\omega^4 + (a_1^2 - 2a_0 a_2)\omega^2 + a_0^2}, \tag{24}$$

$$r_{BS} = \frac{a_3 N_C \omega^5 - a_2 N_S \omega^4 - a_1 N_C \omega^3 + a_0 N_S \omega^2}{a_3^2 \omega^6 + (a_2^2 - 2a_1 a_3)\omega^4 + (a_1^2 - 2a_0 a_2)\omega^2 + a_0^2}. \tag{25}$$

As evident from the derived relations, amplitude of the rotor vibration at locations of the disc and the shaft journals depends on angular speed of the rotor turning (ω) and on controllable damping in the coupling elements (b_B). If damping in the supports remains constant, then relations (21) and (22) represent frequency characteristics of the rotor system.

In the limit case when angular speed of the rotor turning approaches infinity, it holds

$$\lim_{\omega \to \infty} r_B(b_B, \omega) = 0, \tag{26}$$

$$\lim_{\omega \to \infty} r(b_B, \omega) = e_T. \tag{27}$$

The following computational simulations were performed for the system parameters: $m = 500$ kg, $b_P = 10$ Ns/m, $k_H = 26.48$ MN/m, $k_B = 20.0$ MN/m, $e_T = 0.1$ mm and $\psi_O = 0$ rad. The speed of rotation ω and the coefficient of linear damping in the supports b_B were taken from the intervals $0 \div 300$ rad $\cdot$ s^{-1} and $10 \div 1\,000$ kNs/m respectively. The results summarized in Figs. 2 and 3 show that

- dependence of the vibration amplitude on speed of the rotor rotation has one extreme value, which corresponds to the critical angular speed,

- its magnitude depends on amount of damping in the coupling elements,

- maximum value of the critical speed corresponds to infinitely high damping in the rotor supports.

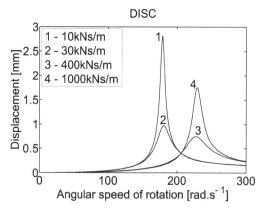

Fig. 2. Resonance characteristic – disc centre

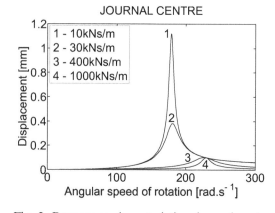

Fig. 3. Resonance characteristic – journal centre

Determination of the minimum maximum amplitude of the disc steady state vibration and corresponding speed of the rotor turning requires solving a set of two nonlinear algebraic equations

$$\frac{\partial r(b_B, \omega)}{\partial b_B} = 0, \tag{28}$$

$$\frac{\partial r(b_B, \omega)}{\partial \omega} = 0. \tag{29}$$

Results of the carried out analysis make it possible to draw several conclusions. If the allowed amplitude of the rotor vibration is higher than its minimum maximum value, then magnitude of damping in the supports can be suitably chosen, the damping coefficient b_B can remain constant and no control of damping in the coupling elements is needed. If the allowed amplitude is lower than the minimum maximum one, then damping in the supports cannot reduce the maximum amplitude of the steady state vibration below the allowed value. But as Figs. 2 and 3 show, the controlled damping in the supports (control of the damping coefficient b_B) can considerably reduce the speed interval, in which amplitude of the vibration obtains the value higher than the allowed one. The higher damping pushes the lower bound of the interval of not allowed rotational speeds to higher values and the lower damping shifts its upper limit to lower velocities. This clearly demonstrates that the controlled damping contributes to increasing the speed interval, in which the rotor can be operated.

3. The investigated rotor system

The investigated technological device, whose scheme is drawn in Fig. 4, is intended for cleaning surfaces of metal sheets MS. The working part is formed by a rotor driven by a direct current electric motor EM. The rotor consisting of a shaft SH and of one disc D with cleaning brushes CB is attached to the motor by a tilting claw coupling CO. The rotor is flexibly supported and the damping devices whose damping effect can be controlled are added to the coupling elements CE. The rotor is loaded by the weight of the disc, by the technological forces induced by the cleaning process, by the disc imbalance and by the driving moment of the motor.

The task is to study the influence of control of damping in the coupling elements on the steady state vibration of the rotor from the point of view of the limit states of deformation and fatigue failure.

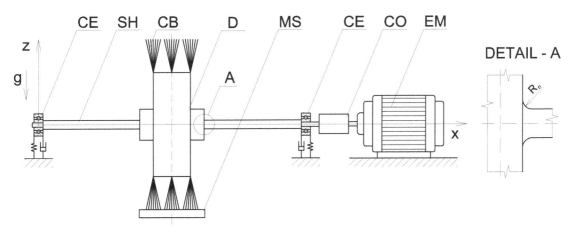

Fig. 4. Scheme of the investigated system

In the computational model the rotating part is represented by a Jeffcott rotor. The coupling and the rotor of the electric motor are considered as absolutely rigid. Interaction between the disc and the metal sheet is modelled by a force coupling. It is assumed that the disc is loaded by two forces acting on it in the horizontal and vertical directions. The forces are repeated, their magnitudes depend on the disc angle of rotation and their angular period is given by the number of the cleaning brushes that are uniformly distributed around the circumference of the disc. Supports of the rotor are flexible, their stiffness and damping are assumed to be linear. Supports of the rotor of the driving motor are absolutely rigid. Dependence of the moment produced by the motor on its angular speed is described by a static characteristic.

Taking into account the symmetry, the model system has six degrees of freedom (translational motion of the disc and of the shaft journal centres in the horizontal and vertical directions, rotation of the disc, rotation of the rotor of the electric motor). The governing equations are derived by means of the Lagrange equations of the second kind

$$\frac{\mathrm{d}}{\mathrm{d}t}\left(\frac{\partial E_K}{\partial \dot{q}_j}\right) - \frac{\partial E_K}{\partial q_j} + \frac{\partial E_P}{\partial q_j} + \frac{\partial R_D}{\partial \dot{q}_j} = Q_j, \tag{30}$$

$$q_j \in \{y, z, y_B, z_B, \phi, \phi_M\}. \tag{31}$$

E_K, E_P, R_D are the kinetic energy, potential energy and Rayleigh dissipative function respectively, q_j, Q_j are the generalized displacements and forces, y, z, y_B, z_B are displacements of the disc and shaft journal centres and ϕ, ϕ_M are rotations of the disc and rotor of the electric motor. Relations for the energy quantities take the form

$$E_K = \frac{1}{2}m\left[\left(\dot{y} - e_T\dot{\phi}\sin\phi\right)^2 + \left(\dot{z} + e_T\dot{\phi}\cos\phi\right)^2\right] + \frac{1}{2}J_T\dot{\phi}^2 + \frac{1}{2}J_M\dot{\phi}_M^2, \tag{32}$$

$$E_P = \frac{1}{2}k_{HO}\left[(y - y_B)^2 + (z - z_B)^2\right] + k_B\left(y_B^2 + z_B^2\right) + \frac{1}{2}k_{HT}\left(\phi - \phi_M\right)^2, \tag{33}$$

$$R_D = \frac{1}{2}b_{PO}\left(\dot{y}^2 + \dot{z}^2\right) + \frac{1}{2}b_{PT}\dot{\phi}^2 + b_B\left(\dot{y}_B^2 + \dot{z}_B^2\right), \tag{34}$$

$$\delta A = F_y\left(\delta y + R\delta\phi\right) + F_z\delta z - mg\delta z + M_M\delta\phi_M. \tag{35}$$

δA is the virtual work of working forces and moments acting in the system, m is the disc mass, J_T, J_M are moments of inertia of the disc and rotor of the motor (relative to their rotational axes going through their centres of gravity), e_T is eccentricity of the disc centre of gravity, k_{HO}, k_{HT} are the bending and torsional stiffnesses of the shaft, k_B is stiffness of each rotor support, b_{PO}, b_{PT} are the coefficients of external damping of the disc related to the rotor bending and torsional vibration respectively, b_B is the coefficient of damping referred to each rotor support (controllable), F_y, F_z are the y and z components of the technological force, R is the disc radius and M_M is moment of the motor, which is defined by the static characteristic

$$M_M = M_{MS} - k_M\dot{\phi}_M. \tag{36}$$

M_{MS}, k_M are the starting moment and slope of the linear dependence of the moment of the motor on its angular velocity. Components of the technological force depend on angular position of the disc ϕ and on the number of brushes n_B

$$F_y = -0.5F_{maxy}\left[1 + \cos(n_B\phi)\right], \tag{37}$$

$$F_z = 0.5F_{maxz}\left[1 + \cos(n_B\phi)\right]. \tag{38}$$

F_{maxy} and F_{maxz} are the maximum values of the y and z components of the technological force.

After performing the appropriate manipulations vibration of the investigated system is described by a set of six differential equations

$$m\ddot{y} - m e_T \ddot{\phi} \sin\phi = -b_{PO}\dot{y} - k_{HO}(y - y_B) + m e_T \dot{\phi}^2 \cos\phi + F_y, \tag{39}$$

$$m\ddot{z} + m e_T \ddot{\phi} \cos\phi = -b_{PO}\dot{z} - k_{HO}(z - z_B) + m e_T \dot{\phi}^2 \sin\phi + F_z - mg, \tag{40}$$

$$0 = -2b_B \dot{y}_B + k_{HO}y - (k_{HO} + 2k_B)y_B, \tag{41}$$

$$0 = -2b_B \dot{z}_B + k_{HO}z - (k_{HO} + 2k_B)z_B, \tag{42}$$

$$-m e_T \ddot{y} \sin\phi + m e_T \ddot{z} \cos\phi +$$
$$\left(J_T + m e_T^2\right)\ddot{\phi} = -b_{PT}\dot{\phi} - k_{HT}(\phi - \phi_M) - RF_y - mg e_T \cos\phi, \tag{43}$$

$$J_M \ddot{\phi}_M = -k_{HT}(\phi_M - \phi) + M_M. \tag{44}$$

A 5th order computational method was used for their solving. Its detailed algorithm is reported in [13].

The solution of differential Eqs. (39)–(44) gives the time histories of displacements of the centres of the disc and of the shaft journals which makes it possible to evaluate behaviour of the rotor system from the point of view of the limit state of deformation.

To analyse the rotor service life one must know the stresses at the individual points of the rotor shaft. The beam state of stress defined by one axial normal stress produced by the shaft bending and by one shear stress in the cross section perpendicular to the shaft centre line caused by the shaft torque is assumed. The bending stress is given by the bending moment having the components in the horizontal and vertical directions and depends also on the shaft rotation

$$\sigma = \frac{k_B y_B + b_B \dot{y}_B}{W_\sigma} x \cos(\phi + \psi_P) + \frac{k_B z_B + b_B \dot{z}_B}{W_\sigma} x \sin(\phi + \psi_P) \quad \text{for} \quad 0 \le x \le \frac{L_B}{2}, \tag{45}$$

$$\tau = -\frac{k_{HT}(\phi - \phi_M)}{W_\tau}. \tag{46}$$

Moduli W_σ and W_τ are expressed

$$W_\sigma = \frac{\pi}{32}\left(d_{H2}^3 - d_{H1}^3\right), \tag{47}$$

$$W_\tau = \frac{\pi}{16}\left(d_{H2}^3 - d_{H1}^3\right). \tag{48}$$

σ, τ are the bending and shear stresses, ψ_P is the position angle of the point on the circumference of the cross section, L_B is the distance between the bearings and x, d_{H1} and d_{H2} denote the axial position and the inner and outer diameters of the shaft cross section at the investigated location.

4. Evaluation of the rotor state of stress from the view point of the fatigue failure

Due to the character of the shaft loading the highest stresses are estimated on its surface. The stress concentration at location of a contact of two cylindrical parts of different diameters is taken into account by the notch coefficients in bending β_σ and torque β_τ [8,11]. Further it is assumed that the courses of the individual stresses are constant or harmonic functions of time whose mean values can be shifted.

The task is to evaluate the service life of the shaft for unlimited number of loading cycles. The fatigue limits in bending and torque (σ_c and τ_c respectively) determined experimentally on testing samples [5] are corrected to be taken into account the size of the shaft, quality of its surface, influence of the environment and the stress concentration caused by the notch. The

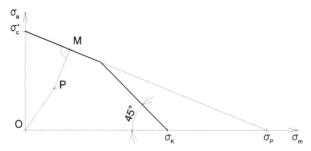

Fig. 5. Haigh diagram

corrected values are denoted σ_c^* and τ_c^* respectively. The safety factors are calculated separately for loadings in bending and torque using the Haigh diagrams (Fig. 5) constructed by means of the Goodman line [12]

$$\sigma_a = \sigma_c^* - \sigma_c^* \frac{\sigma_m}{\sigma_P}, \tag{49}$$

$$\tau_a = \tau_c^* - \tau_c^* \frac{\tau_m}{0.7\sigma_P}. \tag{50}$$

$\sigma_a, \tau_a, \sigma_m, \tau_m$ denote amplitudes and mean values of the time histories of bending and shear stresses respectively, σ_P is the tensile strength and σ_K and τ_K denote the yielding tensile and shear stresses of the shaft material. In the analysis only the cycles with positive (tensile) mean value are evaluated because they have more significant influence on the fatigue failure than the cycles whose mean value is negative (compression).

The safety factors related to the fatigue limits in bending and shear are calculated by means of the Haigh diagrams (Fig. 5) for the worst case of overloading [10]

$$k_{\sigma,\tau} = \frac{\overline{OP} + \overline{PM}}{\overline{OP}}. \tag{51}$$

Consequently, it holds for the resulting safety factor k_C [10]

$$k_C = \frac{k_\sigma k_\tau}{\sqrt{k_\sigma^2 + k_\tau^2}}. \tag{52}$$

This approach assumes that the time histories of the normal and shear stresses produced by the shaft bending and torque have the same phase shifts. If applied for the case when the phase lags are different, the resulting safety factor k_C will be greater than the value obtained by application of relation (52). k_C calculated by means of (52) represents its lower estimation.

5. Results of the computational simulations

The technological parameters of the investigated rotor system are summarized in Table 1. The natural frequency of the undamped bending vibration of the rotor is approximately $105 \, \text{rad} \cdot \text{s}^{-1}$. If dampers placed in the rotor coupling elements are highly overdamped, then the resonance frequency rises to $196 \, \text{rad} \cdot \text{s}^{-1}$. The task is to analyse performance of the damping devices added to the rotor supports. The investigations are carried out for five rotational speeds (80, $125, 165, 210, 250 \, \text{rad} \cdot \text{s}^{-1}$) and three magnitudes of damping (500, 5 000, 500 000 Ns/m). The study is focused on attenuation of the rotor vibration, reducing forces transmitted between the

Table 1. Technological parameters of the investigated system

m	129.53 kg	k_M	5 Nms
J_T	8.476 5 kgm^2	d_{H1}	34 mm
J_M	1 kgm^2	d_{H2} (less)	62 mm
e_T	50 μm	d_{H2} (greater)	160 mm
k_{HO}	4 996 000 Nm^{-1}	L_B	1 100 mm
k_{HT}	193 770 Nm	E	210 GPa
k_B	1 000 000 Nm^{-1}	μ	0.3
b_{PO}	8 Nsm^{-1}	σ_P	340 MPa
b_{PT}	4 Nsm	σ_K	225 MPa
F_{maxy}	56 N	σ_c	170 MPa
F_{maxz}	140 N	σ_c^*	70 MPa
R	0.39 m	τ_K	130 MPa
M_{MS} for 80 rad $\cdot$ s^{-1}	764.84 Nm	τ_c	100 MPa
M_{MS} for 125 rad $\cdot$ s^{-1}	1 141.84 Nm	τ_c^*	54 MPa
M_{MS} for 165 rad $\cdot$ s^{-1}	1 518.84 Nm	α_σ	1.42
M_{MS} for 210 rad $\cdot$ s^{-1}	1 895.84 Nm	α_τ	1.20
M_{MS} for 250 rad $\cdot$ s^{-1}	2 272.84 Nm		

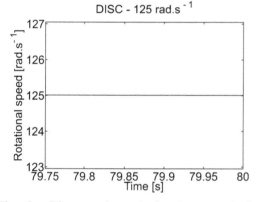

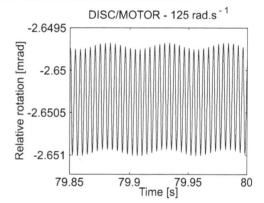

Fig. 6. Disc angular velocity (motor velocity 125 rad $\cdot$ s^{-1})

Fig. 7. Rotation of the disc relative to the motor

rotor and its casing and on evaluation of the influence of the damping elements on the rotor service life. The unlimited number of cycles for all loadings is required.

The technological forces acting on the disc lateral area in the horizontal and vertical directions are of repeated character and have a harmonic course in dependence on the angle of the disc rotation. Their maximum values are 28 N and 70 N respectively and their frequency is given by a product of the rotation angle of the disc and the number of cleaning brushes ($n_B = 12$).

The shaft is made of steel of mark 11 375 whose material constants can be found in Table 1. The critical cross sections are estimated at locations of the notches formed by contact of two shaft steps of different diameters. Their positions are 502 mm from the bearing centres towards the disc.

Results of the carried out computational simulations are summarized in the following figures and tables. The time courses of the steady state angular velocity of the disc and of its rotation relative to the rotor of the electric motor for the mean motor angular speed of rotation of 125 rad $\cdot$ s^{-1} are drawn in Figs. 6 and 7. As evident the disc angular velocity fluctuations are negligible so that its rotation can be considered as uniform.

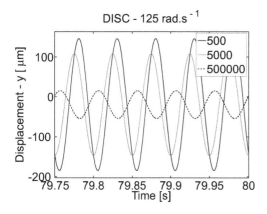

Fig. 8. Displacement y of the disc ($125 \text{ rad} \cdot \text{s}^{-1}$)

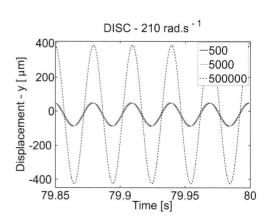

Fig. 9. Displacement y of the disc ($210 \text{ rad} \cdot \text{s}^{-1}$)

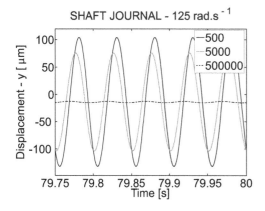

Fig. 10. Displacement y of the shaft journal ($125 \text{ rad} \cdot \text{s}^{-1}$)

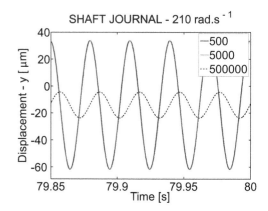

Fig. 11. Displacement y of the shaft journal ($210 \text{ rad} \cdot \text{s}^{-1}$)

Figs. 8–11 show the steady state time histories of the horizontal displacements of the disc and of the shaft journal centres for three magnitudes of damping in the rotor supports. Corresponding courses of the total force transmitted between the rotor and its casing in the horizontal direction are drawn in Figs. 12 and 13.

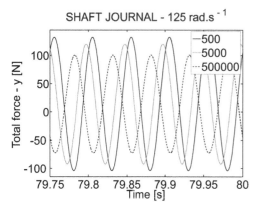

Fig. 12. Force acting on the shaft journal ($125 \text{ rad} \cdot \text{s}^{-1}$)

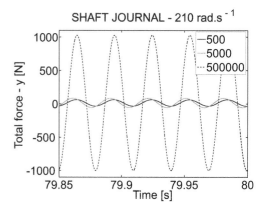

Fig. 13. Force acting on the shaft journal ($210 \text{ rad} \cdot \text{s}^{-1}$)

Table 2. Displacements and transmitted forces for transmitted through the coupling elements

Rotational speed	Peak-to-peak displacement y Disc [μm]			Peak-to-peak displacement y Rotor journal centre [μm]			Maximum force transmitted through the rotor support in y direction [N]		
[rad $\cdot$ s^{-1}]	500 Ns/m	5 000 Ns/m	500 000 Ns/m	500 Ns/m	5 000 Ns/m	500 000 Ns/m	500 Ns/m	5 000 Ns/m	500 000 Ns/m
80	174	128	23	125	91	1	76.4	63.5	42.3
125	330	255	69	236	179	3	132.2	119.8	100.8
165	165	168	267	118	116	8	73.0	89.8	347.2
210	134	139	812	95	95	19	62.0	82.9	1 028.0
250	121	127	257	87	85	5	57.6	82.5	334.3

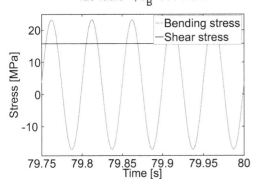

Fig. 14. Stresses time history (125 rad $\cdot$ s^{-1}, 500 Ns/m)

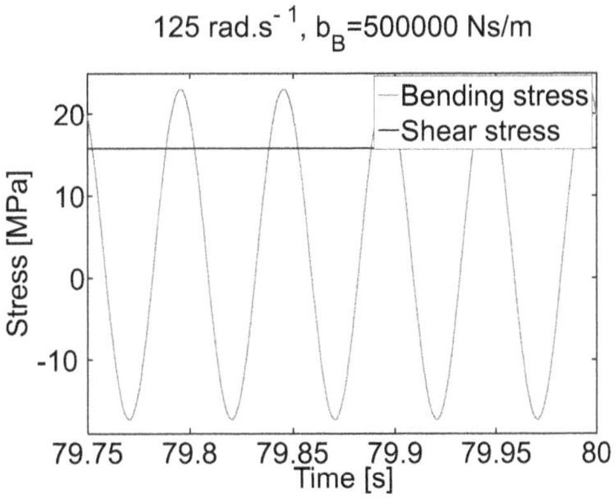

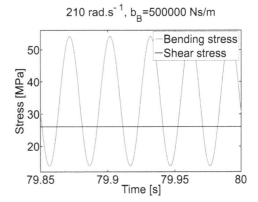

Fig. 15. Stresses time history (125 rad $\cdot$ s^{-1}, 500 000 Ns/m)

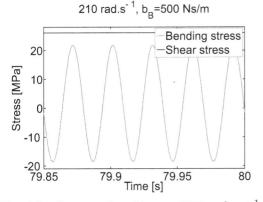

Fig. 16. Stresses time history (210 rad $\cdot$ s^{-1}, 500 Ns/m)

Fig. 17. Stresses time history (210 rad $\cdot$ s^{-1}, 500 000 Ns/m)

Further results related to all investigated speeds and damping magnitudes can be found in Table 2.

The state of stress is analysed in 30 points uniformly distributed around the shaft circumference at location of its critical cross section (location of the notch). Time histories of the bending and shear stresses in one selected point referred to two speeds of the rotor rotation (125, 210 rad $\cdot$ s^{-1}) and two magnitudes of damping in the damping elements (500, 500 000 Ns/m) are drawn in Figs. 14–17. The stresses are related to the smaller diametre from both contacting

Table 3. Stresses referred to the rotor steady state vibration

Rotational speed	Mean bending stress [MPa]			Bending stress amplitude [MPa]			Mean shear stress [MPa]		
	500 Ns/m	5 000 Ns/m	500 000 Ns/m	500 Ns/m	5 000 Ns/m	500 000 Ns/m	500 Ns/m	5 000 Ns/m	500 000 Ns/m
$[rad \cdot s^{-1}]$									
80	2.06	1.63	0.91	20.13	20.15	20.16	10.62	10.62	10.62
125	3.95	3.55	2.82	20.12	20.13	20.13	15.77	15.77	15.77
165	1.96	2.54	11.12	20.12	20.12	20.10	20.92	20.92	20.92
210	1.60	2.30	33.96	20.12	20.12	20.12	26.06	26.06	26.06
250	1.45	2.28	10.71	20.12	20.12	20.11	31.21	31.21	31.21

Table 4. Safety factors

Rotational speed	Safety factor – bending stress [–]			Safety factor – shear stress [–]			Resulting safety factor [–]		
	500 Ns/m	5 000 Ns/m	500 000 Ns/m	500 Ns/m	5 000 Ns/m	500 000 Ns/m	500 Ns/m	5 000 Ns/m	500 000 Ns/m
$[rad \cdot s^{-1}]$									
80	3.39	3.40	3.41	5.86	5.86	5.86	2.94	2.94	2.95
125	3.34	3.35	3.37	4.22	4.22	4.22	2.62	2.63	2.64
165	3.40	3.38	3.03	3.39	3.39	3.39	2.40	2.40	2.26
210	3.41	3.39	2.06	2.89	2.89	2.89	2.20	2.20	1.68
250	3.41	3.39	3.05	2.55	2.55	2.55	2.04	2.04	1.96

shaft steps. It is evident that the time courses of normal stresses can be considered, with sufficient accuracy, as harmonic with nonzero mean value and magnitudes of the shear stresses as constant. This makes it possible to evaluate the limit state of fatigue by means of the procedure described above. The stresses referred to all investigated speeds and damping magnitudes are summarized in Table 3.

The safety factors are determined for all 30 points introduced in the critical cross section and their minimum value is considered as the safety factor referred to the operating conditions (speed of the rotor rotation and magnitude of the damping in the coupling elements), at which the stresses are investigated. The results are given in Table 4.

As evident from Figs. 8–17, frequencies of the time histories of the disc displacements, normal stresses in the investigated points and forces transmitted between the rotor and its casing are equal to the frequency of the rotor rotation, which means that the bending vibrations are influenced by the torsional ones only negligibly.

Results of the carried out computer simulations confirm that to achieve optimum performance of the damping devices their damping effect must be controllable in dependence on the rotor angular velocity. For the speeds of rotation of 80 and 125 $rad \cdot s^{-1}$ the high damping in the support elements is needed. This reduces amplitude of the vibration both of the disc and of the shaft journal in the bearings and leads to minimum force transmitted through the coupling elements into the rotor casing. On the contrary, for a wide interval of rotational velocities around the higher rotor critical speed (196 $rad \cdot s^{-1}$) the damping produced by the damping devices should be as small as possible to achieve low amplitude of the disc vibration and magnitude of the transmitted force. The same conclusions are valid also for behaviour of the rotor from the point of view of the limit state of fatigue strength. Especially for the speed of rotation of $210 \, \text{rad} \cdot \text{s}^{-1}$ the value of the resulting safety factor goes sharply down with increasing damping.

6. Conclusions

Flexural vibration of rotors has a dominant influence on their work. Results of the carried out study show that for some design arrangements and operating conditions appropriate amount of damping in the rotor coupling elements can ensure that parameters of the rotor vibration do not exceed their limit values in the whole extent of the rotor running speed. Then no damping control is needed. But if this is not possible to be achieved due to any reason (e.g. stiffness of the squirrel springs would be very low) then the suitably proposed dependence of damping on the rotor angular velocity enables to reduce the speed interval, in which amplitude of the vibration or magnitude of the transmitted force exceed the allowed values.

This manipulation based on control of damping in the coupling elements placed between the rotor and its casing can be used for a wide class of rotating machines. The advantage is that for its application only the semiactive damping devices that do not require complicated controllers and control systems for their operation can be utilized.

Acknowledgements

Funding of the work reported in this article has come from the research projects P101/10/0209 (Czech Science Foundation), MSM 6198910027 (Ministry of Education of the Czech Republic) and CZ.1.05/1.1.00/02.0070 (framework of the IT4Innovations Centre of Excellence project, Operational Programme "Research and Development for Innovations" funded by Structural Funds of the European Union and state budget of the Czech Republic). Their support is gratefully acknowledged.

References

[1] Carmignani, C., Forte, P., Rustighi, E., Design of a novel magneto-rheological squeeze-film damper, Smart Materials and Structures 15(1) (2006) 164–170.

[2] Carmignani, C., Forte, P., Badalassi, P., Zini, G., Classical control of a magnetorheological squeeze-film damper, Proceedings of the Stability and Control Processes 2005, Saint-Petersburg, Russia, 2005, pp. 1 237–1 246.

[3] Forte, P., Paterno, M., Rustighi, E., A magnetorheological fluid damper for rotor applications, Proceedings of the IFToMM Sixth International Conference on Rotor Dynamics, Sydney, Australia, 2002, pp. 63–70.

[4] Forte, P., Paterno, M., Rustighi, E., A magnetorheological fluid damper for rotor applications, International Journal of Rotating Machinery 10(3) (2004) 175–182.

[5] Fürbacher, I., Lexicon of technical materials with foreign equivalents, Dashöfer, Praha, 2006. (in Czech)

[6] Gasch, R., Nordmann, R., Pfützner, H., Rotor dynamics, Springer-Verlag, Berlin, Heidelberg, New York, 2002. (in German)

[7] Krämer, E., Dynamics of Rotors and Foundations, Springer-Verlag, Berlin, 1993.

[8] Kučera, J., A brief introduction to fracture mechanics — Part II. Material fatigue, VŠB-TUO, Ostrava, 1994. (in Czech)

[9] Mu, C., Darling, J., Burrows, C. R., An appraisal of a proposed active squeeze film damper, ASME Journal of Tribology 113(4) (1991) 750–754.

[10] Ondráček, E., Vrbka, J., Janíček, P., Solid mechanics — mechanics of materials II, Ediční středisko VUT, Brno, 1988. (in Czech)

[11] Řasa, J., Engineering tables for school and practice, Scientia, Praha, 2004. (in Czech)

[12] Růžička, M., Havlíček, V., Calculation of machine parts to fatigue under normal and elevated temperatures I, Dům techniky, Praha, 1988. (in Czech)

[13] Shampine, L. F., Reichelt, M. W., Kierzenka, J. A., Solving Index-1 DAEs in MATLAB and Simulink, SIAM Review 41(3) (1999) 538–552.

[14] Tonoli, A., Dynamic characteristics of eddy current dampers and couplers, Journal of Sound and Vibration 301 (3–5) (2007) 576–591.

[15] Tonoli, A., Silvagni, M., Amati, N., Staples, B., Karpenko, E., Design of electromagnetic damper for aero-engine applications, Proceedings of the IMechE International Conference on Vibration in Rotating Machinery, Exeter, England, 2008, pp. 761–774.

16

Stress analysis of finite length cylinders of layered media

P. Desai[a,*], T. Kant[a]

[a] *Department of Civil Engineering, Indian Institute of Technology Bombay, Powai, Mumbai, 400 076, India*

Abstract

In this paper, we analyze an orthotropic, layered $(0°/90°)$ and $(0°/\text{core}/0°)$ sandwich cylinders under pressurized load with a diaphragm supported boundary conditions which is considered as a two dimensional (2D) plane strain boundary value problem of elasticity in (r, z) direction. A simplified numerical cum analytical approach is used for the analysis. Boundary conditions are satisfied exactly by using an analytical expression in longitudinal (z) direction in terms of Fourier series expansion. Resulting first order simultaneous ordinary differential equations (ODEs) with boundary conditions prescribed at $r = r_i, r_o$ defines a two point boundary value problem (BVP), whose equations are integrated in radial direction through an effective numerical integration technique by first transforming the BVP into a set of initial value problems (IVPs). Numerical solutions are first validated for their accuracy with 1D solution of an infinitely long cylinder. Stresses and displacements in cylinders of finite lengths having various l/R and h/R ratios are presented for future reference.

Keywords: elasticity theory, circular cylinder, numerical integration, boundary value problems, laminated composites

1. Introduction

Composites have seen an ever increasing use in the process industry during the last twenty five years. Their use as a material of choice for pressure vessels and components is due to the fact that they possess longer life in a corrosive environment, low weight but high strength and stiffness, and the capability to tailor directional strength properties to design needs. Composite cylinders are widely used in various engineering applications such as aerospace vehicles, nuclear pressure vessels, piping and many other engineering structures and need accurate analysis of deformations and stresses induced by applied pressure loading. The classic problem of an infinitely long elastic cylinder of an isotropic material under internal and external pressure was analyzed first by Lame in 1847 (given in [16]) for isotropic and by [12] for anisotropic and layered materials. This particular problem has been studied by many during later years. In paper [9], authors obtained stresses and displacements by the use of three dimensional (3D) elasticity theory and several shell theories in a long isotropic circular cylinder subjected to an axisymmetric radial line load and compared results with the shell theories of Love and Flugge. An elasticity solution by using a Love function approach for semi-infinite circular cylindrical shell subjected to a concentrated axisymmetric radial line load at the free end was presented in [3]. The problem of an infinite circular cylindrical shell subjected to periodically spaced band loads using 3D elasticity theory and the shell theories of Love (and Donnell), Flugge, and a theory developed by Reissner and Nagdhi was solved in [10]. An approximate solution to the Navier equations of the 3D elasticity for an axisymmetric orthotropic infinitely long circular cylinder subjected to internal and external pressure, axial loads, and closely spaced periodic radial loads was obtained in [13]. An exact solution for a thick, transversely isotropic, simply supported finite length circular cylindrical shell subjected to axisymmetric load using a transfer matrix approach are obtained in [1]. Clamped-clamped and clamped-simply supported cylindrical shells by a so-called segmentation numerical integration technique was analyzed by [8]. The same technique for elastic analysis of cylindrical pressure vessels with various end closures using Love's classical shell theory used in [15].

In this paper, governing differential equations from theory of 3D anisotropic elasticity, which govern the behaviors of a finite length circular orthotropic cylinder in a state of symmetric plane strain in (r, z) under sinusoidal pressurized loading which is a function of both radial and axial coordinates, are taken. By assuming a global analytical solution in the longitudinal direction (z) which satisfies the two end boundary conditions exactly, dimensional reduction is done with this process, the 2D generalized plane strain problem is reduced to a 1D problem in the radial coordinate. The equations are reformulated to enable application of an efficient and accurate numerical integration technique developed and proposed for the solution of BVP [7].

In addition, one dimensional elasticity equations of an infinitely long symmetric cylinder are utilized to formulate the mathematical model suitable for numerical integration. These equations are summarized in the Appendix. This has been done with a view to check and compares the results of the present formulation of finite length cylinder under uniform internal/external pressure load, when the length of the cylinder tends to infinity.

The basic governing equations

Basic governing equations of a symmetric cylinder which is considered plane strain in (r, z) direction [12] in cylindrical coordinates (Fig. 1) is written as:

Equilibrium equations

$$\frac{\partial \sigma_r}{\partial r} + \frac{\partial \tau_{zr}}{\partial z} + \frac{\sigma_r - \sigma_\theta}{r} = 0,$$
$$\frac{\partial \tau_{zr}}{\partial r} + \frac{\partial \sigma_z}{\partial z} + \frac{\tau_{zr}}{r} = 0. \tag{1a}$$

Strain displacement relations

$$\varepsilon_r = \frac{\partial u}{\partial r}, \quad \varepsilon_\theta = \frac{u}{r}, \quad \varepsilon_z = \frac{\partial w}{\partial z}, \quad \gamma_{zr} = \frac{\partial w}{\partial r} + \frac{\partial u}{\partial z}. \tag{1b}$$

Stress-strains relations for cylindrically orthotropic material

$$\varepsilon_r = \frac{\sigma_r}{E_r} - \nu_{\theta r}\frac{\sigma_\theta}{E_\theta} - \nu_{zr}\frac{\sigma_z}{E_z},$$
$$\varepsilon_\theta = -\nu_{r\theta}\frac{\sigma_r}{E_r} + \frac{\sigma_\theta}{E_\theta} - \nu_{z\theta}\frac{\sigma_z}{E_z}, \tag{1c}$$
$$\varepsilon_z = -\nu_{rz}\frac{\sigma_r}{E_r} - \nu_{\theta z}\frac{\sigma_\theta}{E_\theta} + \frac{\sigma_z}{E_z}, \quad \gamma_{rz} = \frac{\tau_{rz}}{G_{rz}}.$$

Stresses in terms of strains can be written as follows:

$$\begin{Bmatrix} \sigma_r \\ \sigma_\theta \\ \sigma_z \end{Bmatrix} = \begin{bmatrix} C_{11} & C_{12} & C_{13} \\ C_{21} & C_{22} & C_{23} \\ C_{31} & C_{32} & C_{33} \end{bmatrix} \begin{Bmatrix} \varepsilon_r \\ \varepsilon_\theta \\ \varepsilon_z \end{Bmatrix}, \quad \tau_{rz} = G\gamma_{rz} \tag{1d}$$

in which

$$\nu_{r\theta} = \frac{\nu_{\theta r}}{E_\theta} E_r, \quad \nu_{rz} = \frac{\nu_{zr}}{E_z} E_r, \quad \nu_{z\theta} = \frac{\nu_{\theta z}}{E_\theta} E_z,$$

$$C_{11} = \frac{E_r(1 - \nu_{\theta z}\nu_{z\theta})}{\Delta}, \quad C_{12} = \frac{E_r(\nu_{\theta r} + \nu_{zr}\nu_{\theta z})}{\Delta}, \quad C_{13} = \frac{E_r(\nu_{zr} + \nu_{\theta r}\nu_{z\theta})}{\Delta},$$

$$C_{22} = \frac{E_\theta(1 - \nu_{rz}\nu_{zr})}{\Delta}, \quad C_{23} = \frac{E_\theta(\nu_{z\theta} + \nu_{r\theta}\nu_{zr})}{\Delta}, \quad C_{33} = \frac{E_z(1 - \nu_{r\theta}\nu_{\theta r})}{\Delta}, \quad (1e)$$

where $\Delta = (1 - \nu_{r\theta}\nu_{\theta r} - \nu_{\theta z}\nu_{z\theta} - \nu_{zr}\nu_{rz} - 2\nu_{\theta r}\nu_{z\theta}\nu_{rz})$, $C_{21} = C_{12}, C_{32} = C_{23}, C_{31} = C_{13}$.
Stresses in terms of displacement components can be cast as follows:

$$\sigma_r = C_{11}\left(\frac{\partial u}{\partial r}\right) + C_{12}\left(\frac{u}{r}\right) + C_{13}\left(\frac{\partial w}{\partial z}\right), \quad \sigma_\theta = C_{21}\left(\frac{\partial u}{\partial r}\right) + C_{22}\left(\frac{u}{r}\right) + C_{23}\left(\frac{\partial w}{\partial z}\right), (1f)$$

$$\sigma_z = C_{31}\left(\frac{\partial u}{\partial r}\right) + C_{32}\left(\frac{u}{r}\right) + C_{33}\left(\frac{\partial w}{\partial z}\right), \quad \tau_{rz} = G\gamma_{rz} = G\left(\frac{\partial w}{\partial r} + \frac{\partial u}{\partial z}\right)$$

and boundary conditions in the longitudinal and radial directions are

$$u = \sigma_z = 0 \text{ for } z = 0, l; \quad \sigma_r = \tau_{rz} = 0 \text{ for } r = r_i; \quad \sigma_r = -p(z), \quad \tau_{rz} = 0 \text{ for } r = r_o \quad (2)$$

in which l is the length, r_i is the inner radius and r_o is the outer radius of a hollow cylinder.

Load $p(z)$ can be represented in terms of Fourier series in general form as follows:

$$p(z) = \sum_{i=1,3,5,\dots}^{N} p_i \sin\frac{i\pi z}{l} \tag{3a}$$

in which p_i is the Fourier load coefficient which can be determined by using the orthogonality conditions and for sinusoidal loading

$$p(z) = p_0 \sin\frac{\pi z}{l}, \tag{3b}$$

p_0 is the maximum intensity of distributed pressure. The positive coordinates and loadings on a cylinder are shown in Fig. 1a, b.

2. Mathematical formulation

Radial direction r is chosen to be a preferred independent coordinate. Four fundamental dependent variables, displacements u and w and corresponding stresses σ_r and τ_{rz} that occur naturally on a tangent plane $r = \text{constant}$, are chosen in the radial direction. Circumferential stress σ_θ and axial stress σ_z are treated here as auxiliary variables since these are found to be dependent on the chosen fundamental variables [16]. A set of four first order partial differential equations in independent coordinate r which involve only fundamental variables is obtained through algebraic manipulation of Eqs. (1a)–(1f). These are

$$\frac{\partial u}{\partial r} = \frac{\sigma_r}{C_{11}} - \frac{C_{12}}{C_{11}}\left(\frac{u}{r}\right) - \frac{C_{13}}{C_{11}}\left(\frac{\partial w}{\partial z}\right),$$

$$\frac{\partial w}{\partial r} = \frac{1}{G}\tau_{rz} - \frac{\partial u}{\partial z}, \tag{4a}$$

$$\frac{\partial \sigma_r}{\partial r} = -\frac{\partial \tau_{rz}}{\partial z} + \frac{\sigma_r}{r}\left(\frac{C_{21}}{C_{11}} - 1\right) - \left(\frac{C_{21}C_{12}}{C_{11}} - C_{22}\right)\left(\frac{u}{r^2}\right) - \left(\frac{C_{21}C_{13}}{C_{11}} - C_{23}\right)\left(\frac{1}{r}\frac{\partial w}{\partial z}\right),$$

$$\frac{\partial \tau_{rz}}{\partial r} = -\frac{\tau_{rz}}{r} - \frac{C_{31}}{C_{11}}\frac{\partial \sigma_r}{\partial z} - \left(C_{32} - \frac{C_{12}C_{31}}{C_{11}}\right)\frac{\partial}{\partial z}\left(\frac{u}{r}\right) - \left(C_{33} - \frac{C_{13}C_{31}}{C_{11}}\right)\frac{\partial}{\partial z}\left(\frac{\partial w}{\partial z}\right)$$

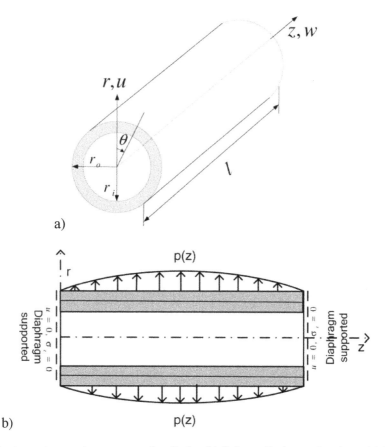

Fig. 1. a) Coordinate system and geometry of cylinder, b) finite cylinder under sinusoidal external pressure loading

and the auxiliary variables

$$\sigma_\theta = C_{21}\left(\frac{\partial u}{\partial r}\right) + C_{22}\left(\frac{u}{r}\right) + C_{23}\left(\frac{\partial w}{\partial z}\right), \quad \sigma_z = C_{31}\left(\frac{\partial u}{\partial r}\right) + C_{32}\left(\frac{u}{r}\right) + C_{33}\left(\frac{\partial w}{\partial z}\right). \quad (4b)$$

Variations of the four fundamental dependent variables which completely satisfy the boundary conditions of simple (diaphragm) supports at $z = 0, l$ can then be assumed as

$$u(r, z) = U(r)\sin\frac{\pi z}{l}, \quad w(r, z) = W(r)\cos\frac{\pi z}{l},$$
$$\sigma_r(r, z) = \sigma(r)\sin\frac{\pi z}{l}, \quad \tau_{rz}(r, z) = \tau(r)\cos\frac{\pi z}{l}. \quad (5)$$

Substitution of Eq. (5) in Eq. (4a) and simplification, resulting from orthogonality conditions of trigonometric functions, leads to the following four simultaneous ordinary differential equations involving only fundamental variables. These are

$$U'(r) = \frac{\sigma(r)}{C_{11}} - \frac{C_{12}}{C_{11}}\left(\frac{U(r)}{r}\right) + \frac{C_{13}}{C_{11}}\left(\frac{\pi}{l}W(r)\right),$$
$$W'(r) = \frac{1}{G}\tau(r) - U(r)\frac{\pi}{l},$$

$$\sigma'(r) = \frac{\pi}{l}\tau(r) + \left(\frac{C_{21}}{C_{11}} - 1\right)\frac{\sigma(r)}{r} - \left(\frac{C_{21} \cdot C_{12}}{C_{11}} - C_{22}\right)\left(\frac{U(r)}{r^2}\right) +$$

$$\left(\frac{C_{21}C_{13}}{C_{11}} - C_{23}\right)\left(\frac{\pi}{l}\frac{W(r)}{r}\right),$$

$$\tau'(r) = -\frac{\tau(r)}{r} - \frac{\pi}{l}\frac{C_{31}}{C_{11}}\sigma(r) - \left(C_{32} - \frac{C_{12}C_{31}}{C_{11}}\right)\left(\frac{\pi}{l}\frac{U(r)}{r}\right) + \qquad (6a)$$

$$\left(C_{33} - \frac{C_{13}C_{31}}{C_{11}}\right)\left(\left(\frac{\pi}{l}\right)^2 W(r)\right)$$

and the auxiliary variables

$$\sigma_\theta = \left[\frac{C_{21}}{C_{11}}\sigma(r) - \left(\frac{C_{21}C_{12}}{C_{11}} - C_{22}\right)\left(\frac{U(r)}{r}\right) + \left(\frac{C_{13}C_{21}}{C_{11}} - C_{23}\right)\left(\frac{\pi}{l}W(r)\right)\right]\sin\frac{\pi z}{l},$$

$$\sigma_z = \left[\frac{C_{31}}{C_{11}}\sigma(r) - \left(\frac{C_{31}C_{12}}{C_{11}} - C_{32}\right)\left(\frac{U(r)}{r}\right) + \left(\frac{C_{13}C_{31}}{C_{11}} - C_{33}\right)\left(\frac{\pi}{l}W(r)\right)\right]\sin\frac{\pi z}{l}. \qquad (6b)$$

3. Solution

The above system of first order simultaneous ordinary differential equations (Eq. (6a)) together with the appropriate boundary conditions at the inner and outer edges of the cylinder (Eq. (2)) forms a two-point BVP. However, a BVP in ODEs cannot be numerically integrated as only a half of the dependent variables (two) are known at the initial edge and numerical integration of an ODE is intrinsically an IVP. It becomes necessary to transform the problem into a set of IVPs. The initial values of the remaining two fundamental variables must be selected so that the complete solution satisfies the two specified conditions at the terminal boundary [7]. This technique has been successfully applied to the solutions of plate's problems [4–6,8,15]. However, this approach was not used for cylindrical problems in that literature. Runge-Kutta fourth order algorithm with modifications suggested by Gill [2] is used for the numerical integration of the IVPs. A computer code in FORTRAN 77 was written to perform the numerical integration.

4. Numerical Results

Nondimensionalized parameters are defined for pressure loading as follows:

$$\bar{r} = \frac{r}{R}, \quad (\bar{u},\bar{w}) = \frac{E_r}{pR}(u,w), \quad (\overline{\sigma_r}, \overline{\sigma_\theta}, \overline{\sigma_z}, \overline{\tau_{rz}}) = \frac{1}{p}(\sigma_r, \sigma_\theta, \sigma_z, \tau_{rz}).$$

Following material properties [11] are taken for orthotropic (0°) for Graphite-epoxy material-fibers are oriented in circumferential direction and layered (0°/90°) cylinders.

Layer-1 (fibers are oriented in circumferential direction 0-degree)

$$E_r = 9.65 \times 10^6, \quad E_\theta = 148 \times 10^6, \quad E_z = 9.65 \times 10^6, \quad G_{zr} = 3.015 \times 10^6,$$
$$\nu_{\theta r} = 0.3, \quad \nu_{zr} = 0.6, \quad \nu_{\theta z} = 0.3.$$

Layer-2 (fibers are oriented in axial direction 90-degree)

$$E_r = 9.65 \times 10^6, \quad E_\theta = 9.65 \times 10^6, \quad E_z = 148 \times 10^6, \quad G_{zr} = 4.55 \times 10^6,$$
$$\nu_{\theta r} = 0.6, \quad \nu_{zr} = 0.3, \quad \nu_{\theta z} = 0.019\,5.$$

Following material properties [14] are taken for the (0°/core/0°) sandwich cylinder:
Face Material properties are

$$E_r = 6.894 \times 10^6, \quad E_\theta = 172.36 \times 10^6, \quad E_z = 6.894 \times 10^6, \quad G_{zr} = 1.378 \times 10^6,$$
$$\nu_{\theta r} = 0.25, \quad \nu_{zr} = 0.25, \quad \nu_{\theta z} = 0.25.$$

Core Material properties are

$$E_r = 3.44 \times 10^6, \quad E_\theta = 0.275 \times 10^6, \quad E_z = 0.275 \times 10^6, \quad G_{zr} = 0.413 \times 10^6,$$
$$\nu_{\theta r} = 0.019\,9, \quad \nu_{zr} = 0.019\,9, \quad \nu_{\theta z} = 0.25.$$

Table 1. Non-dimensional radial stress, radial displacement and hoop stress for simple diaphragm supported orthotropic composite cylinder for $h/R = 1/5, 1/20, 1/50$

Quantity	h/R	$\bar{r}$	Present – Finite Length Cylinder			Analytical elasticity solution Lekhnitskii [12]	Numerical solution for infinitely long cylinder
			l/R				
			1	4	100–200		
$\overline{\sigma}_r\,(z = l/2)$	1/5	0.9	0.000 0	0.000 0	0.000 0	0.000 0	0.000 0
		1	0.526 3	0.532 0	0.532 4	0.537 1	0.537 1
		1.1	1.000 0	1.000 0	1.000 0	1.000 0	1.000 0
	1/20	0.975	0.000 0	0.000 0	0.000 0	0.000 0	0.000 0
		1	0.514 1	0.515 3	0.515 3	0.516 4	0.515 3
		1.02 5	1.000 0	1.000 0	1.000 0	1.000 0	1.000 0
	1/50	0.99	0.000 0	0.000 0	0.000 0	0.000 0	0.000 0
		1	0.506 2	0.506 7	0.506 7	0.507 1	0.507 1
		1.01	1.000 0	1.000 0	1.000 0	1.000 0	1.000 0
$\overline{u}\,(z = l/2)$	1/5	0.9	0.299 5	0.316 2	0.317 0	0.323 1	0.323 1
		1	0.334 4	0.343 5	0.343 8	0.340 5	0.340 5
		1.1	0.384 5	0.400 3	0.401 1	0.406 6	0.406 6
	1/20	0.975	1.321 5	1.326 2	1.326 4	1.327 9	1.326 4
		1	1.322 7	1.325 1	1.325 1	1.324 3	1.325 1
		1.025	1.327 1	1.331 8	1.332 0	1.333 4	1.332 0
	1/50	0.99	3.286 7	3.288 6	3.288 7	3.289 3	3.289 3
		1	3.281 2	3.282 2	3.282 2	3.281 9	3.281 9
		1.01	3.277 1	3.279 0	3.279 0	3.279 6	3.279 6
$\overline{\sigma}_\theta\,(z = l/2)$	1/5	0.9	4.925 3	5.284 6	5.304 3	5.506 2	5.304 3
		1	5.279 7	5.426 6	5.430 6	5.383 4	5.430 6
		1.1	5.831 5	5.973 3	5.978 0	5.969 3	5.978 0
	1/20	0.975	20.598 4	20.764 3	20.773 7	20.887 5	20.773 7
		1	20.439 0	20.476 7	20.477 6	20.465 5	20.477 6
		1.025	20.340 3	20.319 2	20.316 1	20.250 9	20.316 1
	1/50	0.99	50.730 3	50.851 0	50.858 4	50.956 3	50.858 4
		1	50.475 3	50.490 3	50.490 6	50.485 8	50.490 6
		1.01	50.246 5	50.183 8	50.179 0	50.100 5	50.179 0

Table 2. Non-dimensional radial stress, radial displacement and hoop stress for simple diaphragm supported orthotropic laminated $(0°/90°)$ composite cylinder for $h/R = 1/5, 1/20, 1/50$

Quantity	h/R	$\bar{r}$	Present – Finite Length Cylinder			Analytical elasticity solution Lekhnitskii [12]	Numerical solution for infinitely long cylinder
			l/R				
			1	4	100–200		
$\overline{\sigma_r}\,(z=l/2)$	1/5	0.9	0.0000	0.0000	0.0000	0.0000	0.0000
		1	0.9248	0.9833	0.9828	0.9828	0.9801
		1.1	1.0000	1.0000	1.0000	1.0000	1.0000
	1/20	0.975	0.0000	0.0000	0.0000	0.0000	0.0000
		1	0.9474	0.9515	0.9514	0.9514	0.9514
		1.025	1.0000	1.0000	1.0000	1.0000	1.0000
	1/50	0.99	0.0000	0.0000	0.0000	0.0000	0.0000
		1	0.9439	0.9443	0.9443	0.9443	0.9443
		1.01	1.0000	1.0000	1.0000	1.0000	1.0000
$\overline{u}\,(z=l/2)$	1/5	0.9	0.5438	0.5933	0.5930	0.5913	0.5930
		1	0.5773	0.6017	0.5991	0.5914	0.5991
		1.1	0.6051	0.6310	0.6288	0.6203	0.6288
	1/20	0.975	2.4316	2.4358	2.4349	2.4462	2.4349
		1	2.4200	2.4163	2.4148	2.3978	2.4148
		1.025	2.4001	2.3969	2.3955	2.3780	2.3955
	1/50	0.99	6.1038	6.0947	6.0936	6.1245	6.1199
		1	6.0823	6.0700	6.0687	6.0334	6.1062
		1.01	6.0527	6.0407	6.0394	6.0037	6.0760
$\overline{\sigma_\theta}\,(z=l/2)$	1/5	0.9	9.0855	10.1314	10.1444	10.0760	10.1444
		1	1.0635	1.1785	1.1811	1.1811	1.1811
		1.1	1.2205	1.1724	1.1639	1.1639	1.1639
	1/20	0.975	38.1753	38.4852	38.4860	38.4788	38.4860
		1	2.8932	2.9651	2.9685	2.9686	2.9685
		1.025	3.0228	2.9276	2.9199	2.9200	2.9199
	1/50	0.99	94.7739	94.8781	94.8770	94.8796	94.8770
		1	6.5333	6.5957	6.5993	6.6000	6.5993
		1.01	6.6573	6.5512	6.5436	6.5443	6.5436

Radial and hoop quantities are maximum at $z = l/2$ whereas axial quantities are maximum at $z = 0, l$. Analytical solution for radial stress, hoop stress and radial displacement from exact theory of anisotropic elasticity for infinitely long plane strain cylinder is given in Lekhnitskii [12]. These are used to validate and check the present results throughout wherever applicable. Comparisons of the results are given in Tables 1, 2 and 3.

Three sets of numerical results are presented in the above tables, i.e., results from the present finite length cylinder formulation, computations on the analytical formulae available for infinitely long cylinder [12] and numerically integrated values of the BVP of the infinitely long cylinder (see Appendix).

Here, first a long cylinder is subjected to a sinusoidal pressure load; the results within the limited central length zone only are compared with the plane strain one dimensional solutions.

Table 3. Non-dimensional radial stress, radial displacement and hoop stress for simple diaphragm supported sandwich composite cylinder for $h/R = 1/5, 1/20, 1/50$

Quantity	h/R	$\bar{r}$	Present – Finite Length Cylinder			Analytical elasticity solution Lekhnitskii [12]	Numerical solution for infinitely long cylinder
			l/R				
			1	4	100–200		
$\bar{\sigma}_r \,(z = l/2)$	1/5	0.9	0.000 0	0.000 0	0.000 0	0.000 0	0.000 0
		1	0.548 2	0.552 8	0.553 1	0.553 1	0.553 1
		1.1	1.000 0	1.000 0	1.000 0	1.000 0	1.000 0
	1/20	0.975	0.000 0	0.000 0	0.000 0	0.000 0	0.000 0
		1	0.520 2	0.521 3	0.521 4	0.521 4	0.521 4
		1.025	1.000 0	1.000 0	1.000 0	1.000 0	1.000 0
	1/50	0.99	0.000 0	0.000 0	0.000 0	0.000 0	0.000 0
		1	0.508 7	0.509 1	0.509 2	0.509 2	0.509 4
		1.01	1.000 0	1.000 0	1.000 0	1.000 0	1.000 0
$\bar{u} \,(z = l/2)$	1/5	0.9	0.967 4	0.996 4	0.996 7	0.996 2	0.996 5
		1	1.060 6	1.087 0	1.087 0	1.134 0	1.086 9
		1.1	1.149 8	1.179 3	1.179 6	1.179 9	1.180 2
	1/20	0.975	4.040 7	4.049 6	4.049 7	4.049 0	4.049 7
		1	4.057 4	4.065 1	4.065 1	4.109 5	4.065 1
		1.025	4.073 8	4.082 7	4.082 8	4.082 9	4.082 8
	1/50	0.99	10.022 4	10.025 8	10.025 8	10.025 2	10.025 8
		1	10.024 9	10.027 8	10.027 8	10.071 0	10.027 8
		1.01	10.027 4	10.030 8	10.030 8	10.030 9	10.030 8
$\bar{\sigma}_\theta \,(z = l/2)$	1/5	0.9	26.616 4	27.621 1	27.647 8	27.673 7	27.647 8
		1	0.053 9	0.056 2	0.056 2	0.056 2	0.056 2
		1.1	26.599 8	27.088 3	27.081 2	27.068 6	27.081 2
	1/20	0.975	103.329 3	103.791 4	103.809 4	103.826 8	103.809 4
		1	0.173 8	0.174 3	0.174 3	0.174 3	0.174 3
		1.025	99.877 7	99.867 2	99.853 8	99.839 7	99.853 8
	1/50	0.99	252.821 8	253.143 0	253.159 4	253.175 4	253.159 4
		1	0.411 6	0.411 9	0.411 9	0.411 9	0.411 9
		1.01	248.728 1	248.584 2	248.569 6	248.555 1	248.569 6

A good agreement is obtained. It is clearly seen that for long cylinders with higher l/R ratios, the results are close to the elasticity solution given by Lekhnitskii [12], for thick, moderately thick and thin cases.

Figs. 2–4 show the through thickness variation of basic as well as auxiliary hoop quantities for orthotropic cylinder for various l/R and h/R ratios. It is seen from Fig. 2 that radial stress varies linearly through thickness; radial displacement is linear for thick orthotropic cylinder whereas reverse trend is seen in thin cylinder. From Fig. 3 it can be seen that hoop stress varies parabolically in case of thick cylinder whereas it varies linearly in case of thin orthotropic cylinder. Parabolic variation of shear stress is seen in Fig. 4, axial displacement is constant through thickness as seen in Fig. 4. Numerical results for laminated ($0°/90°$) and sandwich ($0°$/core/$0°$) cylinder are presented in Figs. 5–7 and Figs. 8–9.

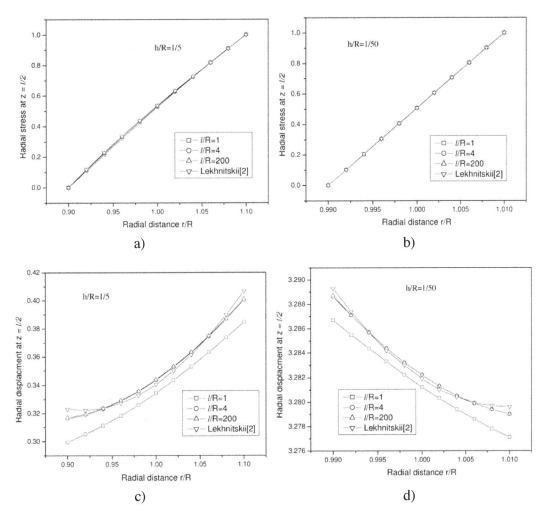

Fig. 2. Distribution of radial stress $\overline{\sigma_r}$ and radial displacement $\overline{u}$ through thickness subjected to sinusoidal loading for orthotropic cylinder

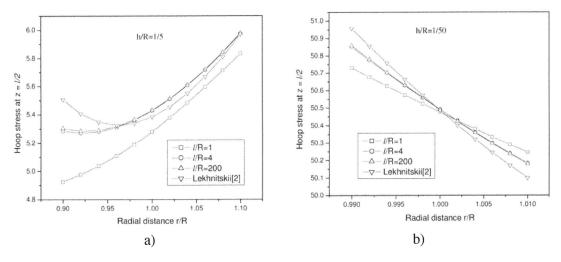

Fig. 3. Distribution of hoop stress $\overline{\sigma_\theta}$ through thickness subjected to sinusoidal loading for orthotropic cylinder

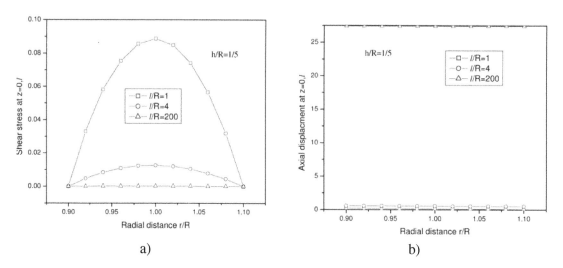

Fig. 4. Distribution of shear stress $\overline{\tau_{rz}}$ and axial displacement $\overline{w}$ through thickness subjected to sinusoidal loading for orthotropic cylinder

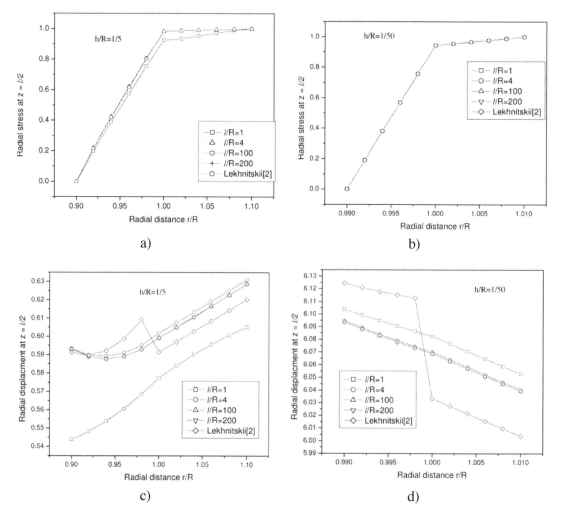

Fig. 5. Distribution of radial stress $\overline{\sigma_r}$ and radial displacement $\overline{u}$ through thickness subjected to sinusoidal loading for layered $(0°/90°)$ cylinder

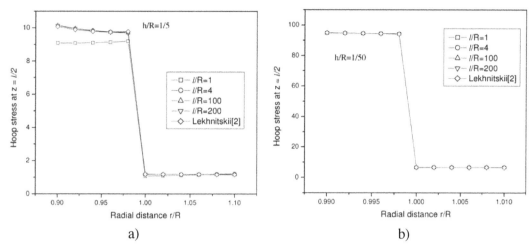

Fig. 6. Distribution of hoop stress $\overline{\sigma_\theta}$ through thickness subjected to sinusoidal loading for layered $(0°/90°)$ cylinder

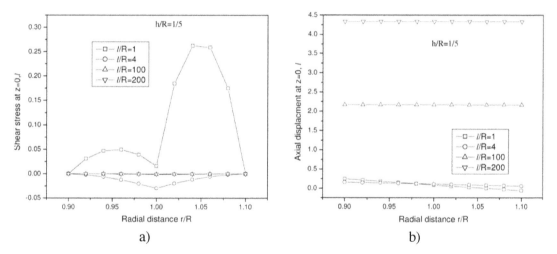

Fig. 7. Distribution of shear stress $\overline{\tau_{rz}}$ and axial displacement $\overline{w}$ through thickness subjected to sinusoidal loading for layered $(0°/90°)$ cylinder

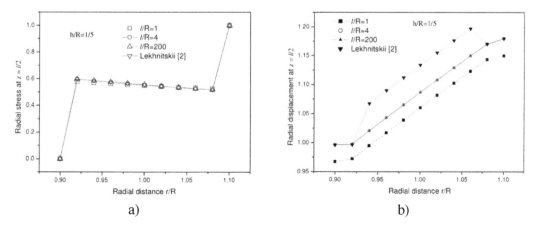

Fig. 8. Distribution of radial stress $\overline{\sigma_r}$ and radial displacement $\overline{u}$ through thickness subjected to sinusoidal loading for sandwich cylinder

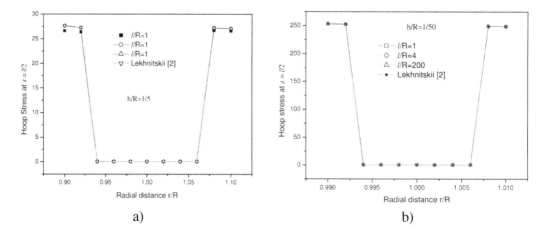

Fig. 9. Distribution of hoop stress $\overline{\sigma}_\theta$ through thickness subjected to sinusoidal loading for sandwich cylinder

5. Conclusions

Numerical analysis of orthotropic, laminated fiber reinforced composite and sandwich cylinders under sinusoidal pressure loading is presented. Homogeneous and anisotropic media are considered under conditions of simply (diaphragm) supported cylinder. Exact analytical solutions are available only for infinitely long cylinders. The present results of cylinders of finite length are not only new but are also very accurate. Proposed numerical technique was found to be efficient, since 1) the derivation involves mixed variables, both displacements and stresses. 2) The continuity conditions between the layers are satisfied automatically while performing the numerical integration in radial coordinate.

Appendix: 1D Formulation for Orthotropic and Layered Cylinder

$$\frac{\mathrm{d}\sigma_r}{\mathrm{d}r} + \frac{1}{r}(\sigma_r - \sigma_\theta) = 0, \quad \varepsilon_r = \frac{\partial u}{\partial r} \quad \varepsilon_\theta = \frac{u}{r}, \tag{A1}$$

$$\sigma_r = C_{11}\varepsilon_r + C_{12}\varepsilon_\theta, \qquad \sigma_r = C_{11}\frac{\mathrm{d}u}{\mathrm{d}r} + C_{12}\frac{u}{r},$$

$$\sigma_\theta = C_{12}\varepsilon_r + C_{22}\varepsilon_\theta, \qquad \sigma_\theta = C_{21}\frac{\mathrm{d}u}{\mathrm{d}r} + C_{22}\frac{u}{r},$$

$$\frac{\mathrm{d}u}{\mathrm{d}r} = \frac{\sigma_r}{C_{11}} - \frac{C_{12}}{C_{11}}\frac{u}{r}, \quad \frac{\mathrm{d}\sigma_r}{\mathrm{d}r} = \frac{\sigma_r}{r}\left(\frac{C_{21}}{C_{11}} - 1\right) + \frac{u}{r^2}\left(C_{22} - \frac{C_{21}C_{12}}{C_{11}}\right), \tag{A2}$$

where

$$\nu_{r\theta} = \frac{\nu_{\theta r}}{E_\theta}E_r, \ C_{11} = \frac{E_r}{(1 - \nu_{r\theta}\nu_{\theta r})}, \ C_{12} = \frac{\nu_{r\theta}E_\theta}{(1 - \nu_{r\theta}\nu_{\theta r})}, \ C_{22} = \frac{E_\theta}{(1 - \nu_{r\theta}\nu_{\theta r})}, \ C_{21} = C_{12}.$$

References

[1] Chandrashekhara, K., Kumar, B. S., Static analysis of a thick laminated circular cylindrical shell subjected to axisymmetric load. Composite Structures (23) (1993) 1–9.

[2] Gill, S., A Process for the step-by-Step integration of differential equations in an automatic digital computing machine. Proceedings of Cambridge Philosophical Society 1 (47) (1951) 96–108.

[3] Iyengar, K. T. S. R., Yogananda, C. V., Comparison of elasticity and shell theory solutions for long circular cylindrical shells, AIAA Journal 12 (4) (1966) 2 090–2 096.

[4] Kant, T., Numerical analysis of elastic plates with two opposite simply supported ends by segmentation method. Computers and Structures (14) (1981) 195–203.

[5] Kant, T., Numerical analysis of thick plates. Computer Methods in Applied Mechanics and Engineering (31) (1982) 1–18.

[6] Kant, T., Hinton, E., Mindlin plate analysis by segmentation method. ASCE Journal of Engineering Mechanics (109) (1983) 537–556.

[7] Kant, T., Ramesh, C. K., Numerical integration of linear boundary value problems in solid mechanics by segmentation method. International Journal of Numerical Methods in Engineering. (17) (1981) 1 233–1 256.

[8] Kant, T., Setlur, A. V., Computer analysis of clamped-clamped and clamped-supported cylindrical shells. Journal of the Aerospace Society of India 2 (25) (1973) 47–55.

[9] Klosner, J. M., Kempner, J., Comparison of elasticity and shell-theory solutions. AIAA Journal, 3 (1) (1963) 627–630.

[10] Klosner, J. M., Levine, H. S., Further comparison of elasticity and shell theory solutions. AIAA Journal 3 (4) (1966) 467–480.

[11] Kollar, L. P., Springer, G. S., Mechanics of Composite Structures. Cambridge University Press, New York, 2003.

[12] Lekhnitskii, S. G., Anisotropic Plates. Gordon and Breach Science, New York, 1968.

[13] Misovec, A. P., Kempner, J., Approximate elasticity solution for orthotropic cylinder under hydrostatic pressure and band loads. ASME Journal of Applied Mechanics 1 (37) (1970) 101–108.

[14] Pagano, N. J., Exact solutions for rectangular bidirectional composites and sandwich plates. Journal of Composite Materials (4) (1970) 21–34.

[15] Ramesh, C. K., Kant, T., Jadhav, V. B., Elastic analysis of cylindrical pressure vessels with various end closures. International Journal of Pressure Vessels and Piping (2) (1974) 143–154.

[16] Timoshenko, S., Goodier, J. N. Theory of Elasticity. McGraw-Hill, New York, 1951.

Nomenclature

r, θ, z	Cylindrical coordinates
u, v, w	Displacement components
$\sigma_r, \sigma_\theta, \sigma_z$	Normal stress components on planes normal to r, θ, and z axes
τ_{zr}	Shearing stress components in cylindrical coordinates
$\varepsilon_r, \varepsilon_\theta, \varepsilon_z$	Unit elongations (normal strains) in cylindrical coordinates
γ_{zr}	Shearing strain component in cylindrical coordinates
C_{ij}	Material constants for orthotropic material
ν	Poisson's ratio
r_i	Inner radius of the cylinder
r_0	Outer radius of the cylinder
l	Length of the cylinder
p	Uniform external pressure

$\overline{u}, \overline{w}$	Nondimensionalized displacement components
$\overline{\sigma_r}, \overline{\sigma_\theta}, \overline{\sigma_z}$	Nondimensionalized normal stress components
$\overline{\tau_{rz}}$	Nondimensionalized shearing stress in cylindrical coordinates
$\overline{r}$	Nondimensionalized radius
R	Mean radius $\frac{(r_0 + r_i)}{2}$

17

Modelling and modal properties of nuclear fuel assembly

V. Zeman[a,*], Z. Hlaváč[a]

[a] *Faculty of Applied Sciences, University of West Bohemia, Univerzitní 22, 306 14 Plzeň, Czech Republic*

Abstract

The paper deals with the modelling and modal analysis of the hexagonal type nuclear fuel assembly. This very complicated mechanical system is created from the many beam type components shaped into spacer grids. The cyclic and central symmetry of the fuel rod package and load-bearing skeleton is advantageous for the fuel assembly decomposition into six identical revolved fuel rod segments, centre tube and skeleton linked by several spacer grids in horizontal planes. The derived mathematical model is used for the modal analysis of the Russian TVSA-T fuel assembly and validated in terms of experimentally determined natural frequencies, modes and static deformations caused by lateral force and torsional couple of forces. The presented model is the first necessary step for modelling of the nuclear fuel assembly vibration caused by different sources of excitation during the nuclear reactor VVER type operation.

Keywords: fuel assembly, modelling of vibration, modal values, decomposition method

1. Introduction

Nuclear fuel assemblies are in term of mechanics very complicated system of beam type, which basic structure is formed from large number of parallel identical fuel rods, some guide thimbles and centre tube, which are linked by transverse spacer grids to each other and with skeleton construction [8]. The spacer grids are placed on several horizontal level spacings between support plates in reactor core [9].

Dynamic properties of nuclear fuel assembly (FA) are usually investigated using global models, whose properties are gained experimentally [4,7]. Eigenfrequencies and eigenvectors, investigated by measurement in the air, serve as initial data for parametric identification of the FA global model considered as one dimensional continuum of beam type [2]. This consideration is acceptable for a mathematical modelling and computer simulation of the whole nuclear reactor vibration caused by seismic excitation [2] and pressure pulsations [10] in terms of FA skeleton deformation. These global models of FA do not enable investigation of dynamic deformations and load of FA components and abrasion of fuel rods coating [5].

The goal of the paper is a development of analytical method for modelling and analysis of the FA modal properties. Motivation of this research work was exchange the American nuclear VVANTAGE 6 FA for Russian TVSA-T FA in NPP Temelín. The newly developed conservative mathematical model and corresponding computer model of the hexagonal type nuclear FA in parametric form enables to analyse modal properties, sensitivity to FA design parameters and parametric identification of FA components on the basis of measured static deformations, eigenfrequencies and eigenvectors. The presented methodology and FA detailed model is the

---

*Corresponding author. e-mail: zemanv@kme.zcu.cz.

first necessary step to modelling the dynamic response caused by forced and kinematics excitation. Dynamic forces between fuel rods and spacer grids will be used for calculation of expected lifetime period of nuclear FA in term of abrasion of fuel rods coating and fatigue live.

2. Mathematical models of FA subsystems

In order to model the fuel assembly, the system is divided into subsystems — identical rod segments (S), centre tube (CT) and load-bearing skeleton (LS) fixed in bottom part in lower piece (Fig. 1).

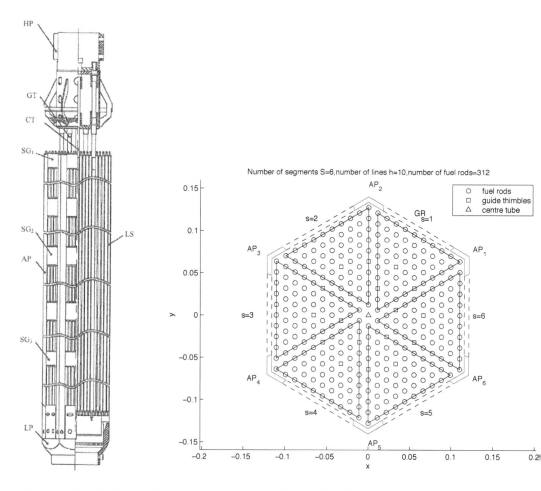

Fig. 1. Scheme of the fuel assembly   Fig. 2. The FA cross-section

2.1. Model of the rod segment

Because of the cyclic and central symmetric package of fuel rods and guide thimbles with respect to centre tube (Fig. 2), the FA decomposition of the identical rod segment $s = 1, \ldots, S$ (on the Fig. 2 for $S = 6$) shall be applied. Each rod segment is composed of R fuel rods with fixed bottom ends in lower piece and guide thimbles (GT) fully restrained in lower and head pieces. The fuel rods and guide thimbles inside the segments are linked by transverse spacer grids of three types ($SG_1 - SG_3$) which elastic properties are expressed by linear springs placed

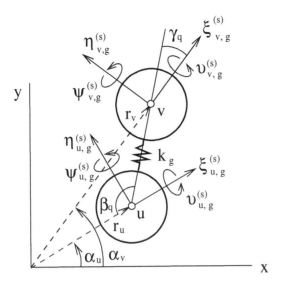

Fig. 3. The spring between two rods replacing the stiffness of spacer grid g

on several level spacings $g = 1, \ldots, G$ (see Fig. 1).The fuel rods are embedded into spacer grids with small initial tension, which wouldn't fall below zero during core operation.

The mathematical model of the rod segments isolated from adjacent segments (without linkages between segments) was derived [11] in the special coordinate system

$$q_s = [q_{1,s}^T, \ldots, q_{r,s}^T, \ldots, q_{R,s}^T]^T \, , \tag{1}$$

where $q_{r,s}$ is vector of nodal point displacements of one rod r (fuel rod or guide thimble) on the level of all spacer grids g in the form

$$q_{r,s} = [\ldots, \xi_{r,g}^{(s)}, \eta_{r,g}^{(s)}, \vartheta_{r,g}^{(s)}, \psi_{r,g}^{(s)}, \ldots]^T \, , \quad g = 1, \ldots, G \, . \tag{2}$$

Lateral displacements $\xi_{r,g}^{(s)}$, $\eta_{r,g}^{(s)}$ in contact nodal points with spacer grid g are mutually perpendicular whereas displacements $\xi_{r,g}^{(s)}$ are radial with respect to vertical central axis of FA. Displacements $\vartheta_{r,g}^{(s)}$, $\psi_{r,g}^{(s)}$ are bending angles of rod cross-section around lateral axes in contact nodal points (see Fig. 3).

The conservative mathematical model of the arbitrary isolated rod segment s was derived on the basis of Rayleigh beam theory in the form [11]

$$M_s \ddot{q}_s + \left(K_S + \sum_{q=1}^{Q} \sum_{g=1}^{G} K_{q,g} \right) q_s = 0 \, , \quad s = 1, \ldots, S \, , \tag{3}$$

where Q is the number of the transverse linear springs of one spacer grid inside one segment and $K_{q,g}$ is stiffness matrix corresponding to the coupling q by means of the spring k_g on the level of spacer grid g between two fuel rods u and v of the segment s. These coupling stiffnesses are determined by polar coordinates r_u, α_u and r_v, α_v of the linked fuel rods [3] and nonzero elements are localized at positions corresponding to displacements $\xi_{u,g}^{(s)}$, $\eta_{u,g}^{(s)}$, $\xi_{v,g}^{(s)}$, $\eta_{v,g}^{(s)}$ in the vector of generalized coordinates q_s in (1). The mass M_s and stiffness K_s matrices of the fuel rods and guide thimbles in one segment are block diagonal and have the form

$$X_s = \mathrm{diag}[X_R, \ldots, X_{GT}, \ldots, X_R] \in R^{4GR}, \quad X = M, K \, , \tag{4}$$

whereas matrices $X_R (X_{GT})$ correspond to one mutually uncoupled fuel rod (guide thimble). All fuel rods and guide thimbles are parallel and have identical boundary conditions (fuel rods have fixed lower ends and guide thimbles are fully restrained).

2.2. Model of the centre tube

The fully restrained centre tube (see Fig. 1 and Fig. 2) is discretized into G nodal points on the level of spacer grids $g = 1, \ldots, G$ by means of $G + 1$ prismatic beam finite elements [6] in the coordinate system

$$q_{CT} = [\ldots, x_g, y_g, \vartheta_g, \psi_g, \ldots]^T, \ g = 1, \ldots, G, \tag{5}$$

where lateral displacements x_g, y_g are oriented into axes x, y (Fig. 3). Mass and stiffness matrices M_{CT}, K_{CT} are symmetric of order $4G$.

2.3. Model of load-bearing skeleton

The load-bearing skeleton (further only skeleton) is created of S (on the Fig. 2 for $S = 6$) angle pieces (AP) coupled by divided grid rim (GR) at all levels of spacer grids (Fig. 4). Each angle piece with fixed bottom ends in lower piece is discretized into nodal points C_g in cross-section centre of gravity on the level of spacer grids $g = 1, \ldots, G$. The mathematical model of the skeleton without of couplings with spacer grids is derived in the coordinate system

$$q_{LS} = [q_{AP_1}^T, \ldots, q_{AP_s}^T, \ldots, q_{AP_S}^T]^T, \tag{6}$$

where q_{AP_s} is vector of nodal points displacements for particular angle piece s on the level of all grid rim g in the form

$$q_{AP_s} = [\ldots, \xi_{AP,g}^{(s)}, \eta_{AP,g}^{(s)}, \varphi_{AP,g}^{(s)}, \vartheta_{AP,g}^{(s)}, \psi_{AP,g}^{(s)}, \ldots]^T, \ g = 1, \ldots, G. \tag{7}$$

Lateral displacements $\xi_{AP,g}^{(s)}$, $\eta_{AP,g}^{(s)}$ of cross-section centre of gravity on the level of spacer grid g are mutually perpendicular whereas displacement $\xi_{AP,g}^{(s)}$ is radial (see Fig. 4). Displacements $\varphi_{AP,g}^{(s)}, \vartheta_{AP,g}^{(s)}, \psi_{AP,g}^{(s)}$ are torsional and bending angles of angle piece cross-section around vertical and lateral axes (in Fig. 4 indexes are let-out).

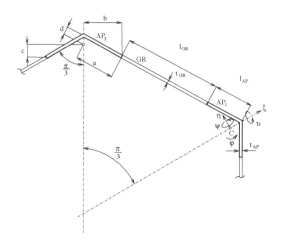

Fig. 4. Scheme of the load-bearing skeleton (part)

Mathematical model of the angle piece beam element between nodal points C_{g-1} and C_g in alternate coordinate system (indexes AP and (s) of coordinates are let-out)

$$q_{AP}^{(e)} = [\xi_{g-1}, \psi_{g-1}, \xi_g, \psi_g, \eta_{g-1}, \vartheta_{g-1}, \eta_g, \vartheta_g, \varphi_{g-1}, \varphi_g]^T \tag{8}$$

is written by mass and stiffness matrices in the form [1]

$$M_{AP}^{(e)} = \rho \begin{bmatrix} S_1^{-T}(AI_\phi + J_\eta I_{\phi'})S_1^{-1} & 0 & 0 \\ 0 & S_2^{-T}(AI_\phi + J_\xi I_{\phi'})S_2^{-1} & 0 \\ 0 & 0 & S_3^{-T} J_p I_\psi S_3^{-1} \end{bmatrix} \in R^{10,10}, \tag{9}$$

$$K_{AP}^{(e)} = \begin{bmatrix} S_1^{-T} E^* J_\eta I_{\phi''} S_1^{-1} & 0 & 0 \\ 0 & S_2^{-T} E^* J_\xi I_{\phi''} S_2^{-1} & 0 \\ 0 & 0 & S_3^{-T} G J_k I_{\psi'} S_3^{-1} \end{bmatrix} \in R^{10,10}, \tag{10}$$

where

$$I_\chi = \int_0^l \chi^T(x)\chi(x)\mathrm{d}x, \ I_{\chi'} = \int_0^l \chi'^T(x)\chi'(x)\mathrm{d}x, \ I_{\chi''} = \int_0^l \chi''^T(x)\chi''(x)\mathrm{d}x, \ \chi = \Phi, \Psi;$$

$$S_1 = \begin{bmatrix} 1 & 0 & 0 & 0 \\ 0 & 1 & 0 & 0 \\ 1 & l & l^2 & l^3 \\ 0 & 1 & 2l & 3l^2 \end{bmatrix}, \ S_2 = \begin{bmatrix} 1 & 0 & 0 & 0 \\ 0 & -1 & 0 & 0 \\ 1 & l & l^2 & l^3 \\ 0 & -1 & -2l & -3l^2 \end{bmatrix}, \ S_3 = \begin{bmatrix} 1 & 0 \\ 0 & l \end{bmatrix}$$

and $\Phi(x) = [1, x, x^2, x^3]$, $\Psi(x) = [1, x]$. Every beam element is determined by parameters ρ (mass density), A (cross-section area), J_ξ, J_η (second moment of the cross-section area to corresponding axes), J_p (polar second moment of area), $J_k \sim \frac{A^4}{40 J_p}$, l (length), E (Young's modulus), G (shear modulus) and $E^* = E \frac{1-\nu}{(1+\nu)(1-2\nu)}$ depends on Poisson's ratio ν.

To transform the model into general coordinates $q_{AP,s}$ defined in (7) by mass and stiffness matrices must be transformed in the form

$$X_e = P^T X^{(e)} P, \ X = M, \ K, \tag{11}$$

where

$$q_{AP}^{(e)} = P\tilde{q}_{AP}^{(e)}, \ \tilde{q}_{AP}^{(e)} = [\xi_{g-1}, \eta_{g-1}, \varphi_{g-1}, \vartheta_{g-1}, \psi_{g-1}, \xi_g, \eta_g, \varphi_g, \vartheta_g, \psi_g]^T.$$

Structure of the mass and stiffness matrices of one angle piece is given by following scheme

$$X_{AP} = \sum_{e=1}^G \mathrm{diag}[0, X_e, 0], \ X = M, \ K \tag{12}$$

with block matrices X_e determined in (11). Matrices of the first (lower) beam element in (12) must be arranged in accordance with angle pieces boundary conditions. The hardening of skeleton by welded grid rims within length l_{GR}, height h_{GR} and thickness t_{GR} (see Fig. 4) is respected by lateral beams fully restrained into adjacent angle pieces on the level of all spacer grids. The total mass and stiffness matrices of the skeleton in the coordinate system (6) have the form

$$M_{LS} = \mathrm{diag}[M_{AP} + \Delta M_{GR}, \ldots, M_{AP} + \Delta M_{GR}] \in R^{5GS},$$

$$\boldsymbol{K}_{LS} = \operatorname{diag}[\boldsymbol{K}_{AP}, \ldots, \boldsymbol{K}_{AP}] + \sum_{s=1}^{S} \sum_{g=1}^{G} \boldsymbol{K}_{GR,g}^{(s,s+1)} \in R^{5GS}, \tag{13}$$

where $\Delta \boldsymbol{M}_{GR}$ expresses additional mass matrix of the grid rims with mass concentrated into ends points of adjacent angle pieces on the level of all spacer grids and $\boldsymbol{K}_{GR,g}^{(s,s+1)}$ is stiffness matrix of one grid rim between adjacent angle pieces AP_s and AP_{s+1} on the level of spacer grid g.

3. Mathematical model of the fuel assembly

3.1. Structure of the fuel assembly model

The subsystems of FA are linked by spacer grids of different types for $g = 1, g = 2, \ldots, G - 1$ and $g = G$. In consequence of radial and orthogonal fuel rods and guide thimbles displacements mathematical models of segments are identical. Therefore the conservative model of the fuel assembly in configuration space

$$\boldsymbol{q} = [\boldsymbol{q}_1^T, \ldots, \boldsymbol{q}_s^T, \ldots, \boldsymbol{q}_S^T, \boldsymbol{q}_{CT}^T, \boldsymbol{q}_{LS}^T]^T \tag{14}$$

of dimension $n = 4GRS + 4G + 5GS$ can be written as

$$\boldsymbol{M}\ddot{\boldsymbol{q}} + (\boldsymbol{K} + \boldsymbol{K}_{S,S} + \boldsymbol{K}_{S,CT} + \boldsymbol{K}_{S,LS})\boldsymbol{q} = \boldsymbol{0}. \tag{15}$$

The mass $\boldsymbol{M}$ and stiffness $\boldsymbol{K}$ matrices correspond to a fictive fuel assembly divided into mutually uncoupled subsystems. Therefore these matrices are block diagonal

$$\boldsymbol{M} = \operatorname{diag}[\boldsymbol{M}_S, \ldots, \boldsymbol{M}_S, \boldsymbol{M}_{CT}, \boldsymbol{M}_{LS}], \quad \boldsymbol{K} = \operatorname{diag}[\boldsymbol{K}_S^*, \ldots, \boldsymbol{K}_S^*, \boldsymbol{K}_{CT}, \boldsymbol{K}_{LS}], \tag{16}$$

where segment stiffness matrix $\boldsymbol{K}_S^*$ includes couplings between all fuel rods and guide thimbles inside the segment. According to equation (3) it holds $\boldsymbol{K}_S^* = \boldsymbol{K}_S + \sum_{q=1}^{Q} \sum_{g=1}^{G} \boldsymbol{K}_{q,g}$.

3.2. Modelling of couplings between FA subsystems

The stiffness matrix $\boldsymbol{K}_{i,j,g}^{(s,s+1)}$ of one coupling by spacer grid g between fuel rod i at segment s and fuel rod j at segment $s+1$ has similar structure as matrix $\boldsymbol{K}_{q,g}$ in (3). Nonzero elements are localized at positions corresponding to displacements $\xi_{i,g}^{(s)}$, $\eta_{i,g}^{(s)}$ and $\xi_{j,g}^{(s+1)}$, $\eta_{j,g}^{(s+1)}$ in the vector of generalized coordinates $\boldsymbol{q}$ in (14). The total *coupling stiffness matrix between all segments* in the case of FA hexagonal type ($s = 1, \ldots, 6$) has structure

$$\boldsymbol{K}_{S,S} = \begin{bmatrix} \boldsymbol{K}_{1,1}^S & \boldsymbol{K}_{1,2}^S & 0 & 0 & 0 & \boldsymbol{K}_{1,6}^S & 0 & 0 \\ \boldsymbol{K}_{2,1}^S & \boldsymbol{K}_{2,2}^S & \boldsymbol{K}_{2,3}^S & 0 & 0 & 0 & 0 & 0 \\ 0 & \boldsymbol{K}_{3,2}^S & \boldsymbol{K}_{3,3}^S & \boldsymbol{K}_{3,4}^S & 0 & 0 & 0 & 0 \\ 0 & 0 & \boldsymbol{K}_{4,3}^S & \boldsymbol{K}_{4,4}^S & \boldsymbol{K}_{4,5}^S & 0 & 0 & 0 \\ 0 & 0 & 0 & \boldsymbol{K}_{5,4}^S & \boldsymbol{K}_{5,5}^S & \boldsymbol{K}_{5,6}^S & 0 & 0 \\ \boldsymbol{K}_{6,1}^S & 0 & 0 & 0 & \boldsymbol{K}_{6,5}^S & \boldsymbol{K}_{6,6}^S & 0 & 0 \\ 0 & 0 & 0 & 0 & 0 & 0 & 0 & 0 \\ 0 & 0 & 0 & 0 & 0 & 0 & 0 & 0 \end{bmatrix}. \tag{17}$$

Elastic properties of the spacer grids between the first fuel rods in all segments $s = 1, \ldots, S$ and the centre tube are expressed by stiffness $k_{S,CT}^{(g)}$ of the transverse springs on all level spacings $g = 1, \ldots, G$. The corresponding coupling stiffness matrix results from identity

$$\frac{\partial E_{S,CT}}{\partial \boldsymbol{q}} = \boldsymbol{K}_{S,CT} \boldsymbol{q} \,. \tag{18}$$

The potential (deformation) energy of these couplings is

$$E_{S,CT} = \sum_{s=1}^{S} \sum_{g=1}^{G} \frac{1}{2} k_{S,CT}^{(g)} (x_g \cos \alpha_s + y_g \sin \alpha_s - \xi_{1,g}^{(s)})^2 \,, \tag{19}$$

where in the case of hexagonal type of FA $\alpha_s = \frac{\pi}{6} + \frac{\pi}{3}(s-1)$ is radius vector angle of the first fuel rod in segment s with respect to x axis. The total *stiffness matrix between all segments and the centre tube* is

$$\boldsymbol{K}_{S,CT} = \sum_{s=1}^{S} \sum_{g=1}^{G} k_{S,CT}^{(g)} \begin{bmatrix} 1 & \cdots & -\cos \alpha_s & -\sin \alpha_s \\ \vdots & & \vdots & \vdots \\ -\cos \alpha_s & \cdots & \cos^2 \alpha_s & \sin \alpha_s \cos \alpha_s \\ -\sin \alpha_s & \cdots & \sin \alpha_s \cos \alpha_s & \sin^2 \alpha_s \end{bmatrix} \,, \tag{20}$$

where the introduced nonzero elements are localized at positions $4GR(s-1) + 4(g-1) + 1$ corresponding to fuel rod coordinates $\xi_{1,g}^{(s)}$ and $4GRS + 4(g-1) + 1 \div 2$ corresponding to centre tube coordinates x_g, y_g in the vector of generalized coordinates $\boldsymbol{q}$ in (14). This matrix for hexagonal type FA has structure

$$\boldsymbol{K}_{S,CT} = \begin{bmatrix} \boldsymbol{K}_{1,1}^{CT} & 0 & 0 & 0 & 0 & 0 & \boldsymbol{K}_{1,CT} & 0 \\ 0 & \boldsymbol{K}_{2,2}^{CT} & 0 & 0 & 0 & 0 & \boldsymbol{K}_{2,CT} & 0 \\ 0 & 0 & \boldsymbol{K}_{3,3}^{CT} & 0 & 0 & 0 & \boldsymbol{K}_{3,CT} & 0 \\ 0 & 0 & 0 & \boldsymbol{K}_{4,4}^{CT} & 0 & 0 & \boldsymbol{K}_{4,CT} & 0 \\ 0 & 0 & 0 & 0 & \boldsymbol{K}_{5,5}^{CT} & 0 & \boldsymbol{K}_{5,CT} & 0 \\ 0 & 0 & 0 & 0 & 0 & \boldsymbol{K}_{6,6}^{CT} & \boldsymbol{K}_{6,CT} & 0 \\ \boldsymbol{K}_{CT,1} & \boldsymbol{K}_{CT,2} & \boldsymbol{K}_{CT,3} & \boldsymbol{K}_{CT,4} & \boldsymbol{K}_{CT,5} & \boldsymbol{K}_{CT,6} & \boldsymbol{K}_{CT,CT} & 0 \\ 0 & 0 & 0 & 0 & 0 & 0 & 0 & 0 \end{bmatrix} \,. \tag{21}$$

Every angle piece AP_s of the skeleton encircles fuel rods 10 and 19 of the segment s and fuel rod 55 of the segment $s - 1$ (see Fig. 5). The potential (deformation) energy of one contact lateral springs $k_{S,AP}^{(g)}$ between single fuel rod r of the segment s and angle pieces AP_s on level spacings g is

$$E_{r,g}^{(s)} = \frac{1}{2} k_{S,AP}^{(g)} [\xi_{r,g}^{(s)} \cos(\delta - \alpha_r) + \eta_{r,g}^{(s)} \sin(\delta - \alpha_r) - \xi_{AP,g}^{(s)} \cos \delta - \eta_{AP,g}^{(s)} \sin \delta + e_r \varphi_{AP,g}^{(s)}]^2 \,. \tag{22}$$

The corresponding coupling stiffness matrix $\boldsymbol{K}_{r,g}^{(s)}$ results from identity

$$\frac{\partial E_{r,g}^{(s)}}{\partial \boldsymbol{q}} = \boldsymbol{K}_{r,g}^{(s)} \boldsymbol{q} \,, \tag{23}$$

whereas

$$
\boldsymbol{K}_{r,g}^{(s)} = k_{S,AP}^{(g)}
\begin{bmatrix}
\vdots & \vdots & & \vdots & \vdots & \vdots & \\
\cdots & C^2 & SC & \cdots & -C\cos\delta & -C\sin\delta & e_r C & \cdots \\
\cdots & CS & S^2 & \cdots & -S\cos\delta & -S\sin\delta & e_r S & \cdots \\
& \vdots & \vdots & & \vdots & \vdots & \vdots & \\
\cdots & -C\cos\delta & -S\cos\delta & \cdots & \cos^2\delta & \sin\delta\cos\delta & -e_r\cos\delta & \cdots \\
\cdots & -C\sin\delta & -S\sin\delta & \cdots & \sin\delta\cos\delta & \sin^2\delta & -e_r\sin\delta & \cdots \\
\cdots & Ce_r & Se_r & \cdots & -e_r\cos\delta & -e_r\sin\delta & e_r^2 & \cdots \\
& \vdots & \vdots & & \vdots & \vdots & \vdots &
\end{bmatrix}
\tag{24}
$$

and $C = \cos(\delta - \alpha_r)$, $S = \sin(\delta - \alpha_r)$. The introduced nonzero elements are localized at positions $4GR(s-1) + 4G(r-1) + 4(g-1) + 1 \div 2$ corresponding to fuel rod r coordinates $\xi_{r,g}^{(s)}$, $\eta_{r,g}^{(s)}$ and $4GRS + 4G + 5G(s-1) + 5(g-1) + 1 \div 3$ corresponding to angle piece AP_s coordinates $\xi_{AP,g}^{(s)}$, $\eta_{AP,g}^{(s)}$, $\varphi_{AP,g}^{(s)}$.

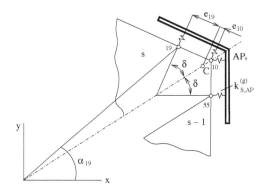

Fig. 5. Couplings between fuel rods of segments s and $s-1$ and the angle piece AP_s

The total stiffness matrix between fuel rods of all segments and all angle pieces (skeleton) is

$$
\boldsymbol{K}_{S,LS} = \sum_{s=1}^{S} \sum_{g=1}^{G} \sum_{r=10,19,55} \boldsymbol{K}_{r,g}^{(s)}.
\tag{25}
$$

This matrix for hexagonal type FA has structure

$$
\boldsymbol{K}_{S,LS} =
\begin{bmatrix}
\boldsymbol{K}_{1,1}^{LS} & 0 & 0 & 0 & 0 & 0 & 0 & \boldsymbol{K}_{1,LS} \\
0 & \boldsymbol{K}_{2,2}^{LS} & 0 & 0 & 0 & 0 & 0 & \boldsymbol{K}_{2,LS} \\
0 & 0 & \boldsymbol{K}_{3,3}^{LS} & 0 & 0 & 0 & 0 & \boldsymbol{K}_{3,LS} \\
0 & 0 & 0 & \boldsymbol{K}_{4,4}^{LS} & 0 & 0 & 0 & \boldsymbol{K}_{4,LS} \\
0 & 0 & 0 & 0 & \boldsymbol{K}_{5,5}^{LS} & 0 & 0 & \boldsymbol{K}_{5,LS} \\
0 & 0 & 0 & 0 & 0 & \boldsymbol{K}_{6,6}^{LS} & 0 & \boldsymbol{K}_{6,LS} \\
0 & 0 & 0 & 0 & 0 & 0 & 0 & 0 \\
\boldsymbol{K}_{LS,1} & \boldsymbol{K}_{LS,2} & \boldsymbol{K}_{LS,3} & \boldsymbol{K}_{LS,4} & \boldsymbol{K}_{LS,5} & \boldsymbol{K}_{LS,6} & 0 & \boldsymbol{K}_{LS,LS}
\end{bmatrix}.
\tag{26}
$$

The coupling stiffness matrices $\boldsymbol{K}_{SS}$, $\boldsymbol{K}_{S,CT}$ and $\boldsymbol{K}_{S,LS}$ express interaction between appropriate subsystems marked in subscripts. Therefore nonzero submatrices correspond to mutually linked subsystems and zero submatrices express an absent of coupling between subsystems.

3.3. Condensed model of the fuel assembly

The FA model (15) has to large DOF number $n = 4GRS + 4G + 5GS$ for calculation of the dynamic response excited by different sources of excitation. Therefore it is necessary to compile the condensed FA model using the modal synthesis method [6]. The *first step* is the modal analysis of the mutually isolated subsystems presented in section two

$$M_S \ddot{q}_s + (K_S + \sum_{q=1}^{Q} \sum_{g=1}^{G} K_{q,g})q_s = 0 \implies \Lambda_S, V_S \in R^{n_S}, n_S = 4GR,$$

$$M_{CT} \ddot{q}_{CT} + K_{CT} q_{CT} = 0 \implies \Lambda_{CT}, V_{CT} \in R^{n_{CT}}, n_{CT} = 4G, \quad (27)$$

$$M_{LS} \ddot{q}_{LS} + K_{LS} q_{LS} = 0 \implies \Lambda_{LS}, V_{LS} \in R^{n_{LS}}, n_{LS} = 5GS,$$

where $\Lambda_X, V_X, X = S, CT, LS$ are spectral and modal matrices of the subsystems, fulfilling the orthonormality conditions $V_X^T M_X V_X = E, V_X^T K_X V_X = \Lambda_X$. Further we choose a set of m_S low-frequency eigenvectors of the rod segment which will be arranged in its modal submatrix ${}^m V_S \in R^{n_S,m_S}$ corresponding to spectral submatrix ${}^m \Lambda_S \in R^{m_S,m_S}$. A set of other rod segment eigenmodes of each segment will be neglected. The *second step* is the transformation of the global vector of generalized coordinates defined in (14) by means of modal matrices (submatrices) of subsystems in the form

$$q = \begin{bmatrix} q_1 \\ q_2 \\ \vdots \\ q_S \\ q_{CT} \\ q_{LS} \end{bmatrix} = \begin{bmatrix} {}^m V_S & 0 & \dots & 0 & 0 & 0 \\ 0 & {}^m V_S & \dots & 0 & 0 & 0 \\ \dots & & & & & \\ 0 & 0 & \dots & {}^m V_S & 0 & 0 \\ 0 & 0 & \dots & 0 & V_{CT} & 0 \\ 0 & 0 & \dots & 0 & 0 & V_{LS} \end{bmatrix} \begin{bmatrix} x_1 \\ x_2 \\ \vdots \\ x_S \\ x_{CT} \\ x_{LS} \end{bmatrix} \quad (28)$$

for short

$$q = {}^m V x, \; {}^m V \in R^{n,m}, \; m = S m_S + 4G + 5GS. \quad (29)$$

The stiffness matrix of all couplings between subsystems is

$$K_C = K_{S,S} + K_{S,CT} + K_{S,LS} \quad (30)$$

and according to (17), (21) and (26) for hexagonal type FA ($S = 6$) has block structure

$$K_C = \begin{bmatrix} K_{1,1} & K_{1,2}^S & 0 & 0 & 0 & K_{1,6}^S & K_{1,CT} & K_{1,LS} \\ K_{2,1}^S & K_{2,2} & K_{2,3}^S & 0 & 0 & 0 & K_{2,CT} & K_{2,LS} \\ 0 & K_{3,2}^S & K_{3,3} & K_{3,4}^S & 0 & 0 & K_{3,CT} & K_{3,LS} \\ 0 & 0 & K_{4,3}^S & K_{4,4} & K_{4,5}^S & 0 & K_{4,CT} & K_{4,LS} \\ 0 & 0 & 0 & K_{5,4}^S & K_{5,5} & K_{5,6}^S & K_{5,CT} & K_{5,LS} \\ K_{6,1}^S & 0 & 0 & 0 & K_{6,5}^S & K_{6,6} & K_{6,CT} & K_{6,LS} \\ K_{CT,1} & K_{CT,2} & K_{CT,3} & K_{CT,4} & K_{CT,5} & K_{CT,6} & K_{CT,CT} & 0 \\ K_{LS,1} & K_{LS,2} & K_{LS,3} & K_{LS,4} & K_{LS,5} & K_{LS,6} & 0 & K_{LS,LS} \end{bmatrix}. \quad (31)$$

The diagonal block matrices are

$$K_{i,i} = K_{i,i}^S + K_{i,i}^{CT} + K_{i,i}^{LS}, \; i = 1, \dots, 6.$$

After the transformation (29) applied to FA model (15) the FA condensed conservative model

$$\ddot{x} + (^{m}\boldsymbol{\Lambda} +^{m}\boldsymbol{V}^{T}\boldsymbol{K}_{C}{}^{m}\boldsymbol{V})x = 0 \tag{32}$$

has $m = Sm_S + 4G + 5GS$ DOF number. Matrices of the condensed model have the block structure corresponding to FA decomposition

$$^{m}\boldsymbol{\Lambda} = \mathrm{diag}[^{m}\boldsymbol{\Lambda}_{S}, \ldots, {}^{m}\boldsymbol{\Lambda}_{S}, \boldsymbol{\Lambda}_{CT}, \boldsymbol{\Lambda}_{LS}] \in R^{m,m} \tag{33}$$

and

$$^{m}\boldsymbol{V}^{T}\boldsymbol{K}_{C}{}^{m}\boldsymbol{V} =
\begin{bmatrix}
\widetilde{\boldsymbol{K}}_{1,1} & \widetilde{\boldsymbol{K}}_{1,2} & 0 & 0 & 0 & \widetilde{\boldsymbol{K}}_{1,6} & \widetilde{\boldsymbol{K}}_{1,CT} & \widetilde{\boldsymbol{K}}_{1,LS} \\
\widetilde{\boldsymbol{K}}_{2,1} & \widetilde{\boldsymbol{K}}_{2,2} & \widetilde{\boldsymbol{K}}_{2,3} & 0 & 0 & 0 & \widetilde{\boldsymbol{K}}_{2,CT} & \widetilde{\boldsymbol{K}}_{2,LS} \\
0 & \widetilde{\boldsymbol{K}}_{3,2} & \widetilde{\boldsymbol{K}}_{3,3} & \widetilde{\boldsymbol{K}}_{3,4} & 0 & 0 & \widetilde{\boldsymbol{K}}_{3,CT} & \widetilde{\boldsymbol{K}}_{3,LS} \\
0 & 0 & \widetilde{\boldsymbol{K}}_{4,3} & \widetilde{\boldsymbol{K}}_{4,4} & \widetilde{\boldsymbol{K}}_{4,5} & 0 & \widetilde{\boldsymbol{K}}_{4,CT} & \widetilde{\boldsymbol{K}}_{4,LS} \\
0 & 0 & 0 & \widetilde{\boldsymbol{K}}_{5,4} & \widetilde{\boldsymbol{K}}_{5,5} & \widetilde{\boldsymbol{K}}_{5,6} & \widetilde{\boldsymbol{K}}_{5,CT} & \widetilde{\boldsymbol{K}}_{5,LS} \\
\widetilde{\boldsymbol{K}}_{6,1} & 0 & 0 & 0 & \widetilde{\boldsymbol{K}}_{6,5} & \widetilde{\boldsymbol{K}}_{6,6} & \widetilde{\boldsymbol{K}}_{6,CT} & \widetilde{\boldsymbol{K}}_{6,LS} \\
\widetilde{\boldsymbol{K}}_{CT,1} & \widetilde{\boldsymbol{K}}_{CT,2} & \widetilde{\boldsymbol{K}}_{CT,3} & \widetilde{\boldsymbol{K}}_{CT,4} & \widetilde{\boldsymbol{K}}_{CT,5} & \widetilde{\boldsymbol{K}}_{CT,6} & \widetilde{\boldsymbol{K}}_{CT} & 0 \\
\widetilde{\boldsymbol{K}}_{LS,1} & \widetilde{\boldsymbol{K}}_{LS,2} & \widetilde{\boldsymbol{K}}_{LS,3} & \widetilde{\boldsymbol{K}}_{LS,4} & \widetilde{\boldsymbol{K}}_{LS,5} & \widetilde{\boldsymbol{K}}_{LS,6} & 0 & \widetilde{\boldsymbol{K}}_{LS}
\end{bmatrix}, \tag{34}$$

where

$$\widetilde{\boldsymbol{K}}_{i,j} =^{m}\boldsymbol{V}_{S}^{T}\boldsymbol{K}_{i,j}{}^{m}\boldsymbol{V}_{S}; \;\; \widetilde{\boldsymbol{K}}_{i,CT} =^{m}\boldsymbol{V}_{S}^{T}\boldsymbol{K}_{i,CT}\boldsymbol{V}_{CT}; \;\; \widetilde{\boldsymbol{K}}_{i,LS} =^{m}\boldsymbol{V}_{S}^{T}\boldsymbol{K}_{i,LS}\boldsymbol{V}_{LS};$$

$$\widetilde{\boldsymbol{K}}_{CT,j} = \boldsymbol{V}_{CT}^{T}\boldsymbol{K}_{CT,j}{}^{m}\boldsymbol{V}_{S}; \;\; \widetilde{\boldsymbol{K}}_{LS,j} = \boldsymbol{V}_{LS}^{T}\boldsymbol{K}_{LS,j}{}^{m}\boldsymbol{V}_{S}; \;\; \widetilde{\boldsymbol{K}}_{CT} = \boldsymbol{V}_{CT}^{T}\boldsymbol{K}_{CT,CT}\boldsymbol{V}_{CT};$$

$$\widetilde{\boldsymbol{K}}_{LS} = \boldsymbol{V}_{LS}^{T}\boldsymbol{K}_{LS,LS}\boldsymbol{V}_{LS}; \;\; i = 1, \ldots, 6; \;\; j = 1, \ldots, 6.$$

Eigenfrequencies Ω_{ν} and eigenvectors

$$\boldsymbol{x}_{\nu} = [\boldsymbol{x}_{1,\nu}^{T}, \ldots, \boldsymbol{x}_{S,\nu}^{T}, \boldsymbol{x}_{CT,\nu}^{T}, \boldsymbol{x}_{LS,\nu}^{T}]^{T}, \; \nu = 1, \ldots, m$$

of FA are obtained from the modal analysis of the condensed model (32). Subvectors $\boldsymbol{x}_{s,\nu}$ ($s = 1, \ldots, S$) corresponding to rod segments, $\boldsymbol{x}_{CT,\nu}$ to centre tube and $\boldsymbol{x}_{LS,\nu}$ to skeleton, can be transformed according to (28) from the space of coordinates of the condensed model (32) to the original configuration space of the generalized coordinates ob subsystems by

$$\boldsymbol{q}_{X,\nu} =^{m}\boldsymbol{V}_{X}\boldsymbol{x}_{X,\nu}, \; X = 1, \ldots, S, CT, LS.$$

The eigenvalues calculated using condensed model (32) must be checked in light of accuracy with respect to noncondensed model (15) for different number m_S of applied rod segment master eigenvectors on the basis of the cumulative relative error of the eigenfrequencies and the normalized cross orthogonality matrix [11].

4. Application

The presented methodology and developed software in Matlab code was tested for the Russian TVSA-T fuel assembly used in nuclear power plant Temelín [8]. This FA of the hexagonal type (Fig. 1 and Fig. 2) has six rod segments ($S = 6$) and eight spacer grids ($G = 8$). Each segment contains 52 fuel rods and 3 guide thimbles ($R = 55$) linked by 135 transverse springs between adjacent rods within stiffnesses $k_1 = 2 \cdot 10^5$, $k_2 = \cdots = k_7 = 1,83 \cdot 10^5$, $k_8 = 2,07 \cdot 10^5$ N/m on particular levels of spacer grids $g = 1, \ldots, 8$. The rod spacing is 12.75 mm. The noncondensed FA model under consideration has $n = 10\,832$ ($n_S = 1\,760$, $n_{CT} = 32$, $n_{LS} = 240$) DOF number. The lowest FA eigenfrequencies are $f_1 = f_2 = 3.43$ Hz at temperature $20\,°C$ and $f_1 = f_2 = 3.09$ Hz at temperature $350\,°C$. Pairs of eigenfrequencies correspond to flexural and breathing mode shapes and single eigenfrequencies correspond to torsion mode shapes. The spectrum of nineteen lowest (up to 20 Hz) eigenfrequencies with the characteristics of corresponding mode shapes is presented in Table 1. For the sake of completeness we introduce measured flexural mode shapes at temperature 20 °C provided by ŠKODA, Nuclear Machinery, Co.Ltd.

Table 1. Eigenfrequencies and characteristics of corresponding natural modes of the FA model

	Eigenfrequencies [Hz]			Characteristics of mode shapes
ν	$t = 350°$	$t = 20°$	Measured	
1	3.09	3.43	3.9	Flexural, 1.mode
2	3.09	3.43		
3	4.13	4.58		Torsional, 1. mode
4	6.24	6.90	6.6	Flexural, 2. mode
5	6.24	6.90		
6	8.71	9.46		Torsional, 2. mode
7	9.49	10.46	9.4	Flexural, 3. mode
8	9.49	10.46		
9	11.74	12.56		Torsional, 3. mode
10	12.88	14.21	12.5	Flexural, 4. mode
11	12.88	14.21		
12	14.24	15.22		Torsional, 4. mode
13	16.47	18.17	18.6	Flexural, 5. mode
14	16.47	18.17		
15	17.23	18.60		Torsional, 5. mode
16	19.26	19.33		Breathing mode
17	19.26	19.33		
18	19.79	19.98		Breathing mode
19	19.79	19.98		

The spectrum of eigenfrequencies is very crowded especially for higher frequencies. The flexural mode shapes (Fig. 6,7) are characterized by inphase deformations of all FA components whereas spacer grids are practically non-deformed. The torsional mode shapes are characterized by maximal deformations of outsider fuel rods (Fig. 8) and spacer grids roll up practically without their deformations. The breathing modes (Fig. 9, 10) corresponding to higher eigenfrequencies approximately from 20 Hz (see Table 1) are characterized by spacer grid deformations and relatively high contact forces between fuel rods and spacer grids. All mode shapes in Fig. 6–10 are vizualized on the FA cross-section on the level of the fourth (central) spacer grid.

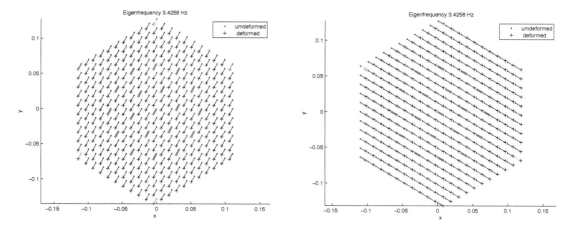

Fig. 6. The first FA mode shape (flexural mode)   Fig. 7. The second FA mode shape (flexural mode)

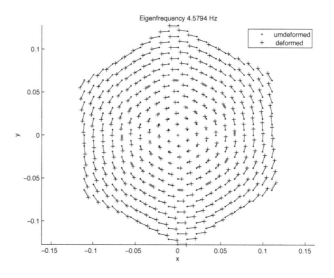

Fig. 8. The third FA mode shape (torsional mode)

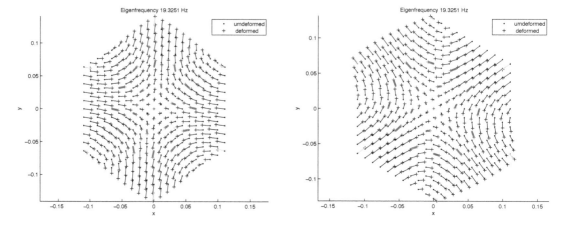

Fig. 9. The 16th FA mode shape (breathing mode)   Fig. 10. The 17th FA mode shape (breathing mode)

5. Conclusion

The described method enables to model effectively the flexural and torsional vibration of nuclear fuel assemblies. The special coordinate system of radial and orthogonal displacements of the fuel assembly components — fuel rods, guide thimbles, centre tube and skeleton angle pieces — enables to separate the system into several identical revolved rod segments characterized by identical mass and stiffness matrices, centre tube and load-bearing skeleton as subsystems. The subsystems are linked by spacer grids of different types on particular levels of spacer grids.

This new approach to modelling based on the system decomposition enables simple including of model particular components with identified parameters into global FA model, to significantly decrease of time demands of computing program assembladge and to save the computer memory. The preliminary results of the modal analysis of the Russian TVSA-T fuel assembly show, that in low-frequency spectrum of excitation (approximately up to 20 Hz) the flexural and torsional mode shapes are employed and in high-frequency spectrum the breathing mode shapes, characterized by spacer grid deformation on all levels, are employed.

The condensed FA model based on modal synthesis method with reduction of rod segment DOF number, will be applied to calculation of forced vibration caused by pressure pulsations and seismic excitation in terms of fuel assembly component deformations and abrasion of fuel rods coating.

Acknowledgements

This work was supported by the research project MSM 4977751303 of the Ministry of Education, Youth and Sports of the Czech Republic.

References

[1] Byrtus, M., Hajžman, M., Zeman, V., Dynamics of rotating systems, University of West Bohemia, Plzeň, 2010 (in Czech).

[2] Hlaváč, Z., Zeman, V., The seismic response affection of the nuclear reactor WWER1000 by nuclear fuel assemblies, Engineering Mechanics, 3/4 (17) (2010) 147–160.

[3] Hlaváč, Z., Zeman, V., Flexural vibration of the package of rods linked by lattices, Proceedings of the 8-th conference Dynamic of rigid and deformable bodies 2010, Ústí nad Labem, 2010 (in Czech).

[4] Lavreňuk, P. I., Obosnovanije sovmestnosti TVSA-T PS CUZ i SVP s projektom AES Temelín, Statement from technical report TEM-GN-01, Sobstvennost' OAO TVEL (inside information of NRI Řež, 2009).

[5] Pečínka, L., Criterion assessment of fuel assemblies behaviour VV6 and TVSA-T at standard operating conditions of ETE V1000/320 type reactor, Research report DITI 300/406, NRI Řež, 2009 (in Czech).

[6] Slavík, J., Stejskal, V., Zeman, V., Elements of dynamics of machines, ČVUT, Praha, 1997 (in Czech).

[7] Smolík, J., and coll., Vvantage 6 Fuel Assembly Mechanical Test, Technical Report No. Ae 18018T, Škoda, Nuclear Machinery, Pilsen, Co.Ltd., 1995.

[8] Sýkora, M., Reactor TVSA-T fuel assembly insertion, part 4, Research report Pp BZ1, 2, ČEZ-ETE, 2009 (in Czech).

[9] Zeman, V., Hlaváč, Z., Modelling of WWER 1000 type reactor by means of decomposition method, Engineering Mechanics 2006, Institute of Theoretical and Applied Mechanics AS CR, Prague 2006, p. 444 (full paper on CD-ROM in Czech).

[10] Zeman, V., Hlaváč, Z., Dynamic response of VVER 1000 type reactor excited by pressure pulsations, Engineering Mechanics 6 (15) (2008) 435–446.

[11] Zeman, V., Hlaváč, Z., Modal properties of the flexural vibrating package of rods linked by spacer grids, Applied and Computational Mechanics 1 (5) (2011) 111–122.

18

Constitutive modelling of an arterial wall supported by microscopic measurements

J. Vychytil[a,*], P. Kochová[a], Z. Tonar[a], J. Kuncová[b], J. Švíglerová[b]

[a]*Faculty of Applied Sciences, University of West Bohemia in Pilsen, Univerzitní 22, 306 14 Pilsen, Czech Republic*
[b]*Faculty of Medicine in Pilsen, Charles University in Prague, Lidická 1, 301 66 Pilsen, Czech Republic*

Abstract

An idealized model of an arterial wall is proposed as a two-layer system. Distinct mechanical response of each layer is taken into account considering two types of strain energy functions in the hyperelasticity framework. The outer layer, considered as a fibre-reinforced composite, is modelled using the structural model of Holzapfel. The inner layer, on the other hand, is represented by a two-scale model mimicking smooth muscle tissue. For this model, material parameters such as shape, volume fraction and orientation of smooth muscle cells are determined using the microscopic measurements. The resulting model of an arterial ring is stretched axially and loaded with inner pressure to simulate the mechanical response of a porcine arterial segment during inflation and axial stretching. Good agreement of the model prediction with experimental data is promising for further progress.

Keywords: arterial wall, hyperelasticity, multi-scale modelling, stereology

1. Introduction

Mechanical response of an arterial wall is often predicted using hyperelastic models in the framework of continuum mechanics. In that case, the stress-strain relationship is defined via the so-called strain energy function representing the density of the Helmholtz free energy. In order to fit experimental data, phenomenological models such as the Mooney-Rivlin, the Ogden or the exponential model of the Fung's type are introduced using the empirical formulae of strain energy functions, see e.g. [3,14]. Although widely used (see e.g. [1,2,18]), their application on the description of arterial wall is limited due to its complex microstructure. Phenomenological models may be suitable for the description of an overall mechanical response, however, they are incapable of providing the insight into the microscopic level. At the same time, physical meaning of the material constants appearing in these models is not always clear. The least-squares fitting instead of direct measurements must be applied for their identification, as it is detailed e.g. in [13].

Therefore, an increasing effort has been made in developing so-called structural models that are able to relate the overall mechanical response to the corresponding effects at microscopic level. An example of the strain energy function, proposed with respect to the underlying microstructure, is shown in [7] and [9]. In both cases, anisotropy and the structure of arterial layers are taken into account via a single parameter representing the angle between preferred fibre directions. The first mentioned model is later improved in [5] by considering the dispersion of fibres orientation. Possible application is shown in [6] for the soft tissue remodeling, i.e. the

*Corresponding author. e-mail: vychytil@kme.zcu.cz.

alignment of collagen fibres along the directions of principal stresses. In [10], an anisotropic hyperelastic model is proposed for fibre-reinforced materials. Taking into account an arbitrary number of fibre families, the strain energy function is proposed as a sum of exponential functions which makes it suitable for the description of soft collagenous tissues. Also, an issue of polyconvexity as a favourable property of strain energy functions is addressed. The model is generalized in [4] and applied for the description of uniaxial and biaxial tension tests with human coronary arteries and abdominal aorta. In this case, five material parameters are determined upon the comparison with experimental data.

Another approach of including the fibrous nature of the soft tissue into a hyperelastic model is demonstrated in [20]. Here the eight chain model is introduced as a representative volume element of the arterial layer. As a result, orthotropic strain energy function of the tissue is obtained respecting the entropy elasticity of individual polymer chains. Especially in this case of the bottom-up approach, all material parameters have a clear physical meaning related to the microstructure. However, as in the case of the angle between fibre orientation in [7], most of these parameters have to be determined by the least-squares fitting as their direct identification is not feasible.

The aim of this work is to contribute to the trend of the bottom-up approach in the constitutive modelling of arteries by employing the two-scale hyperelastic model which is based on the arrangement of soft tissues at the cellular level. The paper is organized as follows. In section 2 an idealized model of an artery is proposed as a two-layer thick walled tube composed of media and adventitia layers. Description of kinematics follows the approach proposed in [7] considering three configurations. Hence, residual stresses are taken into account via a parameter representing the opening angle of the arterial ring at the reference state. Distinct mechanical properties of individual layers are taken into account by considering two types of strain energy functions. Adventitia, the outer layer, is represented by the structural model of Holzapfel et al. proposed in [7]. Media, on the other hand, is described using the so-called "balls and springs" (BS) model introduced in [8].

Direct identification of some of the model parameters is done in section 3. Experimental methods of stereological assessment are performed for the smooth muscle tissue of a porcine abdominal aorta and a gastropod as it is detailed in [16,17]. Thus the structural information is obtained, namely the relative volume fraction of smooth muscle cells within the media, their orientation, their size and their shape. It corresponds directly to several constants appearing in the BS model representing the material of media. Moreover, thickness of both media and adventitia is determined for a carotid artery from a freshly killed domestic pig.

Finally, the model predictions are compared to the experimental data in section 4. The porcine carotid artery is stretched longitudinally and pressurized at the same time following the experimental procedure described in [19]. As a result, pressure-diameter diagram is obtained for a given value of the axial stretch. Theoretical curves predicted by the model are obtained using the combination of both direct measurement of material parameters and the least-squares fitting.

2. Model of an arterial segment

2.1. Description of deformations

An arterial segment is considered as an axisymmetric thick-walled tube consisting of two layers, the media (inner) and the adventitia (outer). Its description follows the approach detailed in [7,19]. The reference configuration is characterized with the length L, inner and outer radii

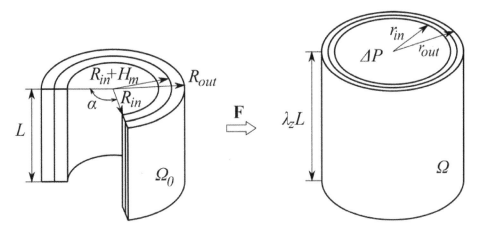

Fig. 1. The reference (left) and the current configuration (right) of an arterial segment loaded with inner pressure and axial stretch

R_{in}, R_{out}, media thickness H_m and the opening angle α, see Fig. 1. Loaded with the inner pressure ΔP and the axial stretch λ_z, the segment occupies the current configuration characterized with the length $\lambda_z L$, and the inner and outer radii r_{in}, r_{out}. Clearly, we assume the segment remaining as an axisymmetric tube also at this configuration, see Fig. 1. Due to the symmetry of the geometry and the loading, the description of kinematics is in fact reduced into a one-dimensional problem of the deformed radius $r(R)$. Applying the assumption of incompressibility, it is given with the formula

$$r(R) = \sqrt{\frac{R^2}{h\lambda_z} + C}. \tag{1}$$

Here, h is a parameter related to the opening angle as

$$h = \frac{2\pi}{2\pi - \alpha}, \tag{2}$$

quantifying in fact the residual stress and C is a constant to be determined from the boundary conditions. Applying the constitutive equations of hyperelasticity,

$$\boldsymbol{\sigma} = \frac{\partial \hat{W}}{\partial \mathbf{F}} \mathbf{F}^T - p\mathbf{I}, \tag{3}$$

we obtain the relationship between ΔP and C,

$$\Delta P = \int_{R_{in}}^{R_{in}+H_m} \frac{r'}{r}\left(\frac{hr}{R}\hat{W}_2^m - r'\hat{W}_1^m\right)\,\mathrm{d}R + \int_{R_{in}+H_m}^{R_{out}} \frac{r'}{r}\left(\frac{hr}{R}\hat{W}_2^a - r'\hat{W}_1^a\right)\,\mathrm{d}R. \tag{4}$$

Here, $\boldsymbol{\sigma}$ is the Cauchy stress tensor, $\hat{W}$ the strain-energy function, $\mathbf{F}$ the deformation gradient, p the hydrostatic pressure and the superscripts m and a refer to the media and adventitia, respectively. The notion $\hat{W}_i$ stands for the partial derivative of $\hat{W}$ with respect to the corresponding diagonal component of the deformation gradient,

$$\hat{W}_i = \frac{\partial \hat{W}}{\partial \lambda_i}, \quad \lambda_i = F_{ii}. \tag{5}$$

2.2. Material models

In agreement with [7], the adventitia is considered as a fibre-reinforced composite, characterized by the strain energy function

$$\hat{W}^a = \frac{\mu}{2}\left(\hat{I}_1 - 3\right) + \frac{\kappa_1}{2\kappa_2}\left[\left(e^{\kappa_2(\hat{I}_4-1)^2} - 1\right) + \left(e^{\kappa_2(\hat{I}_6-1)^2} - 1\right)\right], \tag{6}$$

where μ, κ_1 and κ_2 are material constants, $\hat{I}_1$, $\hat{I}_4$ and $\hat{I}_6$ are invariants and pseudo-invariants, respectively. They are defined as

$$\hat{I}_1 = \operatorname{tr}\hat{\mathbf{C}}, \quad \hat{I}_4 = \hat{\mathbf{C}} : \mathbf{A}_1, \quad \hat{I}_6 = \hat{\mathbf{C}} : \mathbf{A}_2, \quad \mathbf{A}_1 = \boldsymbol{a}_{01} \otimes \boldsymbol{a}_{01}, \quad \mathbf{A}_2 = \boldsymbol{a}_{02} \otimes \boldsymbol{a}_{02}. \tag{7}$$

Here, $\hat{\mathbf{C}}$ is the distortional component of the right Cauchy-Green tensor,

$$\hat{\mathbf{C}} = (\det \mathbf{C})^{-1/3}\mathbf{C}, \qquad \mathbf{C} = \mathbf{F}^T\mathbf{F}. \tag{8}$$

The material is supposed to be reinforced by two families of fibres with predominant directions $\boldsymbol{a}_{01}$ and $\boldsymbol{a}_{02}$.

Note that the strain-energy function (6) is in fact composed of two contributions. The first one corresponds to the isotropic mechanical response of the matrix and is represented by the neo-Hookean term. The second one corresponds to the anisotropic mechanical response due to collagen fibres. The fibres are supposed to form a symmetrical structure so that their direction vectors can be expressed as

$$\boldsymbol{a}_{01} = [0, \cos\beta, \sin\beta]^T, \qquad \boldsymbol{a}_{02} = [0, \cos\beta, -\sin\beta]^T. \tag{9}$$

The angle β between the fibres represents an additional parameter appearing in the material model, see Fig. 2.

Media, on the other hand, is of different structure. Focusing on the contribution of the smooth muscle tissue we use the two-scale hyperelastic model introduced in [8]. Its microstructure is formed of incompressible balls interconnected mutually via linear springs to mimic the arrangement of smooth muscle cells and extracellular matrix. Each ball is also reinforced with

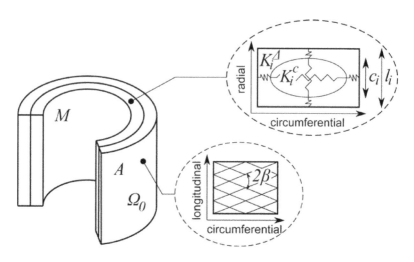

Fig. 2. Representative volume elements of the material models corresponding to the continuum points in the adventitia (A) and the media (M) layers. Two-dimensional sections are depicted for simplicity

linear springs representing cytoskeleton. Hence we call it the "balls and springs" (BS) model. The representative volume element (RVE) of the BS model is depicted in Fig. 2. Here, l_i denotes the sizes of the RVE, c_i the sizes of the ellipsoidal ball, K_i^c the stiffness of the springs reinforcing the ball and K_i^Δ the stiffness of the springs representing the extracellular matrix. The subscript i denotes the spatial direction, $i = 1, 2, 3$. The strain energy function can be expressed in the form

$$W^m = \sum_{\substack{i=1 \\ i \neq j \neq k}}^{3} \frac{K_i}{2} \frac{k_i}{1 + k_i} l_{ij} l_{ik} (\lambda_i - 1)^2 + \min_{\substack{r_i \\ r_1 r_2 r_3 = 1}} \sum_{\substack{i=1 \\ i \neq j \neq k}}^{3} \frac{K_i}{2} (1 + k_i) g_i^2 l_{ij} l_{ik} \left(r_i - \lambda_i^{eff} \right)^2. \quad (10)$$

Here, the effective stretches are introduced,

$$\lambda_i^{eff} = \frac{\lambda_i - 1}{g_i (1 + k_i)} + 1. \quad (11)$$

The material parameters of this model are derived from the microscopic level as

$$K_i = \frac{K_i^\Delta}{l_i}, \qquad k_i = \frac{K_i^c}{K_i^\Delta}, \qquad l_{ij} = \frac{l_i}{l_j}, \qquad g_i = \frac{c_i}{l_i}. \quad (12)$$

Although the strain energy function is formed solely by quadratic contributions of the stretched/compressed springs connected in a simply-looking structure, the resulting formula (10) is rather complex. The reason for this is due to the configuration of inner structure represented by the shape of the ball. Instead of simply assuming the shape of the ball to be governed by the macroscopic deformation gradient (affine micro-deformations), it is "allowed" to occupy the shape which minimizes the overall deformation energy of the RVE. Hence, the strain energy function is split into a part describing an averaged spring elasticity of whole RVE and a part expressing the corresponding minimization problem. The dependence on the shape of the ball is given via dimensionless parameters r_i,

$$r_i = \frac{c_i'}{c_i}, \quad (13)$$

where c_i' represent sizes of the ball at the actual configuration. Clearly, the constraint

$$r_1 r_2 r_3 = 1, \quad (14)$$

stands for the incompressibily of the ball. Notice that the reference configuration of this model coincides with the natural one of zero strain energy as

$$r_i = 1, \qquad \lambda_i = 1 \Rightarrow \lambda_i^{eff} = 1, \qquad W^m = 0. \quad (15)$$

It means that there is neither tension nor compression within springs forming the RVE at the reference state and hence it is stress-free. Unfortunately, the minimization problem (10) has no analytical solution in general. Therefore, there is no analytical formula regarding stress-strain relationship which means that stress components have to be calculated numerically using the definition (3).

It is worth stressing that the proposed model is orthotropic and has a large number of material parameters. However, concerning the smooth muscle tissue within arterial wall, the assumption of the transverse isotropy is employed considering the radial and longitudinal directions to be equivalent. This assumption result in the total number of seven material parameters, namely $K_1, K_2, k_1, k_2, g_1, g_2$ and l_{12}.

3. Identification of model parameters

3.1. Geometry of an arterial ring and the media thickness

A sample of carotid artery was taken from a minipig involved in another experiment performed at the experimental facility of the Faculty of Medicine in Pilsen. The animal received humane care in compliance with the European Convention on Animal Care and the whole project was approved by the Faculty Committee for the Prevention of Cruelty to Animals. The carotid artery was rinsed with Tyrode's solution and placed in ice-cold Tyrode's solution. An approximately 1 cm-long ring was taken from the artery to assess its geometrical parameters, i.e. the inner and outer radii and the opening angle. Using three measurements at each end of the sample, the mean values were obtained as $R_{in} = 1.29$ mm, $R_{out} = 1.49$ mm and $\alpha = 108°$.

The sample was then fixed in 10 % formalin, dehydrated in graded ethanol solutions and embedded in paraffin blocks for histological analysis. The tissue block was cut transversally into 5 μm-thick histological sections, mounted on slides and stained with Verhoeff's haematoxylin and green trichrome [12] in order to show the overall layers of the arterial wall (the intima, the media, and the adventitia) and to distinguish elastin, collagen fibres, and vascular smooth muscle cells (SMC). The cytoplasm of SMC stains reddish with the acid fuchsin. Four micrographs were analyzed using the Ellipse software (ViDiTo, Košice, Slovakia). They were sampled in a systematic uniform random manner round the circumference of the carotid wall. The intima-media thickness was measured using a line tool connecting the surface of the intima with the most external layer of the compact vascular smooth muscle of the media, see Fig. 3. As the intima was represented only by a single layer of endothelium supported with a very thin layer of subendothelial connective tissue, it was neglected in agreement with model assumptions. The measurements thus resulted in a mean value of media thickness, $H_m = 0.89$ mm.

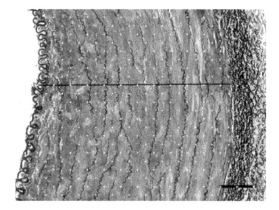

Fig. 3. The intima media thickness (black line) was measured three times per micrograph. Modified green trichrome stain, scale bar 100 μm

3.2. Orientation of smooth muscle tissue within media

Unlike the contours of individual muscle cells, the nuclei are easy to be visualized in routine histological sections. As the smooth muscle cells are spindle shaped and the long axes of their oval nuclei run parallel to the long axis of the muscle cells, describing the orientation of the nuclei is a good estimate for assessing the orientation of the whole smooth muscle cells. For the model presented in this paper, we use the previously published distribution of angles between

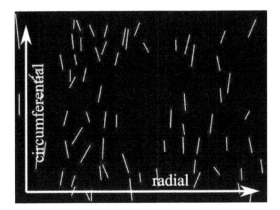

Fig. 4. Orientation of nuclei of smooth muscle cells in a transverse section of tunica media of porcine aorta. Based on a method previously published in [17]

the long axes of smooth muscle cells and the circumferential direction of arterial wall [17]. In the transverse section, the angle ranges within the interval $\langle -38; 35 \rangle^\circ$, i.e. the orientation of nuclei and therefore smooth muscle cells is considered as circumferential, see Fig. 4.

3.3. Model parameters of a single smooth muscle cell

Due to the variability in size and shape of individual biological cells, estimating of their typical morphological parameters relies on stochastic geometry. Morphometric characteristics of vascular smooth muscle cells have not been published so far, therefore we use the values published in gastropod smooth muscle cells, where the mean cell volume is $1\,358\ \mu\mathrm{m}^3$ [11,16] and the maximum transversal diameter is approx. $6\ \mu\mathrm{m}$ [15]. If the shape of smooth muscle cell is approximated by an ellipsoid with the minor axes of the same size, its length is calculated as $72\ \mu\mathrm{m}$.

Concerning the proposed model, the observed orientation of smooth muscle tissue and the parameters of individual cells result in following conclusions. First, the model is considered as transversally isotropic with the sizes of the ball $c_1 = c_3 = 6\ \mu\mathrm{m}, c_2 = 72\ \mu\mathrm{m}$. Second, the main axes of the ball and therefore the orientation of the RVE correspond to the spatial directions of the cylindrical coordinate system as it is depicted in Fig. 2.

3.4. Volume fraction of smooth muscle cells

In order to determine the relative size of the ball within the RVE in our model, the volume fraction of smooth muscle cells is employed. This parameter represents the ratio of the volume of cells and the volume of the whole tunica media. We use the values for the porcine abdominal aorta published in [17]. Here the stereological point-grid method is used, allowing a reliable estimation of area fraction of smooth muscle cells within the media. According to the Cavalieri principle, the volume fraction is estimated with the mean value of $V_{rel} = 0.65$.

In the BS model, let us assume the thickness of the extracellular matrix is expressed with a constant δ in all three spatial directions, i.e.

$$l_i = c_i + \delta. \tag{16}$$

Introducing the dimensionless parameters $\delta^* = \delta/c_1$ and $c^* = c_2/c_1 = 12$, the expression for the relative volume fraction,

$$V_{rel} = \frac{c_1^2 c_2}{l_1^2 l_2}, \tag{17}$$

leads to the equation

$$(\delta^*)^3 + (2 + c^*)(\delta^*)^2 + (2c^* + 1)\delta^* + \left(1 - \frac{1}{V_{rel}}\right)c^* = 0. \tag{18}$$

As a solution of this equation, parameter δ^* is obtained. Finally, using the definitions (12) and (16), three material parameters of the BS model are identified,

$$g_1 = \frac{1}{1 + \delta^*} \doteq 0.81, \qquad g_2 = \frac{c^*}{c^* + \delta^*} \doteq 0.98, \qquad l_{12} = \frac{1 + \delta^*}{c^* + \delta^*} \doteq 0.10. \tag{19}$$

4. Mechanical response of an arterial segment during inflation and axial stretching

4.1. Experimental data

In order to compare the model predictions with experimental data, a sample of porcine carotid artery is investigated. Namely, we use the sample described in section 3.1. The experiment follows the procedure that is detailed in [19]. Therefore, only a brief summary is given in this section. The sample is clamped into a measurement device (Tissue bath MAYFLOWER, Perfusion of tubular organs version, type 813/6, Hugo Sachs Electronik, Germany) with the axial stretch of $\lambda_z = 1.5$ to mimic *in vivo* conditions. After preconditioning, the sample is loaded with inner pressure from 0 up to 200 mm Hg. At the same time the outer diameter is measured at the middle region of the segment. Resulting pressure-diameter diagram is plotted in Fig. 5. Here the units are converted as 1 mm Hg = 133.322 Pa.

4.2. Model predictions

Theoretical pressure-diameter curves are obtained by the model of an arterial segment using the formulae derived in section 2 and the material parameters obtained in section 3. As an axisymmetric tube formed of two layers, its geometry is described with inner radius $R_{in} =$

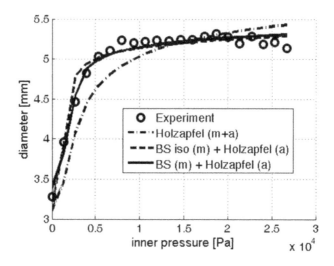

Fig. 5. Experimental pressure-diameter diagram and the theoretical curves

1.29 mm, outer radius $R_{out} = 2.49$ mm, opening angle $\alpha = 108°$ and the media thickness $H_m = 0.89$ mm. Adventitia, the outer layer, is considered as the structural material model of Holzapfel with four unknown material parameters, μ, κ_1, κ_2, β. Concerning the media, described using the BS model, three material parameters related to the microstructure of smooth muscle tissue are determined by the microscopic measurements, namely $g_1 = 0.81$, $g_2 = 0.98$ and $l_{12} = 0.10$. However, four material parameters related to the stiffness of smooth muscle cells and extracellular matrix are unknown, i.e. K_1, K_2, k_1 and k_2. All the unknown parameters (eight in total) are obtained by the comparison of the theoretical pressure-diameter curve with the experimental data using the least-squares fitting. The result is plotted using the solid line in Fig. 5. Material parameters, obtained by the least-squares fitting, are listed in Table 1.

For comparison, two other models are considered. In the first one the isotropic restriction of the BS model is considered for media (dashed line). It corresponds to the cubic shape of the representative volume element containig spherical ball with stifnesses that are identical in all three spatial directions, i.e. $K_i^\Delta = K^\Delta$, $K_i^c = K^c$, $c_i = c$, $l_i = l$, see Fig. 2. Number of material parameters is thus reduced to three for this model, namely $K = K^\Delta/l$, $k = K^c/K^\Delta$, $g = c/l$. The structural parameter is identified using the volume fraction of smooth muscle cells as $g = \sqrt[3]{V_{rel}} \doteq 0.87$. The rest of the material parameters (six in total) is left for the least-squares fitting.

Table 1. The list of material parameters identified by the least-squares fitting

BS (m) + Holzapfel (a)	BS isotropic (m) + Holzapfel (a)	Holzapfel (m+a)
$K_1 = 5.9 \times 10^5$ Pa	$K = 5.9 \times 10^5$ Pa	$\mu^m = 1.8$ kPa
$K_2 = 5.9 \times 10^5$ Pa	$k = 9.8 \times 10^{-3}$	$\kappa_1^m = 1.1$ kPa
$k_1 = 1.8 \times 10^{-2}$		$\kappa_2^m = 0.2$
$k_2 = 2.1 \times 10^{-4}$		$\beta^m = 42°$
$\mu = 82$ Pa	$\mu = 66$ Pa	$\mu^a = 1.3 \times 10^2$ Pa
$\kappa_1 = 1.4 \times 10^2$ Pa	$\kappa_1 = 1.3 \times 10^2$ Pa	$\kappa_1^a = 1.8 \times 10^2$ Pa
$\kappa_2 = 2.2$	$\kappa_2 = 2.4$	$\kappa_2 = 1.5$
$\beta = 2.4°$	$\beta = 18°$	$\beta^m = 26°$

In the second case, the structural material model of Holzapfel is considered for both media and adventitia (dash-dotted line). In fact, it corresponds to the description of arterial wall as it is proposed in [7]. All material parameters (eight in total) must be determined by the least-squares fitting, namely μ^m, κ_1^m, κ_2^m, β^m, μ^a, κ_1^a, κ_2^a, β^a. Here the superscripts m and a refer to the media and the adventitia, respectively.

In all three models, good agreement with experimental data is provided. However, combination of the BS model with the model of Holzapfel seems to be more accurate in comparison to the case when the model of Holzapfel is considered for both media and adventitia. Concerning the material parameters related to the Holzapfel model, the values obtained by the least squares fitting are comparable to those considered for the rabbit carotid artery in [7]. Obviously, there is only a slight difference in the mechanical response of the BS model and its isotropic restriction. This is caused by the influence of adventitia which dominates the mechanical response for higher pressure (due to the exponential character of the model).

5. Conclusion

In this paper, an idealized model of an arterial segment as an axisymmetric tube is proposed. Although real arteries are composed of three layers, intima (the innermost layer) is very thin and makes an insignificant contribution to the overall mechanical response in healthy young arteries. Hence, it is neglected in this model.

Static analysis and description of deformations for the given load (axial stretch and inner pressure) follow the approach detailed in [7]. Due to the symmetry of the loading, the arterial segment keeps the shape of an axisymmetric tube also at the current configuration. During the experiment, on the other hand, the sample is clamped within the device and perfused intraluminarly which results in a more complex shape at the current configuration. However, accuracy of this simplification for the presented experimental setting is confirmed by FE analysis in [19].

Structural model of Holzapfel et al., employed for the description of the mechanical response of adventitia, is characterized with three material parameters related to stiffness and one material parameter related to structure. Although having a transparent physical meaning (angle between preferred fibre directions), direct identification of this parameter has not been done so far. Hence, the least-squares fitting is applied for all material parameters related to adventitia. BS model, representing the material of media, is proposed using the bottom-up approach which leads to the transparent physical meaning of all material parameters. Although this model is orthotropic in general, we use the transverse isotropic restriction upon the fact that smooth muscle cells are of approximately ellipsoidal shape with equal minor axes. Resulting material model is characterized with seven material parameters, three of them are identified directly using the microscopic measurements. It is worth stressing that the measurements take place using also the samples of porcine aorta and the smooth muscle tissue of gastropod even if the model represents an arterial segment of porcine carotid artery. Nevertheless, we assume this inconsistency to be tolerable for the particular parameters (relative volume fraction of smooth muscle cells in porcine aorta and shape of smooth muscle cells in gastropod). Performing these measurements in one sample corresponding to the model is favourable in future work though.

Comparison of the model prediction with the mechanical response of real arterial segment is performed using the porcine carotid artery undergoing inflation test. Theoretical curves exhibit a good agreement with experimental data, however, there are still eight material parameters identified by the least-squares fitting (six material parameters in the case of isotropic restriction). In the case of BS model, the resulting material parameters are of the order $K_i \sim 10^5$ Pa, $k_i \sim 10^{-4} \div 10^{-2}$, meaning the living cells to be much softer compared to the extracellular matrix within smooth muscle tissue. Concerning the Holzapfel model, the resulting material parameters are comparable and of the same order as the parameters published for rabbit carotid artery in [7]. Only the value of the angle β is scattered, however, there is no measurement so far which could be useful in identifying this structural parameter. Moreover, the existence of preferred fibre directions of collagen fibres within adventitia is questionable.

Difference in the mechanical response of the presented model and its isotropic restriction is of small significance namely for higher pressures. This is caused by the exponential character of the mechanical response of adventitia which dominates in the overall mechanical response and acts as a stiff tube reinforcing the arterial segment. This phenomena is in agreement with experimental observations, although undesirable in our research. To be able to study the effect of material parameters related to microscopic level on the overall mechanical response, loading of the media alone is preferable.

Another goal for future studies is to increase the number of material parameters identified directly and not using the least-squares fitting. Also, the presented material model of media is a very drastic simplification of real soft tissue. Improvement of this model using for instance polymer chains instead of linear springs or embedding motor proteins might help in studying some effects related to microscopic level such as prestress or smooth muscle tissue activation.

Acknowledgements

The work has been supported by the grant project GAČR 106/09/0734 and the research project MSM 4977751303.

References

[1] Bathe, M., A fluid-structure interaction finite element analysis of pulsatile blood flow through a compliant stenotic artery, Bachelor thesis, MIT, 1998.

[2] Delfino, A., Stergiopulos, N., Moore, J. E., Meister, J.-J., Residual strain effects on the stress field in a thick wall finite element model of the human carotid bifurcation, Journal of Biomechanics 30(8) (1997) 777–786.

[3] Chuong, C. J., Fung, Y. C., Three-dimensional stress distribution in arteries, Journal of Biomechanical Engineering 105 (1983) 268–274.

[4] Ehret, A. E., Itskov, M., A polyconvex hyperelastic model for fiber-reinforced materials in application to soft tissues, Journal of Materials Science 42 (2007) 8 853–8 863.

[5] Gasser, T. C., Ogden, R. W., Holzapfel, G. A., Hyperelastic modelling of arterial layers with distributed collagen fibre orientation, Journal of the Royal Society Interface 3 (2006) 15–35.

[6] Hariton, I., deBotton, G., Gasser, T. C., Holzapfel, G. A., Stress-driven collagen fiber remodeling in arterial walls, Biomechanics and Modeling in Mechanobiology 6 (2007) 163–175.

[7] Holzapfel, G. A., Gasser, T. C., Ogden, R. W., A new constitutive framework for arterial wall mechanics and a comparative study of material models, Journal of Elasticity 61 (2000) 1–48.

[8] Holeček, M., Moravec, F., Hyperelastic model of a material which microstructure is formed of "balls and springs", International Journal of Solids and Structures 43 (2006) 7 393–7 406.

[9] Horny, L., Zitny, R., Chlup, H., Strain energy function for arterial walls based on limiting fiber extensibility, IFMBE Proceedings 22 (2008) 1 910–1 913.

[10] Itskov, M., Ehret, A. E., Mavrilas, D., A polyconvex anisotropic strain-energy function for soft collagenous tissues, Biomechanics and Modeling in Mechanobiology 5 (2006) 17–26.

[11] Kochova, P., Study of mechanical properties of smooth muscle tissue from cellular level, Ph.D thesis, University of West Bohemia in Pilsen, 2008.

[12] Kočová, J., Overall staining of connective tissue and the muscular layer of vessels. Folia Morphologica 18 (1970) 293–295.

[13] Lally, C., Reid, A. J., Prendergast, P. J., Elastic behavior of porcine coronary artery tissue under uniaxial and equibiaxial tension, Annals of Biomedical Engineering, 32 (10) (2004) 1 355–1 364.

[14] Ogden, R. W., Non-linear elastic deformations, Dover Publications, New York, 1997.

[15] Tonar, Z., Markoš, A., Microscopy and morphometry of integument of the foot of pulmonate gastropods Arion rufus and Helix pomatia. Acta veterinaria Brno 73 (2004) 3–8.

[16] Tonar, Z., Kochová, P., Holeček, M., Janáček, J., Stereological assessment, mechanical measurement and computer modelling of smooth muscle, Materials Science Forum 567–568 (2007) 353–356.

[17] Tonar, Z., Kochová, P., Janáček, J., Orientation, anisotropy, clustering, and volume fraction of smooth muscle cells within the wall of porcine abdominal aorta, Applied and Computational Mechanics 2 (2008) 145–156.

[18] Valencia, A., Solis, F., Blood flow dynamics and arterial wall interaction in a saccular aneurysm model of the basilar artery, Computers and Structures 84 (2006) 1 326–1 337.

[19] Vychytil, J., Moravec, F., Kochová, P., Kuncová, J., Švíglerová, J., Modelling of the mechanical behaviour of the porcine carotid artery undergoing inflation-deflation test, Applied and Computational Mechanics 4 (2010) 251–262.

[20] Zhang, Y., Dunn, M. L., Drexler, E. S., McCowan, C. N., Slifka, A. J., Ivy, D. D., Shandas, R., A microstructural hyperelastic model of pulmonary arteries under normo- and hypertensive conditions, Annals of Biomedical Engineering 33 (8) (2005) 1 042–1 052.

Permissions

List of Contributors

S. Seitl and Z. Knésla
Institute of Physics of Materials, Academy of Sciences of the Czech Republic, v.v.i. Žižkova 22, 616 62 Brno, Czech Republic

V. Veselý and L. Řoutil
Institute of Structural Mechanics, Faculty of Civil Engineering, Brno University of Technology, Veveřì 331/95, 602 00 Brno, Czech Republic

S. Seitla and Z. Knésla
Institute of Physics of Materials, Academy of Sciences of the Czech Republic, v.v.i., Žižkova 22, 616 62 Brno, Czech Republic

Z. Keřsner
Institute of Structural Mechanics, Civil Engineering Faculty, Brno University of Technology, Veveří 331/95, 602 00 Brno, Czech Republic

V. Bílek
ZPSV, a.s., Testing laboratory Brno, Křižìkova 68, 660 90 Brno, Czech Republic

M. Ševčík and L. Náhlík
Institute of Physics of Materials, Czech Academy of Sciences, Žižkova 22, 616 62 Brno, Czech Republic

P. Hutařa
Institute of Solid Mechanics, Mechatronics and Biomechanics, Brno University of Technology, Technická2, 616 69 Brno, Czech Republic

J. Zapoměl and P. Ferfecki
Institute of Thermomechanics, branch Ostrava, Czech Academy of Science, 17. listopadu 15, 708 33 Ostrava, Czech Republic

Očeňašek and J. Voldřich
New Technologies Research Centre, University of West Bohemia, Univerzitní 8, 306 14 Plzn, Czech Republic

R. Melichera
Faculty of Applied Mechanics, Žilina university in Žilina, Univerzitná 1, 010 26 Žilina, Slovak Republic

P. Polach and M. Hajžman
Section of Materials and Mechanical Engineering Research, ŠKODA VÝZKUM, s. r. o., Tylova 1/57, 316 00 Plzeň, Czech Republic

J. Soukup and J. Volek
Department of Mechanics and Machines, Faculty of Production Technology and Management, University of J. E. Purkyně instínad Labem, Na Okraji 1001, 400 96 Uśtínad Labem, Czech Republic

V. Veselý and P. Frantík
Faculty of Civil Engineering, BUT Brno, Veveří331/95, 602 00 Brno, Czech Republic

M. Byrtus, M. Hajžman and V. Zeman
Faculty of Applied Sciences, University of West Bohemia, Univerzitní22, 306 14 Plzeň, Czech Republic

M. Holeček, F.Moravec and J. Vychytil
Faculty of Applied Sciences, University of West Bohemia, Univerzitní22, 306 14 Plzeň, Czech Republic

P. Sváček
Institute of Thermomechanics, Czech Academy of Sciences, Dolejškova 5, 182 00 Praha 8, Czech Republic

Dep. of Technical Mathematics, Faculty of Mechanical Engineering, Czech Technical University in Prague, Karlovo nam. 13, 121 35 Praha 2, Czech Republic

J. Horáček
Institute of Thermomechanics, Czech Academy of Sciences, Dolejškova 5, 182 00 Praha 8, Czech Republic

R. Ghayour, M. Ghayour and S. Ziaei-Rad
Department of Mechanical Engineering, Isfahan University of Technology, Isfahan 84156-83111, Iran

H. Kocková
New Technologies – Research Centre in the West bohemian Region, University of West Bohemia, Univerzitní22, 306 14 Plzeň, Czech Republic

R. Cimrman
Department of Mechanics, Faculty of Applied Sciences, University of West Bohemia, Univerzitní22, 306 14 Plzeň, Czech Republic

V. Kleisner and R. Zemčík
Faculty of Applied Sciences, University of West Bohemia, Univerzitní22, 306 14 Plzeň, Czech Republic

M. Zajíček, V. Adámek and J. Dupal
Faculty of Applied Sciences, University of West Bohemia, Univerzitní22, 306 14 Plzeň, Czech Republic

V. Pelikán, P. Hora and O. Červená
Institute of Thermomechanics of the ASCR, v.v.i., Veleslavínova 11, 301 14 Plzeň, Czech Republic

A. Spielmannová and A. Machová
Institute of Thermomechanics of the ASCR, v.v.i., Dolejškova 5, 182 00 Praha, Czech Republic

V. Zeman and Z. Hlaváč
Faculty of Applied Sciences, University of West Bohemia, Univerzitní22, 306 14 Plzeň, Czech Republic

Index

Printed in the USA
CPSIA information can be obtained
at www.ICGtesting.com
JSHW051407221024
72173JS00006B/1317

9 781632 406033